COLUMBIA

ACCIDENT INVESTIGATION BOARD

REPORT VOLUME I

AUGUST 2003

On the Title Page

This was the crew patch for STS-107. The central element of the patch was the microgravity symbol, µg, flowing into the rays of the Astronaut symbol. The orbital inclination was portrayed by the 39-degree angle of the Earth's horizon to the Astronaut symbol. The sunrise was representative of the numerous science experiments that were the dawn of a new era for continued microgravity research on the International Space Station and beyond. The breadth of science conducted on this mission had widespread benefits to life on Earth and the continued exploration of space, illustrated by the Earth and stars. The constellation Columba (the dove) was chosen to symbolize peace on Earth and the Space Shuttle Columbia. In addition, the seven stars represent the STS-107 crew members, as well as honoring the original Mercury 7 astronauts who paved the way to make research in space possible. The Israeli flag represented the first person from that country to fly on the Space Shuttle.

On the Back Cover

This emblem memorializes the three U.S. human space flight accidents – Apollo 1, Challenger, and Columbia. The words across the top translate to: "To The Stars, Despite Adversity – Always Explore"

This is a reproduction of the first printing of the Columbia Accident Report as it appeared August 2003, subsequent errors corrected by the CAIB have been included up to September 12th 2003. Additional material in this book and on the accompanying CDROM were selected by the editor. Some minor aesthetic changes were made to accomodate the new layout

http://www.apogeebooks.com
Editor: Robert Godwin
ISBN: 1-894959-06-X

We acknowledge the financial support of the Government of Canada through the Book Publishing Industry Development Program for our publishing activities. Published by Collector's Guide Publishing Inc., Box 62034, Burlington, Ontario, Canada, L7R 4K2— Printed and bound in Canada

IN MEMORIAM

Rick D. Husband
Commander

William C. McCool
Pilot

Michael P. Anderson
Payload Commander

David M. Brown
Mission Specialist

Kalpana Chawla
Mission Specialist

Laurel Blair Salton Clark
Mission Specialist

Ilan Ramon
Payload Specialist

Jules F. Mier, Jr.
Debris Search Pilot

Charles Krenek
Debris Search Aviation Specialist

This cause of exploration and discovery is not an option we choose; it is a desire written in the human heart …
We find the best among us, send them forth into unmapped darkness, and pray they will return.
They go in peace for all mankind, and all mankind is in their debt.

– President George W. Bush, February 4, 2003

The quarter moon, photographed from Columbia on January 26, 2003, during the STS-107 mission.

VOLUME I

BOARD STATEMENT

For all those who are inspired by flight, and for the nation where powered flight was first achieved, the year 2003 had long been anticipated as one of celebration – December 17 would mark the centennial of the day the Wright *Flyer* first took to the air. But 2003 began instead on a note of sudden and profound loss. On February 1, Space Shuttle *Columbia* was destroyed in a disaster that claimed the lives of all seven of its crew.

While February 1 was an occasion for mourning, the efforts that ensued can be a source of national pride. NASA publicly and forthrightly informed the nation about the accident and all the associated information that became available. The Columbia Accident Investigation Board was established within two hours of the loss of signal from the returning spacecraft in accordance with procedures established by NASA following the *Challenger* accident 17 years earlier.

The crew members lost that morning were explorers in the finest tradition, and since then, everyone associated with the Board has felt that we were laboring in their legacy. Ours, too, was a journey of discovery: We sought to discover the conditions that produced this tragic outcome and to share those lessons in such a way that this nation's space program will emerge stronger and more sure-footed. If those lessons are truly learned, then *Columbia*'s crew will have made an indelible contribution to the endeavor each one valued so greatly.

After nearly seven months of investigation, the Board has been able to arrive at findings and recommendations aimed at significantly reducing the chances of further accidents. Our aim has been to improve Shuttle safety by multiple means, not just by correcting the specific faults that cost the nation this Orbiter and this crew. With that intent, the Board conducted not only an investigation of what happened to *Columbia*, but also – to determine the conditions that allowed the accident to occur – a safety evaluation of the entire Space Shuttle Program. Most of the Board's efforts were undertaken in a completely open manner. By necessity, the safety evaluation was conducted partially out of the public view, since it included frank, off-the-record statements by a substantial number of people connected with the Shuttle program.

In order to understand the findings and recommendations in this report, it is important to appreciate the way the Board looked at this accident. It is our view that complex systems almost always fail in complex ways, and we believe it would be wrong to reduce the complexities and weaknesses associated with these systems to some simple explanation. Too often, accident investigations blame a failure only on the last step in a complex process, when a more comprehensive understanding of that process could reveal that earlier steps might be equally or even more culpable. In this Board's opinion, unless the technical, organizational, and cultural recommendations made in this report are implemented, little will have been accomplished to lessen the chance that another accident will follow.

From its inception, the Board has considered itself an independent and public institution, accountable to the American public, the White House, Congress, the astronaut corps and their families, and NASA. With the support of these constituents, the Board resolved to broaden the scope of the accident investigation into a far-reaching examination of NASA's operation of the Shuttle fleet. We have explored the impact of NASA's organizational history and practices on Shuttle safety, as well as the roles of public expectations and national policy-making.

In this process, the Board identified a number of pertinent factors, which we have grouped into three distinct categories: 1) physical failures that led directly to *Columbia*'s destruction; 2) underlying weaknesses, revealed in NASA's organization and history, that can pave the way to catastrophic failure; and 3) "other significant observations" made during the course of the investigation, but which may be unrelated to the accident at hand. Left uncorrected, any of these factors could contribute to future Shuttle losses.

To establish the credibility of its findings and recommendations, the Board grounded its examinations in rigorous scientific and engineering principles. We have consulted with leading authorities not only in mechanical systems, but also in organizational theory and practice. These authorities' areas of expertise included risk management, safety engineering, and a review of "best business practices" employed by other high-risk, but apparently reliable enterprises. Among these are nuclear power plants, petrochemical facilities, nuclear weapons production, nuclear submarine operations, and expendable space launch systems.

NASA is a federal agency like no other. Its mission is unique, and its stunning technological accomplishments, a source of pride and inspiration without equal, represent the best in American skill and courage. At times NASA's efforts have riveted the nation, and it is never far from public view and close scrutiny from many quarters. The loss of *Columbia* and her crew represents a turning point, calling for a renewed public policy debate and commitment regarding human space exploration. One of our goals has been to set forth the terms for this debate.

Named for a sloop that was the first American vessel to circumnavigate the Earth more than 200 years ago, in 1981 *Columbia* became the first spacecraft of its type to fly in Earth orbit and successfully completed 27 missions over more than two decades. During the STS-107 mission, *Columbia* and its crew traveled more than six million miles in 16 days.

The Orbiter's destruction, just 16 minutes before scheduled touchdown, shows that space flight is still far from routine. It involves a substantial element of risk, which must be recognized, but never accepted with resignation. The seven *Columbia* astronauts believed that the risk was worth the reward. The Board salutes their courage and dedicates this report to their memory.

Harold W. Gehman, Jr.
Admiral, U.S. Navy (retired)
Chairman

John L. Barry
Major General, U.S. Air Force

Duane W. Deal
Brigadier General, U.S. Air Force

James N. Hallock, Ph.D.
Manager, Aviation Safety Division, DOT/RSPA Volpe Center

Kenneth W. Hess
Major General, U.S. Air Force

G. Scott Hubbard
Director, NASA Ames Research Center

John M. Logsdon, Ph.D.
Professor, George Washington University

Douglas D. Osheroff, Ph.D.
Professor, Stanford University

Sally K. Ride, Ph.D.
Professor, University of California at San Diego

Roger E. Tetrault
Chairman and CEO, McDermott International (retired)

Stephen A. Turcotte
Rear Admiral, U.S. Navy

Steven B. Wallace
Director, FAA Office of Accident Investigation

Sheila E. Widnall, Ph.D
Professor, Massachusetts Institute of Technology

Columbia *inside the Orbiter Processing Facility on November 20, 2002.*

EXECUTIVE SUMMARY

The Columbia Accident Investigation Board's independent investigation into the February 1, 2003, loss of the Space Shuttle *Columbia* and its seven-member crew lasted nearly seven months. A staff of more than 120, along with some 400 NASA engineers, supported the Board's 13 members. Investigators examined more than 30,000 documents, conducted more than 200 formal interviews, heard testimony from dozens of expert witnesses, and reviewed more than 3,000 inputs from the general public. In addition, more than 25,000 searchers combed vast stretches of the Western United States to retrieve the spacecraft's debris. In the process, *Columbia*'s tragedy was compounded when two debris searchers with the U.S. Forest Service perished in a helicopter accident.

The Board recognized early on that the accident was probably not an anomalous, random event, but rather likely rooted to some degree in NASA's history and the human space flight program's culture. Accordingly, the Board broadened its mandate at the outset to include an investigation of a wide range of historical and organizational issues, including political and budgetary considerations, compromises, and changing priorities over the life of the Space Shuttle Program. The Board's conviction regarding the importance of these factors strengthened as the investigation progressed, with the result that this report, in its findings, conclusions, and recommendations, places as much weight on these causal factors as on the more easily understood and corrected physical cause of the accident.

The physical cause of the loss of *Columbia* and its crew was a breach in the Thermal Protection System on the leading edge of the left wing, caused by a piece of insulating foam which separated from the left bipod ramp section of the External Tank at 81.7 seconds after launch, and struck the wing in the vicinity of the lower half of Reinforced Carbon-Carbon panel number 8. During re-entry this breach in the Thermal Protection System allowed superheated air to penetrate through the leading edge insulation and progressively melt the aluminum structure of the left wing, resulting in a weakening of the structure until increasing aerodynamic forces caused loss of control, failure of the wing, and breakup of the Orbiter. This breakup occurred in a flight regime in which, given the current design of the Orbiter, there was no possibility for the crew to survive.

The organizational causes of this accident are rooted in the Space Shuttle Program's history and culture, including the original compromises that were required to gain approval for the Shuttle, subsequent years of resource constraints, fluctuating priorities, schedule pressures, mischaracterization of the Shuttle as operational rather than developmental, and lack of an agreed national vision for human space flight. Cultural traits and organizational practices detrimental to safety were allowed to develop, including: reliance on past success as a substitute for sound engineering practices (such as testing to understand why systems were not performing in accordance with requirements); organizational barriers that prevented effective communication of critical safety information and stifled professional differences of opinion; lack of integrated management across program elements; and the evolution of an informal chain of command and decision-making processes that operated outside the organization's rules.

This report discusses the attributes of an organization that could more safely and reliably operate the inherently risky Space Shuttle, but does not provide a detailed organizational prescription. Among those attributes are: a robust and independent program technical authority that has complete control over specifications and requirements, and waivers to them; an independent safety assurance organization with line authority over all levels of safety oversight; and an organizational culture that reflects the best characteristics of a learning organization.

This report concludes with recommendations, some of which are specifically identified and prefaced as "before return to flight." These recommendations are largely related to the physical cause of the accident, and include preventing the loss of foam, improved imaging of the Space Shuttle stack from liftoff through separation of the External Tank, and on-orbit inspection and repair of the Thermal Protection System. The remaining recommendations, for the most part, stem from the Board's findings on organizational cause factors. While they are not "before return to flight" recommendations, they can be viewed as "continuing to fly" recommendations, as they capture the Board's thinking on what changes are necessary to operate the Shuttle and future spacecraft safely in the mid- to long-term.

These recommendations reflect both the Board's strong support for return to flight at the earliest date consistent with the overriding objective of safety, and the Board's conviction that operation of the Space Shuttle, and all human spaceflight, is a developmental activity with high inherent risks.

A view from inside the Launch Control Center as Columbia rolls out to Launch Complex 39-A on December 9, 2002.

Columbia sits on *Launch Complex 39-A prior to STS-107.*

Report Synopsis

The Columbia Accident Investigation Board's independent investigation into the tragic February 1, 2003, loss of the Space Shuttle *Columbia* and its seven-member crew lasted nearly seven months and involved 13 Board members, approximately 120 Board investigators, and thousands of NASA and support personnel. Because the events that initiated the accident were not apparent for some time, the investigation's depth and breadth were unprecedented in NASA history. Further, the Board determined early in the investigation that it intended to put this accident into context. We considered it unlikely that the accident was a random event; rather, it was likely related in some degree to NASA's budgets, history, and program culture, as well as to the politics, compromises, and changing priorities of the democratic process. We are convinced that the management practices overseeing the Space Shuttle Program were as much a cause of the accident as the foam that struck the left wing. The Board was also influenced by discussions with members of Congress, who suggested that this nation needed a broad examination of NASA's Human Space Flight Program, rather than just an investigation into what physical fault caused *Columbia* to break up during re-entry.

Findings and recommendations are in the relevant chapters and all recommendations are compiled in Chapter 11.

Volume I is organized into four parts: The Accident; Why the Accident Occurred; A Look Ahead; and various appendices. To put this accident in context, Parts One and Two begin with histories, after which the accident is described and then analyzed, leading to findings and recommendations. Part Three contains the Board's views on what is needed to improve the safety of our voyage into space. Part Four is reference material. In addition to this first volume, there will be subsequent volumes that contain technical reports generated by the Columbia Accident Investigation Board and NASA, as well as volumes containing reference documentation and other related material.

Part One: The Accident

Chapter 1 relates the history of the Space Shuttle Program before the *Challenger* accident. With the end looming for the Apollo moon exploration program, NASA unsuccessfully attempted to get approval for an equally ambitious (and expensive) space exploration program. Most of the proposed programs started with space stations in low-Earth orbit and included a reliable, economical, medium-lift vehicle to travel safely to and from low-Earth orbit. After many failed attempts, and finally agreeing to what would be untenable compromises, NASA gained approval from the Nixon Administration to develop, on a fixed budget, only the transport vehicle. Because the Administration did not approve a low-Earth-orbit station, NASA had to create a mission for the vehicle. To satisfy the Administration's requirement that the system be economically justifiable, the vehicle had to capture essentially all space launch business, and to do that, it had to meet wide-ranging requirements. These sometimes-competing requirements resulted in a compromise vehicle that was less than optimal for manned flights. NASA designed and developed a remarkably capable and resilient vehicle, consisting of an Orbiter with three Main Engines, two Solid Rocket Boosters, and an External Tank, but one that has never met any of its original requirements for reliability, cost, ease of turnaround, maintainability, or, regrettably, safety.

Chapter 2 documents the final flight of *Columbia*. As a straightforward record of the event, it contains no findings or recommendations. Designated STS-107, this was the Space Shuttle Program's 113th flight and *Columbia*'s 28th. The flight was close to trouble-free. Unfortunately, there were no indications to either the crew onboard *Columbia* or to engineers in Mission Control that the mission was in trouble as a result of a foam strike during ascent. Mission management failed to detect weak signals that the Orbiter was in trouble and take corrective action.

Columbia was the first space-rated Orbiter. It made the Space Shuttle Program's first four orbital test flights. Because it was the first of its kind, *Columbia* differed slightly from Orbiters *Challenger*, *Discovery*, *Atlantis*, and *Endeavour*. Built to an earlier engineering standard, *Columbia* was slightly heavier, and, although it could reach the high-inclination orbit of the International Space Station, its payload was insufficient to make *Columbia* cost-effective for Space Station missions. Therefore, *Columbia* was not equipped with a Space Station docking system, which freed up space in the payload bay for longer cargos, such as the science modules Spacelab and SPACEHAB. Consequently, *Columbia* generally flew science missions and serviced the Hubble Space Telescope.

STS-107 was an intense science mission that required the seven-member crew to form two teams, enabling round-the-clock shifts. Because the extensive science cargo and its extra power sources required additional checkout time, the launch sequence and countdown were about 24 hours longer than normal. Nevertheless, the countdown proceeded as planned, and *Columbia* was launched from Launch Complex 39-A on January 16, 2003, at 10:39 a.m. Eastern Standard Time (EST).

At 81.7 seconds after launch, when the Shuttle was at about 65,820 feet and traveling at Mach 2.46 (1,650 mph), a large piece of hand-crafted insulating foam came off an area where the Orbiter attaches to the External Tank. At 81.9 seconds, it struck the leading edge of *Columbia*'s left wing. This event was not detected by the crew on board or seen by ground support teams until the next day, during detailed reviews of all launch camera photography and videos. This foam strike had no apparent effect on the daily conduct of the 16-day mission, which met all its objectives.

The de-orbit burn to slow *Columbia* down for re-entry into Earth's atmosphere was normal, and the flight profile throughout re-entry was standard. Time during re-entry is measured in seconds from "Entry Interface," an arbitrarily determined altitude of 400,000 feet where the Orbiter begins to experience the effects of Earth's atmosphere. Entry Interface for STS-107 occurred at 8:44:09 a.m. on February 1. Unknown to the crew or ground personnel, because the data is recorded and stored in the Orbiter instead of being transmitted to Mission Control at Johnson Space Center, the first abnormal indication occurred 270 seconds after Entry Interface. Chapter 2 reconstructs in detail the events leading to the loss of *Columbia* and her crew, and refers to more details in the appendices.

In Chapter 3, the Board analyzes all the information available to conclude that the direct, physical action that initiated the chain of events leading to the loss of *Columbia* and her crew was the foam strike during ascent. This chapter reviews five analytical paths – aerodynamic, thermodynamic, sensor data timeline, debris reconstruction, and imaging evidence – to show that all five independently arrive at the same conclusion. The subsequent impact testing conducted by the Board is also discussed.

That conclusion is that *Columbia* re-entered Earth's atmosphere with a pre-existing breach in the leading edge of its left wing in the vicinity of Reinforced Carbon-Carbon (RCC) panel 8. This breach, caused by the foam strike on ascent, was of sufficient size to allow superheated air (probably exceeding 5,000 degrees Fahrenheit) to penetrate the cavity behind the RCC panel. The breach widened, destroying the insulation protecting the wing's leading edge support structure, and the superheated air eventually melted the thin aluminum wing spar. Once in the interior, the superheated air began to destroy the left wing. This destructive process was carefully reconstructed from the recordings of hundreds of sensors inside the wing, and from analyses of the reactions of the flight control systems to the changes in aerodynamic forces.

By the time *Columbia* passed over the coast of California in the pre-dawn hours of February 1, at Entry Interface plus 555 seconds, amateur videos show that pieces of the Orbiter were shedding. The Orbiter was captured on videotape during most of its quick transit over the Western United States. The Board correlated the events seen in these videos to sensor readings recorded during re-entry. Analysis indicates that the Orbiter continued to fly its pre-planned flight profile, although, still unknown to anyone on the ground or aboard *Columbia*, her control systems were working furiously to maintain that flight profile. Finally, over Texas, just southwest of Dallas-Fort Worth, the increasing aerodynamic forces the Orbiter experienced in the denser levels of the atmosphere overcame the catastrophically damaged left wing, causing the Orbiter to fall out of control at speeds in excess of 10,000 mph.

The chapter details the recovery of about 38 percent of the Orbiter (some 84,000 pieces) and the reconstruction and analysis of this debris. It presents findings and recommendations to make future Space Shuttle operations safer.

Chapter 4 describes the investigation into other possible physical factors that may have contributed to the accident. The chapter opens with the methodology of the fault tree analysis, which is an engineering tool for identifying every conceivable fault, then determining whether that fault could have caused the system in question to fail. In all, more than 3,000 individual elements in the *Columbia* accident fault tree were examined.

In addition, the Board analyzed the more plausible fault scenarios, including the impact of space weather, collisions with micrometeoroids or "space junk," willful damage, flight crew performance, and failure of some critical Shuttle hardware. The Board concludes in Chapter 4 that despite certain fault tree exceptions left "open" because they cannot be conclusively disproved, none of these factors caused or contributed to the accident. This chapter also contains findings and recommendations to make Space Shuttle operations safer.

PART TWO: WHY THE ACCIDENT OCCURRED

Part Two, "Why the Accident Occurred," examines NASA's organizational, historical, and cultural factors, as well as how these factors contributed to the accident.

As in Part One, Part Two begins with history. Chapter 5 examines the post-*Challenger* history of NASA and its Human Space Flight Program. A summary of the relevant portions of the *Challenger* investigation recommendations is presented, followed by a review of NASA budgets to indicate how committed the nation is to supporting human space flight, and within the NASA budget we look at how the Space Shuttle Program has fared. Next, organizational and management history, such as shifting management systems and locations, are reviewed.

Chapter 6 documents management performance related to *Columbia* to establish events analyzed in later chapters. The chapter begins with a review of the history of foam strikes on the Orbiter to determine how Space Shuttle Program managers rationalized the danger from repeated strikes on the Orbiter's Thermal Protection System. Next is an explanation of the intense pressure the program was under to stay on schedule, driven largely by the self-imposed requirement to complete the International Space Station. Chapter 6 then relates in detail the effort by some NASA engineers to obtain additional imagery of *Columbia* to determine if the foam strike had damaged the Orbiter, and how management dealt with that effort.

In Chapter 7, the Board presents its view that NASA's organizational culture had as much to do with this accident as foam did. By examining safety history, organizational theory, best business practices, and current safety failures, the report notes that only significant structural changes to NASA's organizational culture will enable it to succeed.

This chapter measures the Shuttle Program's practices against this organizational context and finds them wanting. The Board concludes that NASA's current organization does not provide effective checks and balances, does not have an independant safety program, and has not demonstrated the characteristics of a learning organization. Chapter 7 provides recommendations for adjustments in organizational culture.

Chapter 8, the final chapter in Part Two, draws from the previous chapters on history, budgets, culture, organization, and safety practices, and analyzes how all these factors contributed to this accident. The chapter opens with "echoes of *Challenger*" that compares the two accidents. This chapter captures the Board's views of the need to adjust management to enhance safety margins in Shuttle operations, and reaffirms the Board's position that without these changes, we have no confidence that other "corrective actions" will improve the safety of Shuttle operations. The changes we recommend will be difficult to accomplish – and will be internally resisted.

PART THREE: A LOOK AHEAD

Part Three summarizes the Board's conclusions on what needs to be done to resume our journey into space, lists significant observations the Board made that are unrelated to the accident but should be recorded, and provides a summary of the Board's recommendations.

In Chapter 9, the Board first reviews its short-term recommendations. These return-to-flight recommendations are the minimum that must be done to essentially fix the problems that were identified by this accident. Next, the report discusses what needs to be done to operate the Shuttle in the mid-term, 3 to 15 years. Based on NASA's history of ignoring external recommendations, or making improvements that atrophy with time, the Board has no confidence that the Space Shuttle can be safely operated for more than a few years based solely on renewed post-accident vigilance.

Chapter 9 then outlines the management system changes the Board feels are necessary to safely operate the Shuttle in the mid-term. These changes separate the management of scheduling and budgets from technical specification authority, build a capability of systems integration, and establish and provide the resources for an independent safety and mission assurance organization that has supervisory authority. The third part of the chapter discusses the poor record this nation has, in the Board's view, of developing either a complement to or a replacement for the Space Shuttle. The report is critical of several bodies in the U.S. government that share responsibility for this situation, and expresses an opinion on how to proceed from here, but does not suggest what the next vehicle should look like.

Chapter 10 contains findings, observations, and recommendations that the Board developed over the course of this extensive investigation that are not directly related to the accident but should prove helpful to NASA.

Chapter 11 is a compilation of all the recommendations in the previous chapters.

PART FOUR: APPENDICES

Part Four of the report by the Columbia Accident Investigation Board contains material relevant to this volume organized in appendices. Additional, stand-alone volumes will contain more reference, background, and analysis materials.

This Earth view of the Sinai Peninsula, Red Sea, Egypt, Nile River, and the Mediterranean was taken from Columbia during STS-107.

AN INTRODUCTION TO THE SPACE SHUTTLE

The Space Shuttle is one of the most complex machines ever devised. Its main elements – the Orbiter, Space Shuttle Main Engines, External Tank, and Solid Rocket Boosters – are assembled from more than 2.5 million parts, 230 miles of wire, 1,060 valves, and 1,440 circuit breakers. Weighing approximately 4.5 million-pounds at launch, the Space Shuttle accelerates to an orbital velocity of 17,500 miles per hour – 25 times faster than the speed of sound – in just over eight minutes. Once on orbit, the Orbiter must protect its crew from the vacuum of space while enabling astronauts to conduct scientific research, deploy and service satellites, and assemble the International Space Station. At the end of its mission, the Shuttle uses the Earth's atmosphere as a brake to decelerate from orbital velocity to a safe landing at 220 miles per hour, dissipating in the process all the energy it gained on its way into orbit.

THE ORBITER

The Orbiter is what is popularly referred to as "the Space Shuttle." About the size of a small commercial airliner, the Orbiter normally carries a crew of seven, including a Commander, Pilot, and five Mission or Payload Specialists. The Orbiter can accommodate a payload the size of a school bus weighing between 38,000 and 56,300 pounds depending on what orbit it is launched into. The Orbiter's upper flight deck is filled with equipment for flying and maneuvering the vehicle and controlling its remote manipulator arm. The mid-deck contains stowage lockers for food, equipment, supplies, and experiments, as well as a toilet, a hatch for entering and exiting the vehicle on the ground, and – in some instances – an airlock for doing so in orbit. During liftoff and landing, four crew members sit on the flight deck and the rest on the mid-deck.

Different parts of the Orbiter are subjected to dramatically different temperatures during re-entry. The nose and leading edges of the wings are exposed to superheated air temperatures of 2,800 to 3,000 degrees Fahrenheit, depending upon re-entry profile. Other portions of the wing and fuselage can reach 2,300 degrees Fahrenheit. Still other areas on top of the fuselage are sufficiently shielded from superheated air that ice sometimes survives through landing.

To protect its thin aluminum structure during re-entry, the Orbiter is covered with various materials collectively referred to as the Thermal Protection System. The three major components of the system are various types of heat-resistant tiles, blankets, and the Reinforced Carbon-Carbon (RCC) panels on the leading edge of the wing and nose cap. The RCC panels most closely resemble a hi-tech fiberglass – layers of special graphite cloth that are molded to the desired shape at very high temperatures. The tiles, which protect most other areas of the Orbiter exposed to medium and high heating, are 90 percent air and 10 percent silica (similar to common sand). One-tenth the weight of ablative heat shields, which are designed to erode during re-entry and therefore can only be used once, the Shuttle's tiles are reusable. They come in varying strengths and sizes, depending on which area of the Orbiter they protect, and are designed to withstand either 1,200 or 2,300 degrees Fahrenheit. In a dramatic demonstration of how little heat the tiles transfer, one can place a blowtorch on one side of a tile and a bare hand on the other. The blankets, capable of withstanding either 700 or 1,200 degrees Fahrenheit, cover regions of the Orbiter that experience only moderate heating.

SPACE SHUTTLE MAIN ENGINES

Each Orbiter has three main engines mounted at the aft fuselage. These engines use the most efficient propellants in the world – oxygen and hydrogen – at a rate of half a ton per second. At 100 percent power, each engine produces 375,000 pounds of thrust, four times that of the largest engine on commercial jets. The large bell-shaped nozzle on each engine can swivel 10.5 degrees up and down and 8.5 degrees left and right to provide steering control during ascent.

EXTERNAL TANK

The three main engines burn propellant at a rate that would drain an average-size swimming pool in 20 seconds. The External Tank accommodates up to 143,351 gallons of liquid oxygen and 385,265 gallons of liquid hydrogen. In order to keep the super-cold propellants from boiling and to prevent ice from forming on the outside of the tank while it is sitting on the launch pad, the External Tank is covered with a one-inch-thick coating of insulating foam. This insulation is so effective that the surface of the External Tank feels only slightly cool to the touch, even though the liquid oxygen is stored at minus 297 degrees Fahrenheit and liquid hydrogen at minus 423 degrees Fahrenheit. This insulating foam also protects the tank's aluminum structure from aerodynamic heating during ascent. Although generally considered the least complex of the Shuttle's main components, in fact the External Tank is a remarkable engineering achievement. In addition to holding over 1.5 million pounds of cryogenic propellants, the 153.8-foot long tank must support the weight of the Orbiter while on the launch pad and absorb the 7.3 million pounds of thrust generated by the Solid Rocket Boosters and Space Shuttle Main Engines during launch and ascent. The External Tanks are manufactured in a plant near New

Orleans and are transported by barge to the Kennedy Space Center in Florida. Unlike the Solid Rocket Boosters, which are reused, the External Tank is discarded during each mission, burning up in the Earth's atmosphere after being jettisoned from the Orbiter.

SOLID ROCKET BOOSTERS

Despite their power, the Space Shuttle Main Engines alone are not sufficient to boost the vehicle to orbit – in fact, they provide only 15 percent of the necessary thrust. Two Solid Rocket Boosters attached to the External Tank generate the remaining 85 percent. Together, these two 149-foot long motors produce over six million pounds of thrust. The largest solid propellant rockets ever flown, these motors use an aluminum powder fuel and ammonium perchlorate oxidizer in a binder that has the feel and consistency of a pencil eraser.

A Solid Rocket Booster (SRB) Demonstration Motor being tested near Brigham City, Utah.

Each of the Solid Rocket Boosters consists of 11 separate segments joined together. The joints between the segments were extensively redesigned after the *Challenger* accident, which occurred when hot gases burned through an O-ring and seal in the aft joint on the left Solid Rocket Booster. The motor segments are shipped from their manufacturer in Utah and assembled at the Kennedy Space Center. Once assembled, each Solid Rocket Booster is connected to the External Tank by bolts weighing 65 pounds each. After the Solid Rocket Boosters burn for just over two minutes, these bolts are separated by pyrotechnic charges and small rockets then push the Solid Rocket Boosters safely away from the rest of the vehicle. As the boosters fall back to Earth, parachutes in their nosecones deploy. After splashing down into the ocean 120 miles downrange from the launch pad, they are recovered for refurbishment and reuse.

THE SHUTTLE STACK

The first step in assembling a Space Shuttle for launch is stacking the Solid Rocket Booster segments on the Mobile Launch Platform. Eight large hold-down bolts at the base of the Solid Rocket Boosters will bear the weight of the entire Space Shuttle stack while it awaits launch. The External Tank is attached to the Solid Rocket Boosters, and the Orbiter is then attached to the External Tank at three points – two at its bottom and a "bipod" attachment near the nose. When the vehicle is ready to move out of the Vehicle Assembly Building, a Crawler-Transporter picks up the entire Mobile Launch Platform and carries it – at one mile per hour – to one of the two launch pads.

An Introduction to NASA

"An Act to provide for research into the problems of flight within and outside the Earth's atmosphere, and for other purposes." With this simple preamble, the Congress and the President of the United States created the National Aeronautics and Space Administration (NASA) on October 1, 1958. Formed in response to the launch of *Sputnik* by the Soviet Union, NASA inherited the research-oriented National Advisory Committee for Aeronautics (NACA) and several other government organizations, and almost immediately began working on options for manned space flight. NASA's first high profile program was Project Mercury, an early effort to learn if humans could survive in space. Project Gemini followed with a more complex series of experiments to increase man's time in space and validate advanced concepts such as rendezvous. The efforts continued with Project Apollo, culminating in 1969 when *Apollo 11* landed the first humans on the Moon. The return from orbit on July 24, 1975, of the crew from the Apollo-Soyuz Test Project began a six-year hiatus of American manned space flight. The launch of the first Space Shuttle in April 1981 brought Americans back into space, continuing today with the assembly and initial operations of the International Space Station.

In addition to the human space flight program, NASA also maintains an active (if small) aeronautics research program, a space science program (including deep space and interplanetary exploration), and an Earth observation program. The agency also conducts basic research activities in a variety of fields.

NASA, like many federal agencies, is a heavily matrixed organization, meaning that the lines of authority are not necessarily straightforward. At the simplest level, there are three major types of entities involved in the Human Space Flight Program: NASA field centers, NASA programs carried out at those centers, and industrial and academic contractors. The centers provide the buildings, facilities, and support services for the various programs. The programs, along with field centers and Headquarters, hire civil servants and contractors from the private sector to support aspects of their enterprises.

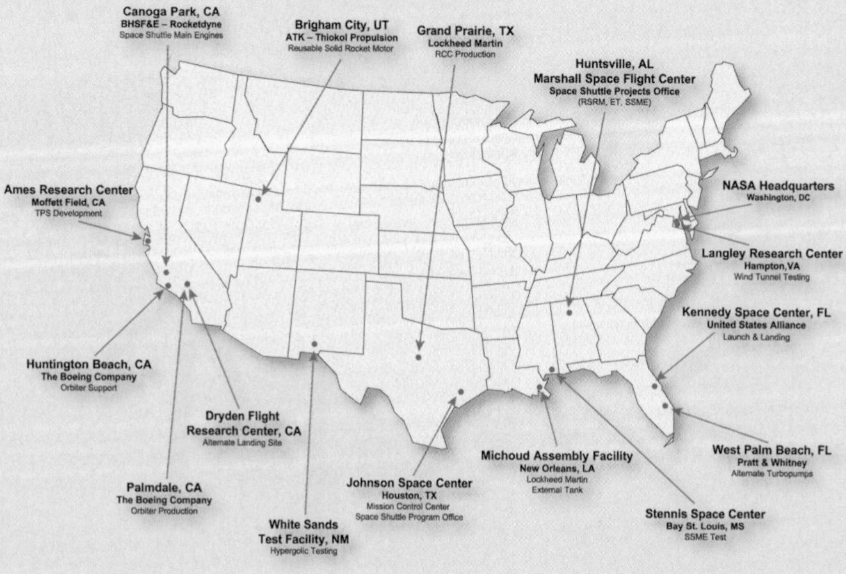

The Locations

NASA Headquarters, located in Washington D.C., is responsible for leadership and management across five strategic enterprises: Aerospace Technology, Biological and Physical Research, Earth Science, Space Science, and Human Exploration and Development of Space. NASA Headquarters also provides strategic management for the Space Shuttle and International Space Station programs.

The Johnson Space Center in Houston, Texas, was established in 1961 as the Manned Spacecraft Center and has led the development of every U.S. manned space flight program. Currently, Johnson is home to both the Space Shuttle and International Space Station Program Offices. The facilities at Johnson include the training, simulation, and mission control centers for the Space Shuttle and Space Station. Johnson also has flight operations at Ellington Field, where the training aircraft for the astronauts and support aircraft for the Space Shuttle Program are stationed, and manages the White Sands Test Facility, New Mexico, where hazardous testing is conducted.

The Kennedy Space Center was created to launch the Apollo missions to the Moon, and currently provides launch and landing facilities for the Space Shuttle. The Center is located on Merritt Island, Florida, adjacent to the Cape Canaveral Air Force Station that also provides support for the Space Shuttle Program (and was the site of the earlier Mercury and Gemini launches). Personnel at Kennedy support maintenance and overhaul services for the Orbiters, assemble and check-out the integrated vehicle prior to launch, and operate the Space Station Processing Facility where components of the orbiting laboratory are packaged for launch aboard the Space Shuttle. The majority of contractor personnel assigned to Kennedy are part of the Space Flight Operations Contract administered by the Space Shuttle Program Office at Johnson.

The Marshall Space Flight Center, near Hunstville, Alabama, is home to most NASA rocket propulsion efforts. The Space Shuttle Projects Office located at Marshall—organizationally part of the Space Shuttle Program Office at Johnson—manages the manufacturing and support contracts to Boeing Rocketdyne for the Space Shuttle Main Engine (SSME), to Lockheed Martin for the External Tank (ET), and to ATK Thiokol Propulsion for the Reusable Solid Rocket Motor (RSRM, the major piece of the Solid Rocket Booster). Marshall is also involved in microgravity research and space product development programs that fly as payloads on the Space Shuttle.

The Stennis Space Center in Bay St. Louis, Mississippi, is the largest rocket propulsion test complex in the United States. Stennis provides all of the testing facilities for the Space

Shuttle Main Engines and External Tank. (The Solid Rocket Boosters are tested at the ATK Thiokol Propulsion facilities in Utah.)

The Ames Research Center at Moffett Field, California, has evolved from its aeronautical research roots to become a Center of Excellence for information technology. The Center's primary importance to the Space Shuttle Program, however, lies in wind tunnel and arc-jet testing, and the development of thermal protection system concepts.

The Langley Research Center, at Hampton, Virginia, is the agency's primary center for structures and materials and supports the Space Shuttle Program in these areas, as well as in basic aerodynamic and thermodynamic research.

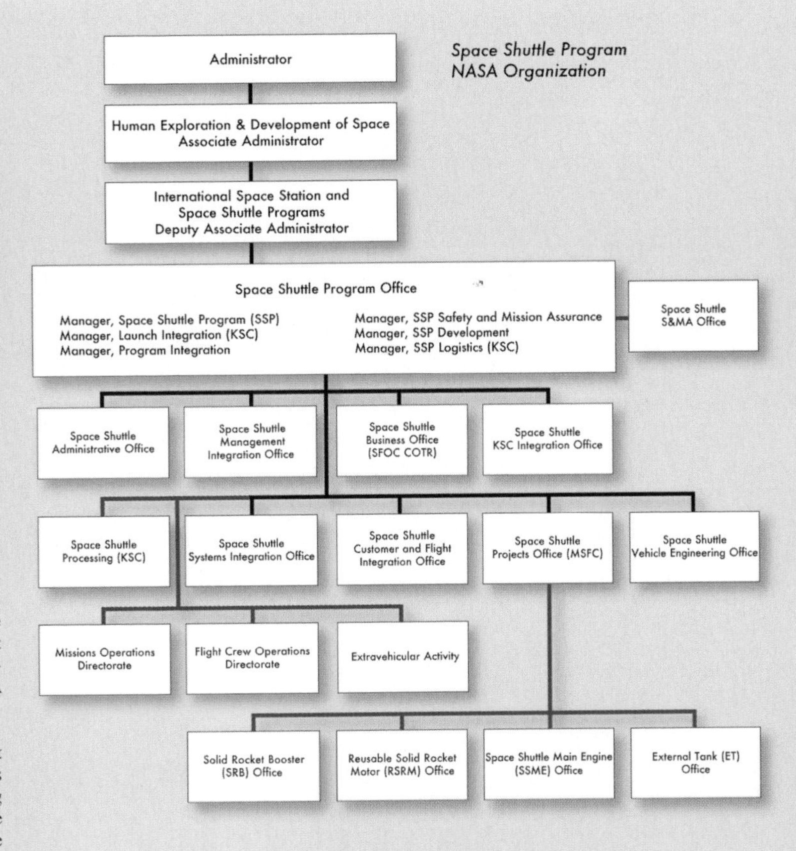

THE PROGRAMS

The two major human space flight efforts within NASA are the Space Shuttle Program and International Space Station Program, both headquartered at Johnson although they report to a Deputy Associate Administrator at NASA Headquarters in Washington, D.C.

The Space Shuttle Program Office at Johnson is responsible for all aspects of developing, supporting, and flying the Space Shuttle. To accomplish these tasks, the program maintains large workforces at the various NASA Centers that host the facilities used by the program. The Space Shuttle Program Office is also responsible for managing the Space Flight Operations Contract with United Space Alliance that provides most of the contractor support at Johnson and Kennedy, as well as a small amount at Marshall.

THE CONTRACTORS

The Space Shuttle Program employs a wide variety of commercial companies to provide services and products. Among these are some of the largest aerospace and defense contractors in the country, including (but not limited to):

United Space Alliance

This is a joint venture between Boeing and Lockheed Martin that was established in 1996 to perform the Space Flight Operations Contract that essentially conducts the day-to-day operation of the Space Shuttle. United Space Alliance is headquartered in Houston, Texas, and employs more than 10,000 people at Johnson, Kennedy, and Marshall. Its contract currently runs through 2005.

The Boeing Company, NASA Systems

The Space Shuttle Orbiter was designed and manufactured by Rockwell International, located primarily in Downey and Palmdale, California. In 1996, The Boeing Company purchased the aerospace assets of Rockwell International, and later moved the Downey operation to Huntington Beach, California, as part of a consolidation of facilities. Boeing is subcontracted to United Space Alliance to provide support to Orbiter modifications and operations, with work performed in California, and at Johnson and Kennedy.

The Boeing Company, Rocketdyne Propulsion & Power

The Rocketdyne Division of Rockwell International was responsible for the development and manufacture of the Space Shuttle Main Engines, and continues to support the engines as a part of The Boeing Company. The Space Shuttle Projects Office at Marshall manages the main engines contract, with most of the work performed in California, Stennis, and Kennedy.

ATK Thiokol Propulsion

ATK Thiokol Propulsion (formerly Morton-Thiokol) in Brigham City, Utah, manufactures the Reusable Solid Rocket Motor segments that are the propellant sections of the Solid Rocket Boosters. The Space Shuttle Projects Office at Marshall manages the Reusable Solid Rocket Motor contract.

Lockheed Martin Space Systems, Michoud Operations

The External Tank was developed and manufactured by Martin Marietta at the NASA Michoud Assembly Facility near New Orleans, Louisiana. Martin Marietta later merged with Lockheed to create Lockheed Martin. The External Tank is the only disposable part of the Space Shuttle system, so new ones are always under construction. The Space Shuttle Projects Office at Marshall manages the External Tank contract.

Lockheed Martin Missiles and Fire Control

The Reinforced Carbon-Carbon (RCC) panels used on the nose and wing leading edges of the Orbiter were manufactured by Ling-Temco-Vought in Grand Prairie, Texas. Lockheed Martin acquired LTV through a series of mergers and acquisitions. The Space Shuttle Program office at Johnson manages the RCC support contract.

The launch of STS-107 on January 16, 2003.

Part One

The Accident

"Building rockets is hard." Part of the problem is that space travel is in its infancy. Although humans have been launching orbital vehicles for almost 50 years now – about half the amount of time we have been flying airplanes – contrast the numbers. Since *Sputnik*, humans have launched just over 4,500 rockets towards orbit (not counting suborbital flights and small sounding rockets). During the first 50 years of aviation, there were over one million aircraft built. Almost all of the rockets were used only once; most of the airplanes were used more often.

There is also the issue of performance. Airplanes slowly built their performance from the tens of miles per hour the Wright Brothers initially managed to the 4,520 mph that Major William J. Knight flew in the X-15A-2 research airplane during 1967. Aircraft designers and pilots would slightly push the envelope, stop and get comfortable with where they were, then push on. Orbital rockets, by contrast, must have all of their performance on the first (and often, only) flight. Physics dictates this – to reach orbit, without falling back to Earth, you have to exceed about 17,500 mph. If you cannot vary performance, then the only thing left to change is the amount of payload – the rocket designers began with small payloads and worked their way up.

Rockets, by their very nature, are complex and unforgiving vehicles. They must be as light as possible, yet attain outstanding performance to get to orbit. Mankind is, however, getting better at building them. In the early days as often as not the vehicle exploded on or near the launch pad; that seldom happens any longer. It was not that different from early airplanes, which tended to crash about as often as they flew. Aircraft seldom crash these days, but rockets still fail between two-and-five percent of the time. This is true of just about any launch vehicle – Atlas, Delta, Soyuz, Shuttle – regardless of what nation builds it or what basic configuration is used; they all fail about the same amount of the time. Building and launching rockets is still a very dangerous business, and will continue to be so for the foreseeable future while we gain experience at it. It is unlikely that launching a space vehicle will ever be as routine an undertaking as commercial air travel – certainly not in the lifetime of anybody who reads this. The scientists and engineers continually work on better ways, but if we want to continue going into outer space, we must continue to accept the risks.

Part One of the report of the Columbia Accident Investigation Board is organized into four chapters. In order to set the background for further discussion, Chapter 1 relates the history of the Space Shuttle Program before the *Challenger* accident. The events leading to the original approval of the Space Shuttle Program are recounted, as well as an examination of some of the promises made in order to gain that approval. In retrospect, many of these promises could never have been achieved. Chapter 2 documents the final flight of *Columbia*. As a straightforward record of the event, it contains no findings or recommendations. Chapter 3 reviews five analytical paths – aerodynamic, thermodynamic, sensor data timeline, debris reconstruction, and imaging evidence – to show that all five independently arrive at the same conclusion. Chapter 4 describes the investigation into other possible physical factors that might have contributed to the accident, but were subsequently dismissed as possible causes.

Sunrise aboard Columbia on Flight Day 7.

The launch of STS-107 on January 16, 2003.

CHAPTER 1

The Evolution of the Space Shuttle Program

More than two decades after its first flight, the Space Shuttle remains the only reusable spacecraft in the world capable of simultaneously putting multiple-person crews and heavy cargo into orbit, of deploying, servicing, and retrieving satellites, and of returning the products of on-orbit research to Earth. These capabilities are an important asset for the United States and its international partners in space. Current plans call for the Space Shuttle to play a central role in the U.S. human space flight program for years to come.

The Space Shuttle Program's remarkable successes, however, come with high costs and tremendous risks. The February 1 disintegration of *Columbia* during re-entry, 17 years after *Challenger* was destroyed on ascent, is the most recent reminder that sending people into orbit and returning them safely to Earth remains a difficult and perilous endeavor.

It is the view of the Columbia Accident Investigation Board that the *Columbia* accident is not a random event, but rather a product of the Space Shuttle Program's history and current management processes. Fully understanding how it happened requires an exploration of that history and management. This chapter charts how the Shuttle emerged from a series of political compromises that produced unreasonable expectations – even myths – about its performance, how the *Challenger* accident shattered those myths several years after NASA began acting upon them as fact, and how, in retrospect, the Shuttle's technically ambitious design resulted in an inherently vulnerable vehicle, the safe operation of which exceeded NASA's organizational capabilities as they existed at the time of the *Columbia* accident. The Board's investigation of what caused the *Columbia* accident thus begins in the fields of East Texas but reaches more than 30 years into the past, to a series of economically and politically driven decisions that cast the Shuttle program in a role that its nascent technology could not support. To understand the cause of the *Columbia* accident is to understand how a program promising reliability and cost efficiency resulted instead in a developmental vehicle that never achieved the fully operational status NASA and the nation accorded it.

1.1 GENESIS OF THE SPACE TRANSPORTATION SYSTEM

The origins of the Space Shuttle Program date to discussions on what should follow Project Apollo, the dramatic U.S. missions to the moon.[1] NASA centered its post-Apollo plans on developing increasingly larger outposts in Earth orbit that would be launched atop Apollo's immense Saturn V booster. The space agency hoped to construct a 12-person space station by 1975; subsequent stations would support 50, then 100 people. Other stations would be placed in orbit around the moon and then be constructed on the lunar surface. In parallel, NASA would develop the capability for the manned exploration of Mars. The concept of a vehicle – or Space Shuttle – to take crews and supplies to and from low-Earth orbit arose as part of this grand vision (see Figure 1.1-1). To keep the costs of these trips to a minimum, NASA intended to develop a fully reusable vehicle.[2]

Figure 1.1-1. Early concepts for the Space Shuttle envisioned a reusable two-stage vehicle with the reliability and versatility of a commercial airliner.

NASA's vision of a constellation of space stations and journeying to Mars had little connection with political realities of the time. In his final year in office, President Lyndon Johnson gave highest priority to his Great Society programs and to dealing with the costs and domestic turmoil associated with the Vietnam war. Johnson's successor, President Richard Nixon, also had no appetite for another large, expensive, Apollo-like space commitment. Nixon rejected NASA's ambitions with little hesitation and directed that the agency's budget be cut as much as was politically feasible. With NASA's space station plans deferred and further production of the Saturn V launch vehicle cancelled, the Space Shuttle was the only manned space flight program that the space agency could hope to undertake. But without space stations to service, NASA needed a new rationale for the Shuttle. That rationale emerged from an intense three-year process of technical studies and political and budgetary negotiations that attempted to reconcile the conflicting interests of NASA, the Department of Defense, and the White House.[3]

1.2 Merging Conflicting Interests

During 1970, NASA's leaders hoped to secure White House approval for developing a fully reusable vehicle to provide routine and low cost manned access to space. However, the staff of the White House Office of Management and Budget, charged by Nixon with reducing NASA's budget, was skeptical of the value of manned space flight, especially given its high costs. To overcome these objections, NASA turned to justifying the Space Shuttle on economic grounds. If the same vehicle, NASA argued, launched all government and private sector payloads and if that vehicle were reusable, then the total costs of launching and maintaining satellites could be dramatically reduced. Such an economic argument, however, hinged on the willingness of the Department of Defense to use the Shuttle to place national security payloads in orbit. When combined, commercial, scientific, and national security payloads would require 50 Space Shuttle missions per year. This was enough to justify – at least on paper – investing in the Shuttle.

Meeting the military's perceived needs while also keeping the cost of missions low posed tremendous technological hurdles. The Department of Defense wanted the Shuttle to carry a 40,000-pound payload in a 60-foot-long payload bay and, on some missions, launch and return to a West Coast launch site after a single polar orbit. Since the Earth's surface – including the runway on which the Shuttle was to land – would rotate during that orbit, the Shuttle would need to maneuver 1,100 miles to the east during re-entry. This "cross-range" requirement meant the Orbiter required large delta-shaped wings and a more robust thermal protection system to shield it from the heat of re-entry.

Developing a vehicle that could conduct a wide variety of missions, and do so cost-effectively, demanded a revolution in space technology. The Space Shuttle would be the first reusable spacecraft, the first to have wings, and the first with a reusable thermal protection system. Further, the Shuttle would be the first to fly with reusable, high-pressure hydrogen/oxygen engines, and the first winged vehicle to transition from orbital speed to a hypersonic glide during re-entry.

Even as the design grew in technical complexity, the Office of Management and Budget forced NASA to keep – or at least promise to keep – the Shuttle's development and operating costs low. In May 1971, NASA was told that it could count on a maximum of $5 billion spread over five years for any new development program. This budget ceiling forced NASA to give up its hope of building a fully reusable two-stage vehicle and kicked off an intense six-month search for an alternate design. In the course of selling the Space Shuttle Program within these budget limitations, and therefore guaranteeing itself a viable post-Apollo future, NASA made bold claims about the expected savings to be derived from revolutionary technologies not yet developed. At the start of 1972, NASA leaders told the White House that for $5.15 billion they could develop a Space Shuttle that would meet all performance requirements, have a lifetime of 100 missions per vehicle, and cost $7.7 million per flight.[4] All the while, many people, particularly those at the White House Office of Management and Budget, knew NASA's in-house and external economic studies were overly optimistic.[5]

Those in favor of the Shuttle program eventually won the day. On January 5, 1972, President Nixon announced that the Shuttle would be "designed to help transform the space frontier of the 1970s into familiar territory, easily accessible for human endeavor in the 1980s and 90s. This system will center on a space vehicle that can shuttle repeatedly from Earth to orbit and back. *It will revolutionize transportation into near space, by routinizing it.* [emphasis added]"[6] Somewhat ironically, the President based his decision on grounds very different from those vigorously debated by NASA and the White House budget and science offices. Rather than focusing on the intricacies of cost/benefit projections, Nixon was swayed by the political benefits of increasing employment in key states by initiating a major new aerospace program in the 1972 election year, and by a geopolitical calculation articulated most clearly by NASA Administrator James Fletcher. One month before the decision, Fletcher wrote a memo to the White House stating, "For the U.S. not to be in space, while others do have men in space, is unthinkable, and a position which America cannot accept."[7]

The cost projections Nixon had ignored were not forgotten by his budget aides, or by Congress. A $5.5 billion ceiling imposed by the Office of Management and Budget led NASA to make a number of tradeoffs that achieved savings in the short term but produced a vehicle that had higher operational costs and greater risks than promised. One example was the question of whether the "strap-on" boosters would use liquid or solid propellants. Even though they had higher projected operational costs, solid-rocket boosters were chosen largely because they were less expensive to develop, making the Shuttle the first piloted spacecraft to use solid boosters. And since NASA believed that the Space Shuttle would be far safer than any other spacecraft, the agency accepted a design with no crew escape system (see Chapter 10.)

The commitments NASA made during the policy process drove a design aimed at satisfying conflicting requirements: large payloads and cross-range capability, but also low development costs and the even lower operating costs of a "routine" system. Over the past 22 years, the resulting vehicle has proved difficult and costly to operate, riskier than expected, and, on two occasions, deadly.

It is the Board's view that, in retrospect, the increased complexity of a Shuttle designed to be all things to all people created inherently greater risks than if more realistic technical goals had been set at the start. Designing a reusable spacecraft that is also cost-effective is a daunting engineering challenge; doing so on a tightly constrained budget is even more difficult. Nevertheless, the remarkable system we have today is a reflection of the tremendous engineering expertise and dedication of the workforce that designed and built the Space Shuttle within the constraints it was given.

In the end, the greatest compromise NASA made was not so much with any particular element of the technical design, but rather with the premise of the vehicle itself. NASA promised it could develop a Shuttle that would be launched almost on demand and would fly many missions each year. Throughout the history of the program, a gap has persisted between the rhetoric NASA has used to market the Space Shuttle and operational reality, leading to an enduring image of the Shuttle as capable of safely and routinely carrying out missions with little risk.

1.3 Shuttle Development, Testing, and Qualification

The Space Shuttle was subjected to a variety of tests before its first flight. However, NASA conducted these tests somewhat differently than it had for previous spacecraft.[8] The Space Shuttle Program philosophy was to ground-test key hardware elements such as the main engines, Solid Rocket Boosters, External Tank, and Orbiter separately and to use analytical models, not flight testing, to certify the integrated Space Shuttle system. During the Approach and Landing Tests (see Figure 1.3-1), crews verified that the Orbiter could successfully fly at low speeds and land safely; however, the Space Shuttle was not flown on an unmanned orbital test flight prior to its first mission – a significant change in philosophy compared to that of earlier American spacecraft.

Figure 1.3-1. The first Orbiter was Enterprise, shown here being released from the Boeing 747 Shuttle Carrier Aircraft during the Approach and Landing Tests at Edwards Air Force Base.

The significant advances in technology that the Shuttle's design depended on led its development to run behind schedule. The date for the first Space Shuttle launch slipped from March 1978 to 1979, then to 1980, and finally to the spring of 1981. One historian has attributed one year of this delay "to budget cuts, a second year to problems with the main engines, and a third year to problems with the thermal protection tiles."[9] Because of these difficulties, in 1979 the program underwent an exhaustive White House review. The program was thought to be a billion dollars over budget, and President Jimmy Carter wanted to make sure that it was worth continuing. A key factor in the White House's final assessment was that the Shuttle was needed to launch the intelligence satellites required for verification of the SALT II arms control treaty, a top Carter Administration priority. The review reaffirmed the need for the Space Shuttle, and with continued White House and Congressional support, the path was clear for its transition from development to flight. NASA ultimately completed Shuttle development for only 15 percent more than its projected cost, a comparatively small cost overrun for so complex a program.[10]

The Orbiter that was destined to be the first to fly into space was *Columbia*. In early 1979, NASA was beginning to feel the pressure of being behind schedule. Despite the fact that only 24,000 of the 30,000 Thermal Protection System tiles had been installed, NASA decided to fly *Columbia* from the manufacturing plant in Palmdale, California, to the Kennedy Space Center in March 1979. The rest of the tiles would be installed in Florida, thus allowing NASA to maintain the appearance of *Columbia*'s scheduled launch date. Problems with the main engines and the tiles were to leave *Columbia* grounded for two more years.

1.4 The Shuttle Becomes "Operational"

On the first Space Shuttle mission, STS-1,[11] *Columbia* carried John W. Young and Robert L. Crippen to orbit on April 12, 1981, and returned them safely two days later to Edwards Air Force Base in California (see Figure 1.4-1). After three years of policy debate and nine years of development, the Shuttle returned U.S. astronauts to space for the first time since the Apollo-Soyuz Test Project flew in July 1975. Post-flight inspection showed that *Columbia* suffered slight damage from excess Solid Rocket Booster ignition pressure and lost 16 tiles, with 148 others sustaining some damage. Over the following 15 months, *Columbia* was launched three more times. At the end of its fourth mission, on July 4, 1982, *Columbia* landed at Edwards where President Ronald Reagan declared to a nation celebrating Independence Day that "beginning with the next flight, the *Columbia* and her sister ships will be *fully operational*, ready to provide *economical and routine access to space* for scientific exploration, commercial ventures, and for tasks related to the national security" [emphasis added].[12]

There were two reasons for declaring the Space Shuttle "operational" so early in its flight program. One was NASA's hope for quick Presidential approval of its next manned space flight program, a space station, which would not move forward while the Shuttle was still considered developmental. The second reason was that the nation was sud-

Figure 1.4-1. The April 12, 1981, launch of STS-1, just seconds past 7 a.m., carried astronauts John Young and Robert Crippen into an Earth orbital mission that lasted 54 hours.

denly facing a foreign challenger in launching commercial satellites. The European Space Agency decided in 1973 to develop Ariane, an expendable launch vehicle. Ariane first flew in December 1979 and by 1982 was actively competing with the Space Shuttle for commercial launch contracts. At this point, NASA still hoped that revenue from commercial launches would offset some or all of the Shuttle's operating costs. In an effort to attract commercial launch contracts, NASA heavily subsidized commercial launches by offering services for $42 million per launch, when actual costs were more than triple that figure.[13] A 1983 NASA brochure titled *We Deliver* touted the Shuttle as "the most reliable, flexible, and cost-effective launch system in the world."[14]

Figure 1.4-2. The crew of STS-5 successfully deployed two commercial communications satellites during the first "operational" mission of the Space Shuttle.

Between 1982 and early 1986, the Shuttle demonstrated its capabilities for space operations, retrieving two communications satellites that had suffered upper-stage misfires after launch, repairing another communications satellite on-orbit, and flying science missions with the pressurized European-built Spacelab module in its payload bay. The Shuttle took into space not only U.S. astronauts, but also citizens of Germany, Mexico, Canada, Saudi Arabia, France, the Netherlands, two payload specialists from commercial enterprises, and two U.S. legislators, Senator Jake Garn and Representative Bill Nelson. In 1985, when four Orbiters were in operation, the vehicles flew nine missions, the most launched in a single calendar year. By the end of 1985, the Shuttle had launched 24 communications satellites (see Figure 1.4-2) and had a backlog of 44 orders for future commercial launches.

On the surface, the program seemed to be progressing well. But those close to it realized that there were numerous problems. The system was proving difficult to operate, with more maintenance required between flights than had been expected. Rather than needing the 10 working days projected in 1975 to process a returned Orbiter for its next flight, by the end of 1985 an average of 67 days elapsed before the Shuttle was ready for launch.[15]

Though assigned an operational role by NASA, during this period the Shuttle was in reality still in its early flight-test stage. As with any other first-generation technology, operators were learning more about its strengths and weaknesses from each flight, and making what changes they could, while still attempting to ramp up to the ambitious flight schedule NASA set forth years earlier. Already, the goal of launching 50 flights a year had given way to a goal of 24 flights per year by 1989. The per-mission cost was more than $140 million, a figure that when adjusted for inflation was seven times greater than what NASA projected over a decade earlier.[16] More troubling, the pressure of maintaining the flight schedule created a management atmosphere that increasingly accepted less-than-specification performance of various components and systems, on the grounds that such deviations had not interfered with the success of previous flights.[17]

1.5 The *Challenger* Accident

The illusion that the Space Shuttle was an operational system, safe enough to carry legislators and a high-school teacher into orbit, was abruptly and tragically shattered on the morning of January 28, 1986, when *Challenger* was destroyed 73 seconds after launch during the 25th mission (see Figure 1.5-1). The seven-member crew perished.

To investigate, President Reagan appointed the 13-member Presidential Commission on the Space Shuttle Challenger Accident, which soon became known as the Rogers Commission, after its chairman, former Secretary of State William P. Rogers.[18] Early in its investigation, the Commission identified the mechanical cause of the accident to be the failure of the joint of one of the Solid Rocket Boosters. The Commission found that the design was not well understood by the engineers that operated it and that it had not been adequately tested.

Figure 1.5-1. the Space Shuttle Challenger *was lost during ascent on January 28, 1986, when an O-ring and seal in the right Solid Rocket Booster failed.*

When the Rogers Commission discovered that, on the eve of the launch, NASA and a contractor had vigorously debated the wisdom of operating the Shuttle in the cold temperatures predicted for the next day, and that more senior NASA managers were unaware of this debate, the Commission shifted the focus of its investigation to "NASA management practices, Center-Headquarters relationships, and the chain of command for launch commit decisions."[19] As the investigation continued, it revealed a NASA culture that had gradually begun to accept escalating risk, and a NASA safety program that was largely silent and ineffective.

The Rogers Commission report, issued on June 6, 1986, recommended a redesign and recertification of the Solid Rocket Motor joint and seal and urged that an independent body oversee its qualification and testing. The report concluded that the drive to declare the Shuttle operational had put enormous pressures on the system and stretched its resources to the limit. Faulting NASA safety practices, the Commission also called for the creation of an independent NASA Office of Safety, Reliability, and Quality Assurance, reporting directly to the NASA Administrator, as well as structural changes in program management.[20] (The Rogers Commission findings and recommendations are discussed in more detail in Chapter 5.) It would take NASA 32 months before the next Space Shuttle mission was launched. During this time, NASA initiated a series of longer-term vehicle upgrades, began the construction of the Orbiter *Endeavour* to replace *Challenger*, made significant organizational changes, and revised the Shuttle manifest to reflect a more realistic flight rate.

The *Challenger* accident also prompted policy changes. On August 15, 1986, President Reagan announced that the Shuttle would no longer launch commercial satellites. As a result of the accident, the Department of Defense made a decision to launch all future military payloads on expendable launch vehicles, except the few remaining satellites that required the Shuttle's unique capabilities.

In the seventeen years between the *Challenger* and *Columbia* accidents, the Space Shuttle Program achieved significant successes and also underwent organizational and managerial changes. The program had successfully launched several important research satellites and was providing most of the "heavy lifting" of components necessary to build the International Space Station (see Figure 1.5-2). But as the Board subsequently learned, things were not necessarily as they appeared. (The post-*Challenger* history of the Space Shuttle Program is the topic of Chapter 5.)

Figure 1.5-2. The International Space Station as seen from an approaching Space Shuttle.

1.6 CONCLUDING THOUGHTS

The Orbiter that carried the STS-107 crew to orbit 22 years after its first flight reflects the history of the Space Shuttle Program. When *Columbia* lifted off from Launch Complex 39-A at Kennedy Space Center on January 16, 2003, it superficially resembled the Orbiter that had first flown in 1981, and indeed many elements of its airframe dated back to its first flight. More than 44 percent of its tiles, and 41 of the 44 wing leading edge Reinforced Carbon-Carbon (RCC) panels were original equipment. But there were also many new systems in *Columbia*, from a modern "glass" cockpit to second-generation main engines.

Although an engineering marvel that enables a wide-variety of on-orbit operations, including the assembly of the International Space Station, the Shuttle has few of the mission capabilities that NASA originally promised. It cannot be launched on demand, does not recoup its costs, no longer carries national security payloads, and is not cost-effective enough, nor allowed by law, to carry commercial satellites. Despite efforts to improve its safety, the Shuttle remains a complex and risky system that remains central to U.S. ambitions in space. *Columbia*'s failure to return home is a harsh reminder that the Space Shuttle is a developmental vehicle that operates not in routine flight but in the realm of dangerous exploration.

ENDNOTES FOR CHAPTER 1

The citations that contain a reference to "CAIB document" with CAB or CTF followed by seven to eleven digits, such as CAB001-0010, refer to a document in the Columbia Accident Investigation Board database maintained by the Department of Justice and archived at the National Archives.

[1] George Mueller, Associate Administrator for Manned Space Flight, NASA, "Honorary Fellowship Acceptance," address delivered to the British Interplanetary Society, University College, London, England, August 10, 1968, contained in John M. Logsdon, Ray A. Williamson, Roger D. Launius, Russell J. Acker, Stephen J. Garber, and Jonathan L. Friedman, editors, Exploring the Unknown: Selected Documents in the History of the U.S. Civil Space Program Volume IV: Accessing Space, NASA SP-4407 (Washington: Government Printing Office, 1999), pp. 202-205.

[2] For detailed discussions of the origins of the Space Shuttle, see Dennis R. Jenkins, Space Shuttle: The History of the National Space Transportation System – The First 100 Missions (Cape Canaveral, FL: Specialty Press, 2001); T. A. Heppenheimer, The Space Shuttle Decision: NASA's Search for a Reusable Space Vehicle, NASA SP-4221 (Washington: Government Printing Office, 1999; also published by the Smithsonian Institution Press, 2002); and T. A. Heppenheimer, Development of the Space Shuttle, 1972-1981 (Washington: Smithsonian Institution Press, 2002). Much of the discussion in this section is based on these studies.

[3] See John M. Logsdon, "The Space Shuttle Program: A Policy Failure?" *Science*, May 30, 1986 (Vol. 232), pp. 1099-1105 for an account of this decision process. Most of the information and quotes in this section are taken from this article.

[4] See also comments by Robert F. Thompson, Columbia Accident Investigation Board Public Hearing, April 23, 2003, in Appendix G.

[5] Heppenheimer, *The Space Shuttle Decision*, pp. 278-289, and Roger A. Pielke, Jr., "The Space Shuttle Program: 'Performance vs. Promise,'" Center for Space and Geosciences Policy, University of Colorado, August 31, 1991; Logsdon, "The Space Shuttle Program: A Policy Failure?" pp. 1099-1105.

[6] Quoted in Jenkins, *Space Shuttle*, p. 171.

[7] Memorandum from J. Fletcher to J. Rose, Special Assistant to the President, November 22, 1971; Logsdon, John, "The Space Shuttle Program: A Policy Failure?" *Science*, May 30, 1986, Volume 232, pp. 1099-1105.

[8] The only actual flight tests conducted of the Orbiter were a series of Approach and Landing Tests where *Enterprise* (OV-101) was dropped from its Boeing 747 Shuttle Carrier Aircraft while flying at 25,000 feet. These tests – with crews aboard – demonstrated the low-speed handling capabilities of the Orbiter and allowed an evaluation of the vehicle's landing characteristics. See Jenkins, *Space Shuttle*, pp. 205-212 for more information.

[9] Heppenheimer, *Development of the Space Shuttle*, p. 355.

[10] As Howard McCurdy, a historian of NASA, has noted: "With the now-familiar Shuttle configuration, NASA officials came close to meeting their cost estimate of $5.15 billion for phase one of the Shuttle program. NASA actually spent $9.9 billion in real year dollars to take the Shuttle through design, development and initial testing. This sum, when converted to fixed year 1971 dollars using the aerospace price deflator, equals $5.9 billion, or a 15 percent cost overrun on the original estimate for phase one. Compared to other complex development programs, this was not a large cost overrun." See Howard McCurdy, "The Cost of Space Flight," *Space Policy* 10 (4) p. 280. For a program budget summary, see Jenkins, *Space Shuttle*, p. 256.

[11] STS stands for Space Transportation System. Although in the years just before the 1986 *Challenger* accident NASA adopted an alternate Space Shuttle mission numbering scheme, this report uses the original STS flight designations.

[12] President Reagan's quote is contained in President Ronald Reagan, "Remarks on the Completion of the Fourth Mission of the Space Shuttle *Columbia*," July 4, 1982, p. 870, in *Public Papers of the Presidents of the United States: Ronald Reagan* (Washington: Government Printing Office, 1982-1991). The emphasis noted is the Board's.

[13] "Pricing Options for the Space Shuttle," Congressional Budget Office Report, 1985.

[14] The quote is from page 2 of the *We Deliver* brochure, reproduced in *Exploring the Unknown Volume IV*, p. 423.

[15] NASA Johnson Space Center, "Technology Influences on the Space Shuttle Development," June 8, 1986, p. 1-7.

[16] The 1971 cost-per-flight estimate was $7.7 million; $140.5 million dollars in 1985 when adjusted for inflation becomes $52.9 million in 1971 dollars or nearly seven times the 1971 estimate. "Pricing Options for the Space Shuttle."

[17] See Diane Vaughan, *The Challenger Launch Decision: Risky Technology, Culture, and Deviance at NASA* (Chicago: The University of Chicago Press, 1996).

[18] See John M. Logsdon, "Return to Flight: Richard H. Truly and the Recovery from the Challenger Accident," in Pamela E. Mack, editor, *From Engineering to Big Science: The NACA and NASA Collier Trophy Research Project Winners*, NASA SP-4219 (Washington: Government Printing Office, 1998) for an account of the aftermath of the accident. Much of the account in this section is drawn from this source.

[19] Logsdon, "Return to Flight," p. 348.

[20] *Presidential Commission on the Space Shuttle Challenger Accident* (Washington: Government Printing Office, June 6, 1986).

CHAPTER 2

Columbia's Final Flight

Space Shuttle missions are not necessarily launched in the same order they are planned (or "manifested," as NASA calls the process). A variety of scheduling, funding, technical, and – occasionally – political reasons can cause the shuffling of missions over the course of the two to three years it takes to plan and launch a flight. This explains why the 113th mission of the Space Shuttle Program was called STS-107. It would be the 28th flight of *Columbia*.

While the STS-107 mission will likely be remembered most for the way it ended, there was a great deal more to the dedicated science mission than its tragic conclusion. The planned microgravity research spanned life sciences, physical sciences, space and earth sciences, and education. More than 70 scientists were involved in the research that was conducted by *Columbia*'s seven-member crew over 16 days. This chapter outlines the history of STS-107 from its mission objectives and their rationale through the accident and its initial aftermath. The analysis of the accident's causes follows in Chapter 3 and subsequent chapters.

2.1 MISSION OBJECTIVES AND THEIR RATIONALES

Throughout the 1990s, NASA flew a number of dedicated science missions, usually aboard *Columbia* because it was equipped for extended-duration missions and was not being used for Shuttle-Mir docking missions or the assembly of the International Space Station. On many of these missions, *Columbia* carried pressurized Spacelab or SPACEHAB modules that extended the habitable experiment space available and were intended as facilities for life sciences and microgravity research.

In June 1997, the Flight Assignment Working Group at Johnson Space Center in Houston designated STS-107, tentatively scheduled for launch in the third quarter of Fiscal Year 2000, a "research module" flight. In July 1997, several committees of the National Academy of Science's Space Studies Board sent a letter to NASA Administrator Daniel Goldin recommending that NASA dedicate several future Shuttle missions to microgravity and life sciences. The purpose would be to train scientists to take full advantage of the International Space Station's research capabilities once it became operational, and to reduce the gap between the last planned Shuttle science mission and the start of science research aboard the Space Station.[1] In March 1998, Goldin announced that STS-107, tentatively scheduled for launch in May 2000, would be a multi-disciplinary science mission modeled after STS-90, the Neurolab mission scheduled later in 1998.[2] In October 1998, the Veterans Affairs and Housing and Urban Development and Independent Agencies Appropriations Conference Report expressed Congress' concern about the lack of Shuttle-based science missions in Fiscal Year 1999, and added $15 million to NASA's budget for STS-107. The following year the Conference Report reserved $40 million for a second science mission. NASA cancelled the second science mission in October 2002 and used the money for STS-107.

In addition to a variety of U.S. experiments assigned to STS-107, a joint U.S./Israeli space experiment – the Mediterranean-Israeli Dust Experiment, or MEIDEX – was added to STS-107 to be accompanied by an Israeli astronaut as part of an international cooperative effort aboard the Shuttle similar to those NASA had begun in the early 1980s. *Triana*, a deployable Earth-observing satellite, was also added to the mission to save NASA from having to buy a commercial launch to place the satellite in orbit. Political disagreements between Congress and the White House delayed *Triana*, and the satellite was replaced by the Fast Reaction Experiments Enabling Science, Technology, Applications, and Research (FREESTAR) payload, which was mounted behind the SPACEHAB Research Double Module.[3]

Figure 2.1-1. Columbia, at the launch pad on January 15, 2003.

Schedule Slippage

STS-107 was finally scheduled for launch on January 11, 2001. After 13 delays over two years, due mainly to other missions taking priority, *Columbia* was launched on January 16, 2003 (see Figure 2.1-1). Delays may take several forms. When any delay is mentioned, most people think of a Space Shuttle sitting on the launch pad waiting for launch. But most delays actually occur long before the Shuttle is configured for a mission. This was the case for STS-107 – of the 13 delays, only a few occurred after the Orbiter was configured for flight; most happened earlier in the planning process. Three specific events caused delays for STS-107:

- Removal of *Triana*: This Earth-observing satellite was replaced with the FREESTAR payload.

- Orbiter Maintenance Down Period: *Columbia*'s depot-level maintenance took six months longer than originally planned, primarily to correct problems encountered with Kapton wiring (see Chapter 4). This resulted in the STS-109 Hubble Space Telescope service mission being launched before STS-107 because it was considered more urgent.

- Flowliner cracks: About one month before the planned July 19, 2002 launch date for STS-107, concerns about cracks in the Space Shuttle Main Engine propellant system flowliners caused a four-month grounding of the Orbiter fleet. (The flowliner, which is in the main propellant feed lines, mitigates turbulence across the flexible bellows to smooth the flow of propellant into the main engine low-pressure turbopump. It also protects the bellows from flow-induced vibration.) First discovered on *Atlantis*, the cracks were eventually discovered on each Orbiter; they were fixed by welding and polishing. The grounding delayed the exchange of the Expedition 5 International Space Station crew with the Expedition 6 crew, which was scheduled for STS-113. To maintain the International Space Station assembly sequence while minimizing the delay in returning the Expedition 5 crew, both STS-112 and STS-113 were launched before STS-107.

COLUMBIA

Columbia was named after a Boston-based sloop commanded by Captain Robert Gray, who noted while sailing to the Pacific Northwest a flow of muddy water fanning from the shore, and decided to explore what he deemed the "Great River of the West." On May 11, 1792, Gray and his crew maneuvered the *Columbia* past the treacherous sand bar and named the river after his ship. After a week or so of trading with the local tribes, Gray left without investigating where the river led. Instead, Gray led the *Columbia* and its crew on the first U.S. circumnavigation of the globe, carrying otter skins to Canton, China, before returning to Boston in 1793.

In addition to *Columbia* (OV-102), which first flew in 1981, *Challenger* (OV-099) first flew in 1983, *Discovery* (OV-103) in 1984, and *Atlantis* (OV-104) in 1985. *Endeavour* (OV-105), which replaced *Challenger*, first flew in 1992. At the time of the launch of STS-107, *Columbia* was unique since it was the last remaining Orbiter to have an internal airlock on the mid-deck. (All the Orbiters originally had internal airlocks, but all excepting *Columbia* were modified to provide an external docking mechanism for flights to *Mir* and the International Space Station.) Because the airlock was not located in the payload bay, *Columbia* could carry longer payloads such as the *Chandra* space telescope, which used the full length of the payload bay. The internal airlock made the mid-deck more cramped than those of other Orbiters, but this was less of a problem when one of the laboratory modules was installed in the payload bay to provide additional habitable volume.

Columbia had been manufactured to an early structural standard that resulted in the airframe being heavier than the later Orbiters. Coupled with a more-forward center of gravity because of the internal airlock, *Columbia* could not carry as much payload weight into orbit as the other Orbiters. This made *Columbia* less desirable for missions to the International Space Station, although planning was nevertheless underway to modify *Columbia* for an International Space Station flight sometime after STS-107.

The Crew

The STS-107 crew selection process followed standard procedures. The Space Shuttle Program provided the Astronaut Office with mission requirements calling for a crew of seven. There were no special requirements for a rendezvous, extravehicular activity (spacewalking), or use of the remote manipulator arm. The Chief of the Astronaut Office announced the crew in July 2000. To maximize the amount of science research that could be performed, the crew formed two teams, Red and Blue, to support around-the-clock operations.

Crew Training

The Columbia Accident Investigation Board thoroughly reviewed all pre-mission training (see Figure 2.1-2) for the STS-107 crew, Houston Mission Controllers, and the Ken-

Figure 2.1-2. Ilan Ramon (left), Laurel Clark, and Michael Anderson during a training exercise at the Johnson Space Center.

Left to right: David Brown, Rick Husband, Laurel Clark, Kalpana Chawla, Michael Anderson, William McCool, Ilan Ramon.

Rick Husband, Commander. Husband, 45, was a Colonel in the U.S. Air Force, a test pilot, and a veteran of STS-96. He received a B.S. in Mechanical Engineering from Texas Tech University and a M.S. in Mechanical Engineering from California State University, Fresno. He was a member of the Red Team, working on experiments including the European Research In Space and Terrestrial Osteoporosis and the Shuttle Ozone Limb Sounding Experiment.

William C. McCool, Pilot. McCool, 41, was a Commander in the U.S. Navy and a test pilot. He received a B.S. in Applied Science from the U.S. Naval Academy, a M.S. in Computer Science from the University of Maryland, and a M.S. in Aeronautical Engineering from the U.S. Naval Postgraduate School. A member of the Blue Team, McCool worked on experiments including the Advanced Respiratory Monitoring System, Biopack, and Mediterranean Israeli Dust Experiment.

Michael P. Anderson, Payload Commander and Mission Specialist. Anderson, 43, was a Lieutenant Colonel in the U.S. Air Force, a former instructor pilot and tactical officer, and a veteran of STS-89. He received a B.S. in Physics/Astronomy from the University of Washington, and a M.S. in Physics from Creighton University. A member of the Blue Team, Anderson worked with experiments including the Advanced Respiratory Monitoring System, Water Mist Fire Suppression, and Structures of Flame Balls at Low Lewis-number.

David M. Brown, Mission Specialist. Brown, 46, was a Captain in the U.S. Navy, a naval aviator, and a naval flight surgeon. He received a B.S. in Biology from the College of William and Mary and a M.D. from Eastern Virginia Medical School. A member of the Blue Team, Brown worked on the Laminar Soot Processes, Structures of Flame Balls at Low Lewis-number, and Water Mist Fire Suppression experiments.

Kalpana Chawla, Flight Engineer and Mission Specialist. Chawla, 41, was an aerospace engineer, a FAA Certified Flight Instructor, and a veteran of STS-87. She received a B.S. in Aeronautical Engineering from Punjab Engineering College, India, a M.S. in Aerospace Engineering from the University of Texas, Arlington, and a Ph.D. in Aerospace Engineering from the University of Colorado, Boulder. A member of the Red Team, Chawla worked with experiments on Astroculture, Advanced Protein Crystal Facility, Mechanics of Granular Materials, and the Zeolite Crystal Growth Furnace.

Laurel Clark, Mission Specialist. Clark, 41, was a Commander (Captain-Select) in the U.S. Navy and a naval flight surgeon. She received both a B.S. in Zoology and a M.D. from the University of Wisconsin, Madison. A member of the Red Team, Clark worked on experiments including the Closed Equilibrated Biological Aquatic System, Sleep-Wake Actigraphy and Light Exposure During Spaceflight, and the Vapor Compression Distillation Flight Experiment.

Ilan Ramon, Payload Specialist. Ramon, 48, was a Colonel in the Israeli Air Force, a fighter pilot, and Israel's first astronaut. Ramon received a B.S. in Electronics and Computer Engineering from the University of Tel Aviv, Israel. As a member of the Red Team, Ramon was the primary crew member responsible for the Mediterranean Israeli Dust Experiment (MEIDEX). He also worked on the Water Mist Fire Suppression and the Microbial Physiology Flight Experiments Team experiments, among others.

The Crew

nedy Space Center Launch Control Team. Mission training for the STS-107 crew comprised 4,811 hours, with an additional 3,500 hours of payload-specific training. The Ascent/Entry Flight Control Team began training with the STS-107 crew on October 22, 2002, and participated in 16 integrated ascent or entry simulations. The Orbiter Flight Control team began training with the crew on April 23, 2002, participating in six joint integrated simulations with the crew and payload customers. Seventy-seven Flight Control Room operators were assigned to four shifts for the STS-107 mission. All had prior certifications and had worked missions in the past.

The STS-107 Launch Readiness Review was held on December 18, 2002, at the Kennedy Space Center. Neither NASA nor United Space Alliance noted any training issues for launch controllers. The Mission Operations Directorate noted no crew or flight controller training issues during the January 9, 2003, STS-107 Flight Readiness Review. According to documentation, all personnel were trained and certified, or would be trained and certified before the flight. Appendix D.1 contains a detailed STS-107 Training Report.

Orbiter Preparation

Board investigators reviewed *Columbia*'s maintenance, or "flow" records, including the recovery from STS-109 and preparation for STS-107, and relevant areas in NASA's Problem Reporting and Corrective Action database, which contained 16,500 Work Authorization Documents consisting of 600,000 pages and 3.9 million steps. This database maintains critical information on all maintenance and modification work done on the Orbiters (as required by the Orbiter Maintenance Requirements and Specifications Document). It also maintains Corrective Action Reports that document problems discovered and resolved, the Lost/Found item database, and the Launch Readiness Review and Flight Readiness Review documentation (see Chapter 7).

The Board placed emphasis on maintenance done in areas of particular concern to the investigation. Specifically, records for the left main landing gear and door assembly and left wing leading edge were analyzed for any potential contributing factors, but nothing relevant to the cause of the accident was discovered. A review of Thermal Protection System tile maintenance records revealed some "non-conformances" and repairs made after *Columbia*'s last flight, but these were eventually dismissed as not relevant to the investigation. Additionally, the Launch Readiness Review and Flight Readiness Review records relating to those systems and the Lost/Found item records were reviewed, and no relevance was found. During the Launch Readiness Review and Flight Readiness Review processes, NASA teams analyzed 18 lost items and deemed them inconsequential. (Although this incident was not considered significant by the Board, a further discussion of foreign object debris may be found in Chapter 4.)

Payload Preparation

The payload bay configuration for STS-107 included the SPACEHAB access tunnel, SPACEHAB Research Double Module (RDM), the FREESTAR payload, the Orbital Acceleration Research Experiment, and an Extended Duration Orbiter pallet to accommodate the long flight time needed to conduct all the experiments. Additional experiments were stowed in the Orbiter mid-deck and on the SPACEHAB roof (see Figures 2.1-3 and 2.1-4). The total liftoff payload weight for STS-107 was 24,536 pounds. Details on STS-107 payload preparations and on-orbit operations are in Appendix D.2.

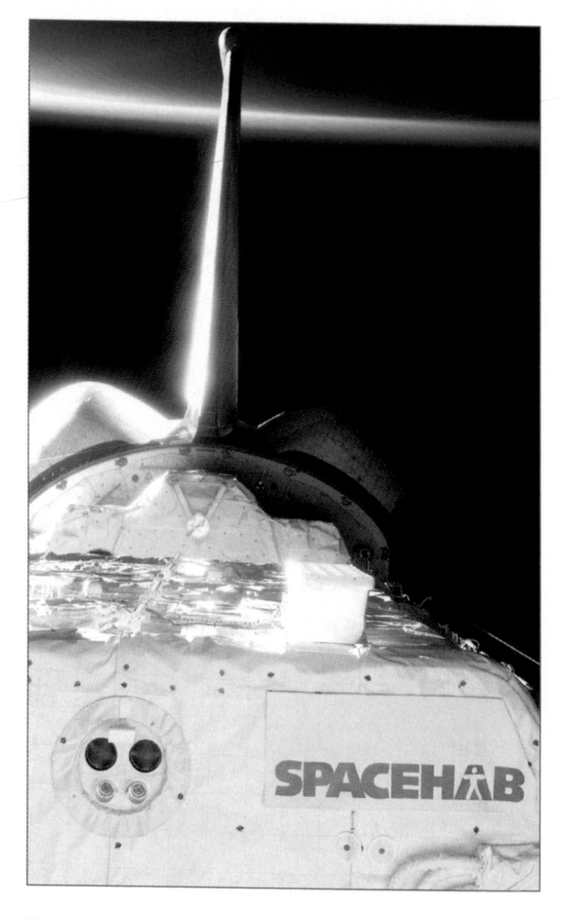

Figure 2.1-3. The SPACEHAB Research Double Module as seen from the aft flight deck windows of Columbia during STS-107. A thin slice of Earth's horizon is visible behind the vertical stabilizer.

Payload readiness reviews for STS-107 began in May 2002, with no significant abnormalities reported throughout the processing. The final Payload Safety Review Panel meeting prior to the mission was held on January 8, 2003, at the Kennedy Space Center, where the Integrated Safety Assessments conducted for the SPACEHAB and FREESTAR payloads were presented for final approval. All payload physical stresses on the Orbiter were reported within acceptable limits. The Extended Duration Orbiter pallet was loaded into the aft section of the payload bay in High Bay 3 of the Orbiter Processing Facility on April 25, 2002. The SPACEHAB

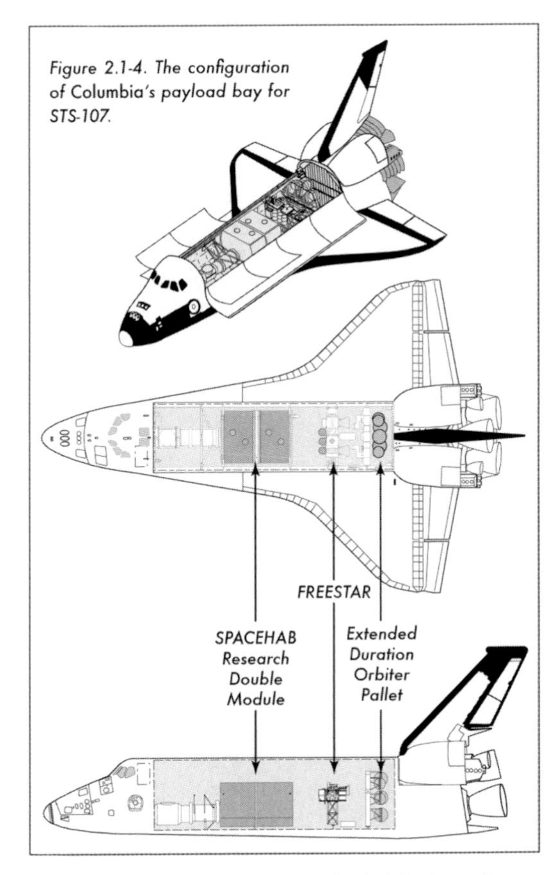

Figure 2.1-4. The configuration of Columbia's payload bay for STS-107.

and FREESTAR payloads were loaded horizontally on March 24, with an Integration Verification Test on June 6. The payload bay doors were closed on October 31 and were not opened prior to launch. (All late stow activities at the launch pad were accomplished in the vertical position using the normal crew entry hatch and SPACEHAB access tunnel.) Rollover of the Orbiter to the Vehicle Assembly Building for mating to the Solid Rocket Boosters and External Tank occurred on November 18. Mating took place two days later, and rollout to Launch Complex 39-A was on December 9.

Unprecedented security precautions were in place at Kennedy Space Center prior to and during the launch of STS-107 because of prevailing national security concerns and the inclusion of an Israeli crew member.

SPACEHAB was powered up at Launch minus 51 (L–51) hours (January 14) to prepare for the late stowing of time-critical experiments. The stowing of material in SPACEHAB once it was positioned vertically took place at L–46 hours and was completed by L–31 hours. Late middeck payload stowage, required for the experiments involving plants and insects, was performed at the launch pad. Flight crew equipment loading started at L–22.5 hours, while middeck experiment loading took place from Launch minus 19 to 16 hours. Fourteen experiments, four of which were powered, were loaded, all without incident.

2.2 FLIGHT PREPARATION

NASA senior management conducts a complex series of reviews and readiness polls to monitor a mission's progress toward flight readiness and eventual launch. Each step requires written certification. At the final review, called the Flight Readiness Review, NASA and its contractors certify that the necessary analyses, verification activities, and data products associated with the endorsement have been accomplished and "indicate a high probability for mission success." The review establishes the rationale for accepting any remaining identifiable risk; by signing the Certificate of Flight Readiness, NASA senior managers agree that they have accomplished all preliminary items and that they agree to accept that risk. The Launch Integration Manager oversees the flight preparation process.

STS-107 Flight Preparation Process

The flight preparation process reviews progress toward flight readiness at various junctures and ensures the organization is ready for the next operational phase. This process includes Project Milestone Reviews, three Program Milestone Reviews, and the Flight Readiness Review, where the Certification of Flight Readiness is endorsed.

The **Launch Readiness Review** is conducted within one month of the launch to certify that Certification of Launch Readiness items from NSTS-08117, Appendices H and Q, Flight Preparation Process Plan, have been reviewed and acted upon. The STS-107 Launch Readiness Review was held at Kennedy Space Center on December 18, 2002. The Kennedy Space Center Director of Shuttle Processing chaired the review and approved continued preparations for a January 16, 2003, launch. Onboard payload and experimental status and late stowage activity were reviewed.

A **Flight Readiness Review**, which is chaired by the Office of Space Flight Associate Administrator, usually occurs about two weeks before launch and provides senior NASA management with a summary of the certification and verification of the Space Shuttle vehicle, flight crew, payloads, and rationales for accepting residual risk. In cases where the Flight Preparation Process has not been successfully completed, Certification of Flight Readiness exceptions will be made, and presented at the Pre-Launch Mission Management Team Review for disposition. The final Flight Readiness Review for STS-107 was held on January 9, 2003, a week prior to launch. Representatives of all organizations except Flight Crew, Ferry Readiness, and Department of Defense Space Shuttle Support made presentations. Safety, Reliability & Quality Assurance summarized the work performed on the Ball Strut Tie Rod Assembly crack, defective booster connector pin, booster separation motor propellant paint chip contamination, and STS-113 Main Engine 1 nozzle leak (see Appendix E.1 for the briefing charts). None of the work performed on these items affected the launch.

Certificate of Flight Readiness: No actions were assigned during the Flight Readiness Review. One exception was included in the Certificate of Flight Readiness pending the completion of testing on the Ball Strut Tie Rod Assembly.

Testing was to be completed on January 15. This exception was to be closed with final flight rationale at the STS-107 Pre-launch Mission Management Team meeting. All principal managers and organizations indicated their readiness to support the mission.

Normally, a Mission Management Team – consisting of managers from Engineering, System Integration, the Space Flight Operations Contract Office, the Shuttle Safety Office, and the Johnson Space Center directors of flight crew operations, mission operations, and space and life sciences – convenes two days before launch and is maintained until the Orbiter safely lands. The Mission Management Team Chair reports directly to the Shuttle Program Manager.

The Mission Management Team resolves outstanding problems outside the responsibility or authority of the Launch and Flight Directors. During pre-launch, the Mission Management Team is chaired by the Launch Integration Manager at Kennedy Space Center, and during flight by the Space Shuttle Program Integration Manager at Johnson Space Center. The guiding document for Mission Management operations is NSTS 07700, Volume VIII.

A **Pre-launch Mission Management Team Meeting** occurs one or two days before launch to assess any open items or changes since the Flight Readiness Review, provide a GO/NO-GO decision on continuing the countdown, and approve changes to the Launch Commit Criteria. Simultaneously, the Mission Management Team is activated to evaluate the countdown and address any issues remaining from the Flight Readiness Review. STS-107's Pre-launch Mission Management Team meeting, chaired by the Acting Manager of Launch Integration, was held on January 14, some 48 hours prior to launch, at the Kennedy Space Center. In addition to the standard topics, such as weather and range support, the Pre-Launch Mission Management Team was updated on the status of the Ball Strut Tie Rod Assembly testing. The exception would remain open pending the presentation of additional test data at the Delta Pre-Launch Mission Management Team review the next day.

The **Delta Pre-Launch Mission Management Team Meeting** was also chaired by the Acting Manager of Launch Integration and met at 9:00 a.m. EST on January 15 at the Kennedy Space Center. The major issues addressed concerned the Ball Strut Tie Rod Assembly and potential strontium chromate contamination found during routine inspection of a (non-STS-107) spacesuit on January 14. The contamination concern was addressed and a toxicology analysis determined there was no risk to the STS-107 crew. A poll of the principal managers and organizations indicated all were ready to support STS-107.

A **Pre-Tanking Mission Management Team Meeting** was also chaired by the Acting Manager of Launch Integration. This meeting was held at 12:10 a.m. on January 16. A problem with the Solid Rocket Booster External Tank Attachment ring was addressed for the first time. Recent mission life capability testing of the material in the ring plates revealed static strength properties below minimum requirements. There were concerns that, assuming worst-case flight environments, the ring plate would not meet the safety factor requirement of 1.4 – that is, able to withstand 1.4 times the maximum load expected in operation. Based on analysis of the anticipated flight environment for STS-107, the need to meet the safety factor requirement of 1.4 was waived (see Chapter 10). No Launch Commit Criteria violations were noted, and the STS-107 final countdown began. The loading of propellants into the External Tank was delayed by some 70 minutes, until seven hours and 20 minutes before launch, due to an extended fuel cell calibration, a liquid oxygen replenish valve problem, and a Launch Processing System reconfiguration. The countdown continued normally, and at T–9 minutes the Launch Mission Management Team was polled for a GO/NO-GO launch decision. All members reported GO, and the Acting Manager of Launch Integration gave the final GO launch decision.

Once the Orbiter clears the launch pad, responsibility passes from the Launch Director at the Kennedy Space Center to the Flight Director at Johnson Space Center. During flight, the mission is also evaluated from an engineering perspective in the Mission Evaluation Room, which is managed by Vehicle Engineering Office personnel. Any engineering analysis conducted during a mission is coordinated through and first presented to the Mission Evaluation Room, and is then presented by the Mission Evaluation Room manager to the Mission Management Team.

NASA Times

Like most engineering or technical operations, NASA generally uses Coordinated Universal Time (UTC, formerly called Greenwich Mean Time) as the standard reference for activities. This is, for convenience, often converted to local time in either Florida or Texas – this report uses Eastern Standard Time (EST) unless otherwise noted. In addition to the normal 24-hour clock, NASA tells time via several other methods, all tied to specific events. The most recognizable of these is "T minus (T–)" time that counts down to every launch in hours, minutes, and seconds. NASA also uses a less precise "L minus" (L–) time that tags events that happens days or weeks prior to launch. Later in this report there are references to "Entry Interface plus (EI+)" time that counts, in seconds, from when an Orbiter begins reentry. In all cases, if the time is "minus" then the event being counted toward has not happened yet; if the time is "plus" then the event has already occurred.

2.3 Launch Sequence

The STS-107 launch countdown was scheduled to be about 24 hours longer than usual, primarily because of the extra time required to load cryogens for generating electricity and water into the Extended Duration Orbiter pallet, and for final stowage of plants, insects, and other unique science payloads. SPACEHAB stowage activities were about 90 minutes behind schedule, but the overall launch countdown was back on schedule when the communication system check was completed at L–24 hours.

At 7 hours and 20 minutes prior to the scheduled launch on January 16, 2003, ground crews began filling the External Tank with over 1,500,000 pounds of cryogenic propellants. At about 6:15 a.m., the Final Inspection Team began its visual and photographic check of the launch pad and vehicle. Frost had been noted during earlier inspections, but it had dissipated by 7:15 a.m., when the Ice Team completed its inspection.

Heavy rain had fallen on Kennedy Space Center while the Shuttle stack was on the pad. The launch-day weather was 65 degrees Fahrenheit with 68 percent relative humidity, dew point 59 degrees, calm winds, scattered clouds at 4,000 feet, and visibility of seven statute miles. The forecast weather for Kennedy Space Center and the Transoceanic Abort Landing sites in Spain and Morocco was within launch criteria limits.

At about 7:30 a.m. the crew was driven from their quarters in the Kennedy Space Center Industrial Area to Launch Complex 39-A. Commander Rick Husband was the first crew member to enter *Columbia*, at the 195-foot level of the launch tower at 7:53 a.m. Mission Specialist Kalpana Chawla was the last to enter, at 8:45 a.m. The hatch was closed and locked at 9:17 a.m.

The countdown clock executed the planned hold at the T–20 minute-mark at 10:10 a.m. The primary ascent computer software was switched over to the launch-ready configuration, communications checks were completed with all crew members, and all non-essential personnel were cleared from the launch area at 10:16 a.m. Fifteen minutes later the countdown clock came out of the planned hold at the T–9 minutes, and at 10:35 a.m., the GO was given for Auxiliary Power Unit start. STS-107 began at 10:39 a.m. with ignition of the Solid Rocket Boosters (see Figure 2.3-1).

Wind Shear

Before a launch, balloons are released to determine the direction and speed of the winds up to 50,000 to 60,000 feet. Various Doppler sounders are also used to get a wind profile, which, for STS-107, was unremarkable and relatively constant at the lower altitudes.

Columbia encountered a wind shear about 57 seconds after launch during the period of maximum dynamic pressure (max-q). As the Shuttle passed through 32,000 feet, it experienced a rapid change in the out-of-plane wind speed of minus 37.7 feet per second over a 1,200-foot altitude range. Immediately after the vehicle flew through this altitude range, its sideslip (beta) angle began to increase in the negative direction, reaching a value of minus 1.75 degrees at 60 seconds.

A negative beta angle means that the wind vector was on the left side of the vehicle, pushing the nose to the right and increasing the aerodynamic force on the External Tank bipod strut attachment. Several studies have indicated that the aerodynamic loads on the External Tank forward attach bipod, and also the interacting aerodynamic loads between the External Tank and the Orbiter, were larger than normal but within design limits.

Predicted and Actual I-Loads

On launch day, the General-Purpose Computers on the Orbiter are updated with information based on the latest observations of weather and the physical properties of the vehicle. These "I-loads" are initializing data sets that contain elements specific to each mission, such as measured winds, atmospheric data, and Shuttle configuration. The I-loads output target angle of attack, angle of sideslip, and dynamic pressure

Figure 2.3-1. The launch of Columbia on STS-107.

as a function of Mach number to ensure that the structural loads the Shuttle experiences during ascent are acceptable.

After the accident, investigators analyzed *Columbia*'s ascent loads using a reconstruction of the ascent trajectory. The wing loads measurement used a flexible body structural loads assessment that was validated by data from the Modular Auxiliary Data System recorder, which was recovered from the accident debris. The wing loads assessment included crosswind effects, angle of attack (alpha) effects, angle of sideslip (beta) effects, normal acceleration (g), and dynamic pressure (q) that could produce stresses and strains on the Orbiter's wings during ascent. This assessment showed that all Orbiter wing loads were approximately 70 percent of their design limit or less throughout the ascent, including the previously mentioned wind shear.

The wind shear at 57 seconds after launch and the Shuttle stack's reaction to it appears to have initiated a very low frequency oscillation, caused by liquid oxygen sloshing inside the External Tank,[4] that peaked in amplitude 75 seconds after launch and continued through Solid Rocket Booster separation at 127 seconds after launch. A small oscillation is not unusual during ascent, but on STS-107 the amplitude was larger than normal and lasted longer. Less severe wind shears at 95 and 105 seconds after launch contributed to the continuing oscillation.

An analysis of the External Tank/Orbiter interface loads, using simulated wind shear, crosswind, beta effects, and liquid oxygen slosh effects, showed that the loads on the External Tank forward attachment were only 70 percent of the design certification limit. The External Tank slosh study confirmed that the flight control system provided adequate stability throughout ascent.

The aerodynamic loads on the External Tank forward attach bipod were analyzed using a Computational Fluid Dynamics simulation, that yielded axial, side-force, and radial loads, and indicated that the external air loads were well below the design limit during the period of maximum dynamic pressure and also when the bipod foam separated.

Nozzle Deflections

Both Solid Rocket Boosters and each of the Space Shuttle Main Engines have exhaust nozzles that deflect ("gimbal") in response to flight control system commands. Review of the STS-107 ascent data revealed that the Solid Rocket Booster and Space Shuttle Main Engine nozzle positions twice exceeded deflections seen on previous flights by a factor of 1.24 to 1.33 and 1.06, respectively. The center and right main engine yaw deflections first exceeded those on previous flights during the period of maximum dynamic pressure, immediately following the wind shear. The deflections were the flight control system's reaction to the wind shear, and the motion of the nozzles was well within the design margins of the flight control system.

Approximately 115 seconds after launch, as booster thrust diminished, the Solid Rocket Booster and Space Shuttle Main Engine exhaust nozzle pitch and yaw deflections exceeded those seen previously by a factor of 1.4 and 1.06 to 1.6, respectively. These deflections were caused by lower than expected Reusable Solid Rocket Motor performance, indicated by a low burn rate; a thrust mismatch between the left and right boosters caused by lower-than-normal thrust on the right Solid Rocket Booster; a small built-in adjustment that favored the left Solid Rocket Booster pitch actuator; and flight control trim characteristics unique to the Performance Enhancements flight profile for STS-107.[5]

The Solid Rocket Booster burn rate is temperature-dependent, and behaved as predicted for the launch day weather conditions. No two boosters burn exactly the same, and a minor thrust mismatch has been experienced on almost every Space Shuttle mission. The booster thrust mismatch on STS-107 was well within the design margin of the flight control system.

Debris Strike

Post-launch photographic analysis showed that one large piece and at least two smaller pieces of insulating foam separated from the External Tank left bipod (–Y) ramp area at 81.7 seconds after launch. Later analysis showed that the larger piece struck *Columbia* on the underside of the left wing, around Reinforced Carbon-Carbon (RCC) panels 5 through 9, at 81.9 seconds after launch (see Figure 2.3-2). Further photographic analysis conducted the day after launch revealed that the large foam piece was approximately 21 to 27 inches long and 12 to 18 inches wide, tumbling at a minimum of 18 times per second, and moving at a relative velocity to the Shuttle Stack of 625 to 840 feet per second (416 to 573 miles per hour) at the time of impact.

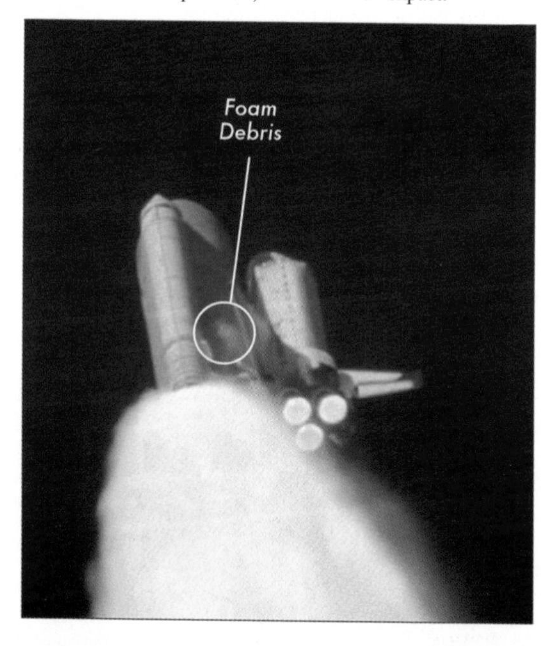

Figure 2.3-2. A shower of foam debris after the impact on Columbia's left wing. The event was not observed in real time.

Arrival on Orbit

Two minutes and seven seconds after launch, the Solid Rocket Boosters separated from the External Tank. They made a normal splashdown in the Atlantic Ocean and were subsequently recovered and returned to the Kennedy Space Center for inspection and refurbishment. Approximately eight and a half minutes after launch, the Space Shuttle Main Engines shut down normally, followed by the separation of the External Tank. At 11:20 a.m., a two-minute burn of the Orbital Maneuvering System engines began to position *Columbia* in its proper orbit, inclined 39 degrees to the equator and approximately 175 miles above Earth.

2.4 ON-ORBIT EVENTS

By 11:39 a.m. EST, one hour after launch, *Columbia* was in orbit and crew members entered the "post-insertion timeline." The crew immediately began to configure onboard systems for their 16-day stay in space.

Flight Day 1, Thursday, January 16

The payload bay doors were opened at 12:36 p.m. and the radiator was deployed for cooling. Crew members activated the Extended Duration Orbiter pallet (containing extra propellants for power and water production) and FREESTAR, and they began to set up the SPACEHAB module (see Figure 2.4-1). The crew then ran two experiments with the Advanced Respiratory Monitoring System stationary bicycle in SPACEHAB.

The crew also set up the Bioreactor Demonstration System, Space Technology and Research Students Bootes, Osteoporosis Experiment in Orbit, Closed Equilibrated Biological Aquatic System, Miniature Satellite Threat Reporting System, and Biopack, and performed Low Power Transceiver communication tests.

Flight Day 2, Friday, January 17

The Ozone Limb Sounding Experiment 2 began measuring the ozone layer, while the Mediterranean Israeli Dust Experiment (MEIDEX) was set to measure atmospheric aerosols over the Mediterranean Sea and the Sahara Desert. The Critical Viscosity of Xenon 2 experiment began studying the fluid properties of Xenon.

The crew activated the SPACEHAB Centralized Experiment Water Loop in preparation for the Combustion Module 2 and Vapor Compression Distillation Flight Experiment and also activated the Facility for Absorption and Surface Tension, Zeolite Crystal Growth, Astroculture, Mechanics of Granular Materials, Combined Two Phase Loop Experiment, European Research In Space and Terrestrial Osteoporosis, Biological Research in Canisters, centrifuge configurations, Enhanced Orbiter Refrigerator/Freezer Operations, and Microbial Physiological Flight Experiment.

Not known to Mission Control, the *Columbia* crew, or anyone else, between 10:30 and 11:00 a.m. on Flight Day 2, an object drifted away from the Orbiter. This object, which subsequent analysis suggests may have been related to the debris strike, had a departure velocity between 0.7 and 3.4 miles per hour, remained in a degraded orbit for approximately two and a half days, and re-entered the atmosphere between 8:45 and 11:45 p.m. on January 19. This object was discovered after the accident when Air Force Space Command reviewed its radar tracking data. (See Chapter 3 for additional discussion.)

Flight Day 3, Saturday, January 18

The crew conducted its first on-orbit press conference. Because of heavy cloud cover over the Middle East, MEIDEX objectives could not be accomplished. Crew members began an experiment to track metabolic changes in their calcium levels. The crew resolved a discrepancy in the SPACEHAB Video Switching Unit, provided body fluid samples for the Physiology and Biochemistry experiment, and activated the Vapor Compression Distillation Flight Experiment.

Figure 2.4-1. The tunnel linking the SPACEHAB module to the Columbia crew compartment provides a view of Kalpana Chawla working in SPACEHAB.

Flight Day 4, Sunday, January 19

Husband, Chawla, Clark, and Ramon completed the first experiments with the Combustion Module 2 in SPACEHAB, which were the Laminar Soot Processes, Water Mist Fire suppression, and Structure of Flame Balls at Low Lewis number. The latter studied combustion at the limits of flammability, producing the weakest flame ever to burn: each flame produced one watt of thermal power (a birthday-cake candle, by comparison, produces 50 watts).

Experiments on the human body's response to microgravity continued, with a focus on protein manufacturing, bone and calcium production, renal stone formation, and saliva and urine changes due to viruses. Brown captured the first ever images of upper-atmosphere "sprites" and "elves," which are produced by intense cloud-to-ground electromagnetic impulses radiated by heavy lightning discharges and are associated with storms near the Earth's surface.

The crew reported about a cup of water under the SPACEHAB module sub-floor and significant amounts clinging to the Water Separator Assembly and Aft Power Distribution Unit. The water was mopped up and Mission Control switched power from Rotary Separator 1 to 2.

Flight Day 5, Monday, January 20

Mission Control saw indications of an electrical short on Rotary Separator 2 in SPACEHAB; the separator was powered down and isolated from the electrical bus. To reduce condensation with both Rotary Separators off, the crew had to reduce the flow in one of *Columbia*'s Freon loops to SPACEHAB in order to keep the water temperature above the dew point and prevent condensation from forming in the Condensing Heat Exchanger. However, warmer water could lead to higher SPACEHAB cabin temperatures; fortunately, the crew was able to keep SPACEHAB temperatures acceptable and avoid condensation in the heat exchanger.

Flight Day 6, Tuesday, January 21

The temperature in the SPACEHAB module reached 81 degrees Fahrenheit. The crew reset the temperature to acceptable levels, and Mission Control developed a contingency plan to re-establish SPACEHAB humidity and temperature control if further degradation occurred. The Miniature Satellite Threat Reporting System, which detects ground-based radio frequency sources, experienced minor command and telemetry problems.

Flight Day 7, Wednesday, January 22

Both teams took a half day off. MEIDEX tracked thunderstorms over central Africa and captured images of four sprites and two elves as well as two rare images of meteoroids entering Earth's atmosphere. Payload experiments continued in SPACEHAB, with no further temperature complications.

Flight Day 8, Thursday, January 23

Eleven educational events were completed using the low-power transceiver to transfer data files to and from schools in Maryland and Massachusetts. The Mechanics of Granular Materials experiment completed the sixth of nine tests. Biopack shut down, and attempts to recycle the power were unsuccessful; ground teams began developing a repair plan.

Mission Control e-mailed Husband and McCool that post-launch photo analysis showed foam from the External Tank had struck the Orbiter's left wing during ascent. Mission Control relayed that there was "no concern for RCC or tile damage" and because the phenomenon had been seen before, there was "absolutely no concern for entry." Mission Control also e-mailed a short video clip of the debris strike, which Husband forwarded to the rest of the crew.

Flight Day 9, Friday, January 24

Crew members conducted the mission's longest combustion test. Spiral moss growth experiments continued, as well as Astroculture experiments that harvested samples of oils from roses and rice flowers. Experiments in the combustion chamber continued. Although the temperature in SPACEHAB was maintained, Mission Control estimated that about a half-gallon of water was unaccounted for, and began planning in-flight maintenance for the Water Separator Assembly.

David Brown stabilizes a digital video camera prior to a press conference in the SPACEHAB Research Double Module aboard Columbia during STS-107.

Flight Day 10, Saturday, January 25

Experiments with bone cells, prostate cancer, bacteria growth, thermal heating, and surface tension continued. MEIDEX captured images of plumes of dust off the coasts of Nigeria, Mauritania, and Mali. Images of sprites were captured over storms in Perth, Australia. Biopack power could not be restored, so all subsequent Biopack sampling was performed at ambient temperatures.

Flight Day 11, Sunday, January 26

Vapor Compression Distillation Flight Experiment operations were complete; SPACEHAB temperature was allowed to drop to 73 degrees Fahrenheit. Scientists received the first live Xybion digital downlink images from MEIDEX and confirmed significant dust in the Middle East. The STARS experiment hatched a fish in the aquatic habitat and a silk moth from its cocoon.

Flight Day 12, Monday, January 27

Combustion and granular materials experiments concluded. The combustion module was configured for the Water Mist experiment, which developed a leak. The Microbial Physiology Flight Experiment expended its final set of samples in yeast and bacteria growth. The crew made a joint observation using MEIDEX and the Ozone Limb Sounding Experiment. MEIDEX captured images of dust over the Atlantic Ocean for the first time.

Flight Day 13, Tuesday, January 28

The crew took another half day off. The Bioreactor experiment produced a bone and prostate cancer tumor tissue sample the size of a golf ball, the largest ever grown in space. The crew, along with ground support personnel, observed a moment of silence to honor the memory of the men and women of *Apollo 1* and *Challenger*. MEIDEX was prepared to monitor smoke trails from research aircraft and bonfires in Brazil. Water Mist runs began after the leak was stopped.

Flight Day 14, Wednesday, January 29

Ramon reported a giant dust storm over the Atlantic Ocean that provided three days of MEIDEX observations. Ground teams confirmed predicted weather and climate effects and found a huge smoke plume in a large cumulus cloud over the Amazon jungle. BIOTUBE experiment ground teams reported growth rates and root curvatures in plant and flax roots different from anything seen in normal gravity on Earth. The crew received procedures from Mission Control for vacuum cleanup and taping of the Water Separator Assembly prior to re-entry. Temperatures in two Biopack culture chambers were too high for normal cell growth, so several Biopack experiments were terminated.

Flight Day 15, Thursday, January 30

Final samples and readings were taken for the Physiology and Biochemistry team experiments. Husband, McCool, and Chawla ran landing simulations on the computer training system. Husband found no excess water in the SPACEHAB sub-floor, but as a precaution, he covered several holes in the Water Separator Assembly.

Flight Day 16, Friday, January 31

The Water Mist Experiment concluded and the combustion module was closed. MEIDEX made final observations of dust concentrations, sprites, and elves. Husband, McCool, and Chawla completed their second computer-based landing simulation. A flight control system checkout was performed satisfactorily using Auxiliary Power Unit 1, with a run time of 5 minutes, 27 seconds.

After the flight control system checkout, a Reaction Control System "hot-fire" was performed during which all thrusters were fired for at least 240 milliseconds. The Ku-band antenna and the radiator on the left payload bay door were stowed.

Flight Day 17, Saturday, February 1

All onboard experiments were concluded and stowed, and payload doors and covers were closed. Preparations were completed for de-orbit, re-entry, and landing at the Kennedy Space Center. Suit checks confirmed that proper pressure would be maintained during re-entry and landing. The payload bay doors were closed. Husband and McCool configured the onboard computers with the re-entry software, and placed *Columbia* in the proper attitude for the de-orbit burn.

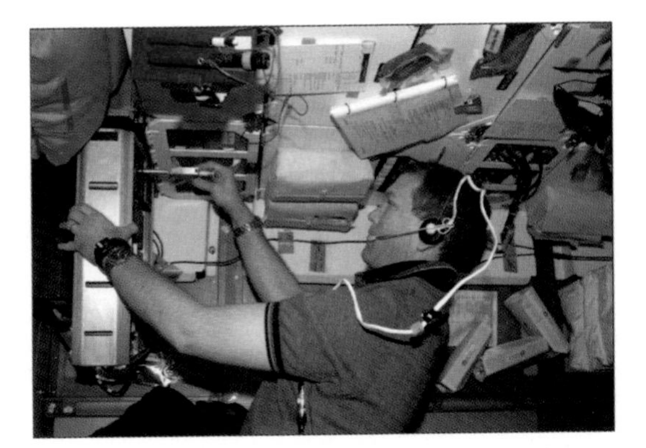

Rick Husband works with the Biological Research in Canister experiment on Columbia's mid-deck.

2.5 Debris Strike Analysis and Requests for Imagery

As is done after every launch, within two hours of the lift-off the Intercenter Photo Working Group examined video from tracking cameras. An initial review did not reveal any unusual events. The next day, when the Intercenter Photo Working Group personnel received much higher resolution film that had been processed overnight, they noticed a debris strike at 81.9 seconds after launch.

A large object from the left bipod area of the External Tank struck the Orbiter, apparently impacting the underside of the left wing near RCC panels 5 through 9. The object's large size and the apparent momentum transfer concerned Intercenter Photo Working Group personnel, who were worried that *Columbia* had sustained damage not detectable in the limited number of views their tracking cameras captured. This concern led the Intercenter Photo Working Group Chair to request, in anticipation of analysts' needs, that a high-resolution image of the Orbiter on-orbit be obtained by the Department of Defense. By the Board's count, this would be the first of three distinct requests to image *Columbia* on-orbit. The exact chain of events and circumstances surrounding the movement of each of these requests through Shuttle Program Management, as well as the ultimate denial of these requests, is a topic of Chapter 6.

After discovering the strike, the Intercenter Photo Working Group prepared a report with a video clip of the impact and sent it to the Mission Management Team, the Mission Evaluation Room, and engineers at United Space Alliance and Boeing. In accordance with NASA guidelines, these contractor and NASA engineers began an assessment of potential impact damage to *Columbia*'s left wing, and soon formed a Debris Assessment Team to conduct a formal review.

The first formal Debris Assessment Team meeting was held on January 21, five days into the mission. It ended with the highest-ranking NASA engineer on the team agreeing to bring the team's request for imaging of the wing on-orbit, which would provide better information on which to base their analysis, to the Johnson Space Center Engineering Management Directorate, with the expectation the request would go forward to Space Shuttle Program managers. Debris Assessment Team members subsequently learned that these managers declined to image *Columbia*.

Without on-orbit pictures of *Columbia*, the Debris Assessment Team was restricted to using a mathematical modeling tool called Crater to assess damage, although it had not been designed with this type of impact in mind. Team members concluded over the next six days that some localized heating damage would most likely occur during re-entry, but they could not definitively state that structural damage would result. On January 24, the Debris Assessment Team made a presentation of these results to the Mission Evaluation Room, whose manager gave a verbal summary (with no data) of that presentation to the Mission Management Team the same day. The Mission Management Team declared the debris strike a "turnaround" issue and did not pursue a request for imagery.

Even after the Debris Assessment Team's conclusion had been reported to the Mission Management Team, engineers throughout NASA and Mission Control continued to exchange e-mails and discuss possible damage. These messages and discussions were generally sent only to people within the senders' area of expertise and level of seniority.

William McCool talks to Mission Control from the aft flight deck of Columbia during STS-107.

2.6 De-Orbit Burn and Re-Entry Events

At 2:30 a.m. EST on February 1, 2003, the Entry Flight Control Team began duty in the Mission Control Center. The Flight Control Team was not working any issues or problems related to the planned de-orbit and re-entry of *Columbia*. In particular, the team indicated no concerns about the debris impact to the left wing during ascent, and treated the re-entry like any other.

The team worked through the de-orbit preparation checklist and re-entry checklist procedures. Weather forecasters, with the help of pilots in the Shuttle Training Aircraft, evaluated landing site weather conditions at the Kennedy Space Center. At the time of the de-orbit decision, about 20 minutes before the initiation of the de-orbit burn, all weather observations and forecasts were within guidelines set by the flight rules, and all systems were normal.

Shortly after 8:00 a.m., the Mission Control Center Entry Flight Director polled the Mission Control room for a GO/NO-GO decision for the de-orbit burn, and at 8:10 a.m., the Capsule Communicator notified the crew they were GO for de-orbit burn.

As the Orbiter flew upside down and tail-first over the Indian Ocean at an altitude of 175 statute miles, Commander Husband and Pilot McCool executed the de-orbit burn at 8:15:30 a.m. using *Columbia*'s two Orbital Maneuvering System engines. The de-orbit maneuver was performed on the 255th orbit, and the 2-minute, 38-second burn slowed the Orbiter from 17,500 mph to begin its re-entry into the atmosphere. During the de-orbit burn, the crew felt about 10 percent of the effects of gravity. There were no problems during the burn, after which Husband maneuvered *Columbia* into a right-side-up, forward-facing position, with the Orbiter's nose pitched up.

Entry Interface, arbitrarily defined as the point at which the Orbiter enters the discernible atmosphere at 400,000 feet, occurred at 8:44:09 a.m. (Entry Interface plus 000 seconds, written EI+000) over the Pacific Ocean. As *Columbia* descended from space into the atmosphere, the heat produced by air molecules colliding with the Orbiter typically caused wing leading-edge temperatures to rise steadily, reaching an estimated 2,500 degrees Fahrenheit during the next six minutes. As superheated air molecules discharged light, astronauts on the flight deck saw bright flashes envelop the Orbiter, a normal phenomenon.

At 8:48:39 a.m. (EI+270), a sensor on the left wing leading edge spar showed strains higher than those seen on previous *Columbia* re-entries. This was recorded only on the Modular Auxiliary Data System, and was not telemetered to ground controllers or displayed to the crew (see Figure 2.6-1).

At 8:49:32 a.m. (EI+323), traveling at approximately Mach 24.5, *Columbia* executed a roll to the right, beginning a preplanned banking turn to manage lift, and therefore limit the Orbiter's rate of descent and heating.

At 8:50:53 a.m. (EI+404), traveling at Mach 24.1 and at approximately 243,000 feet, *Columbia* entered a 10-minute period of peak heating, during which the thermal stresses were at their maximum. By 8:52:00 a.m. (EI+471), nearly eight minutes after entering the atmosphere and some 300 miles west of the California coastline, the wing leading-edge temperatures usually reached 2,650 degrees Fahrenheit. *Columbia* crossed the California coast west of Sacramento at 8:53:26 a.m. (EI+557). Traveling at Mach 23 and 231,600 feet, the Orbiter's wing leading edge typically reached more than an estimated 2,800 degrees Fahrenheit.

Columbia streaking over the Owens Valley Radio Observatory in Big Pine, California.

Now crossing California, the Orbiter appeared to observers on the ground as a bright spot of light moving rapidly across the sky. Signs of debris being shed were sighted at 8:53:46 a.m. (EI+577), when the superheated air surrounding the Orbiter suddenly brightened, causing a noticeable streak in the Orbiter's luminescent trail. Observers witnessed another four similar events during the following 23 seconds, and a bright flash just seconds after *Columbia* crossed from California into Nevada airspace at 8:54:25 a.m. (EI+614), when the Orbiter was traveling at Mach 22.5 and 227,400 feet. Witnesses observed another 18 similar events in the next four minutes as *Columbia* streaked over Utah, Arizona, New Mexico, and Texas.

In Mission Control, re-entry appeared normal until 8:54:24 a.m. (EI+613), when the Maintenance, Mechanical, and Crew Systems (MMACS) officer informed the Flight Director that four hydraulic sensors in the left wing were indicating "off-scale low," a reading that falls below the minimum capability of the sensor. As the seconds passed, the Entry Team continued to discuss the four failed indicators.

At 8:55:00 a.m. (EI+651), nearly 11 minutes after *Columbia* had re-entered the atmosphere, wing leading edge temperatures normally reached nearly 3,000 degrees Fahrenheit. At 8:55:32 a.m. (EI+683), *Columbia* crossed from Nevada into Utah while traveling at Mach 21.8 and 223,400 ft. Twenty seconds later, the Orbiter crossed from Utah into Arizona.

At 8:56:30 a.m. (EI+741), *Columbia* initiated a roll reversal, turning from right to left over Arizona. Traveling at Mach 20.9 and 219,000 feet, *Columbia* crossed the Arizona-New Mexico state line at 8:56:45 (EI+756), and passed just north of Albuquerque at 8:57:24 (EI+795).

Around 8:58:00 a.m. (EI+831), wing leading edge temperatures typically decreased to 2,880 degrees Fahrenheit. At 8:58:20 a.m. (EI+851), traveling at 209,800 feet and Mach 19.5, *Columbia* crossed from New Mexico into Texas, and about this time shed a Thermal Protection System tile, which was the most westerly piece of debris that has been recovered. Searchers found the tile in a field in Littlefield, Texas, just northwest of Lubbock. At 8:59:15 a.m. (EI+906), MMACS informed the Flight Director that pressure readings had been lost on both left main landing gear tires. The Flight Director then told the Capsule Communicator (CAPCOM) to let the crew know that Mission Control saw the messages and was evaluating the indications, and added that the Flight Control Team did not understand the crew's last transmission.

At 8:59:32 a.m. (EI+923), a broken response from the mission commander was recorded: "Roger, [cut off in mid-word] …" It was the last communication from the crew and the last telemetry signal received in Mission Control. Videos made by observers on the ground at 9:00:18 a.m. (EI+969) revealed that the Orbiter was disintegrating.

2.7 EVENTS IMMEDIATELY FOLLOWING THE ACCIDENT

A series of events occurred immediately after the accident that would set the stage for the subsequent investigation.

NASA Emergency Response

Shortly after the scheduled landing time of 9:16 a.m. EST, NASA declared a "Shuttle Contingency" and executed the Contingency Action Plan that had been established after the *Challenger* accident. As part of that plan, NASA Administrator Sean O'Keefe activated the International Space Station and Space Shuttle Mishap Interagency Investigation Board at 10:30 a.m. and named Admiral Harold W. Gehman Jr., U.S. Navy, retired, as its chair.

Senior members of the NASA leadership met as part of the Headquarters Contingency Action Team and quickly notified astronaut families, the President, and members of Congress. President Bush telephoned Israeli Prime Minster Ariel Sharon to inform him of the loss of *Columbia* crew member Ilan Ramon, Israel's first astronaut. Several hours later, President Bush addressed the nation, saying, "The *Columbia* is lost. There are no survivors."

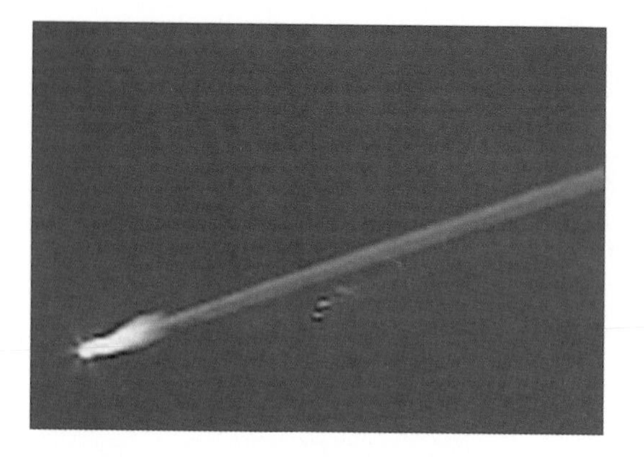

The Orbiter has a large glowing field surrounding it in this view taken from Mesquite, Texas, looking south.

Taken at the same time as the photo at left, but from Hewitt, Texas, looking north.

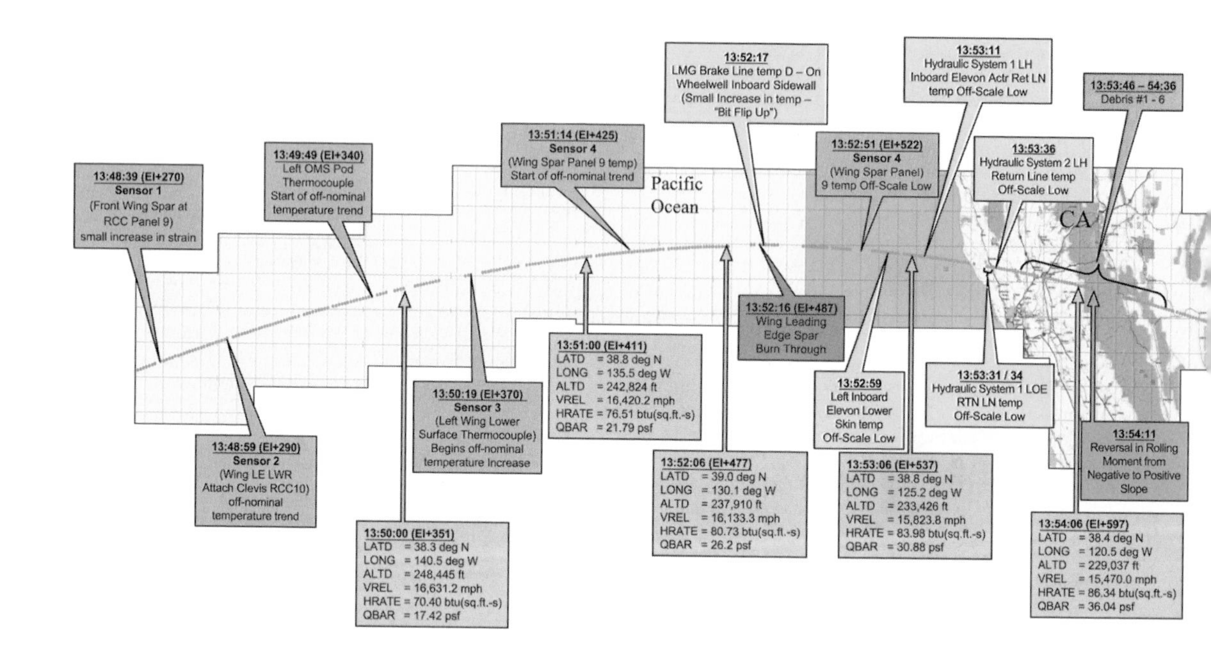

Figure 2.6-1. This simplified timeline shows the re-entry path of Columbia on February 1, 2003. The information presented here is a composite of sensor data telemetered to the ground combined with data from the Modular Auxiliary Data System recorder recovered after the accident. Note that the first off-nominal reading was a small increase in a strain gauge at the front wing spar behind RCC panel 9-left. The chart is color-coded: blue boxes contain position, attitude, and velocity information; orange boxes indicate when debris was shed from the Orbiter; green boxes are significant aerodynamic control events; gray boxes contain sensor information from the Modular Auxiliary Data System; and yellow boxes contain telemetered sensor information. The red boxes indicate other significant events.

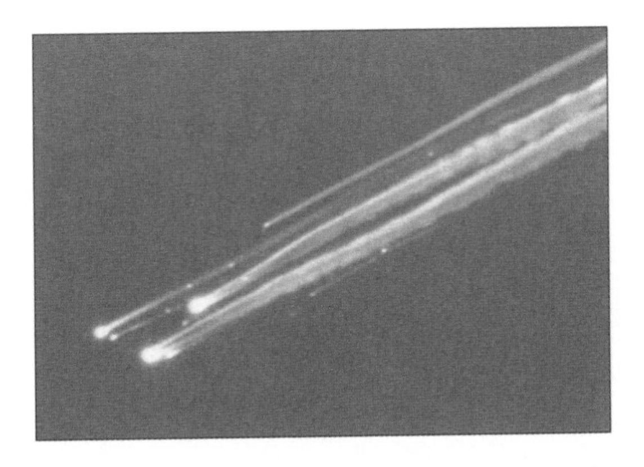

This view was taken from Dallas. (Robert McCullough/© 2003 The Dallas Morning News)

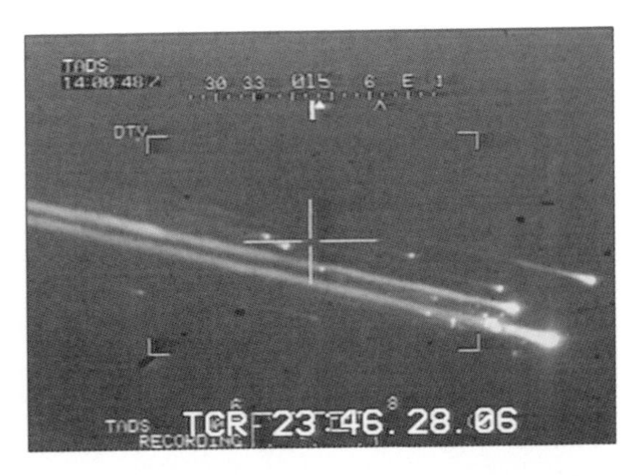

This video was captured by a Danish crew operating an AH-64 Apache helicopter near Fort Hood, Texas.

STS-107 Re-entry Trajectory and Timeline
(First Off-Nominal Event to Loss of Signal)

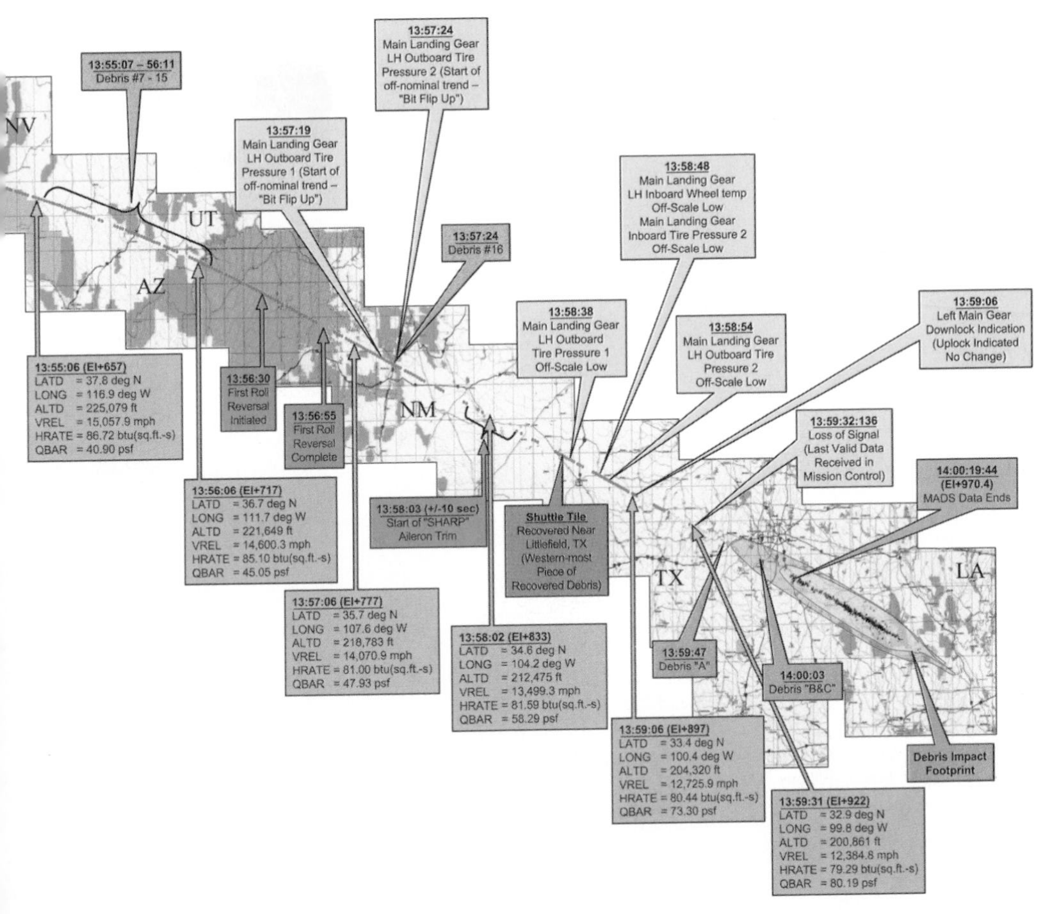

Mission Control Center Communications

At 8:49 a.m. Eastern Standard Time (EI+289), the Orbiter's flight control system began steering a precise course, or drag profile, with the initial roll command occurring about 30 seconds later. At 8:49:38 a.m., the Mission Control Guidance and Procedures officer called the Flight Director and indicated that the "closed-loop" guidance system had been initiated.

The Maintenance, Mechanical, and Crew Systems (MMACS) officer and the Flight Director (Flight) had the following exchange beginning at 8:54:24 a.m. (EI+613).

MMACS: "Flight – MMACS."
Flight: "Go ahead, MMACS."
MMACS: "FYI, I've just lost four separate temperature transducers on the left side of the vehicle, hydraulic return temperatures. Two of them on system one and one in each of systems two and three."
Flight: "Four hyd [hydraulic] return temps?"
MMACS: "To the left outboard and left inboard elevon."
Flight: "Okay, is there anything common to them? DSC [discrete signal conditioner] or MDM [multiplexer-demultiplexer] or anything? I mean, you're telling me you lost them all at exactly the same time?"
MMACS: "No, not exactly. They were within probably four or five seconds of each other."
Flight: "Okay, where are those, where is that instrumentation located?"
MMACS: "All four of them are located in the aft part of the left wing, right in front of the elevons, elevon actuators. And there is no commonality."
Flight: "No commonality."

At 8:56:02 a.m. (EI+713), the conversation between the Flight Director and the MMACS officer continues:

Flight: "MMACS, tell me again which systems they're for."
MMACS: "That's all three hydraulic systems. It's ... two of them are to the left outboard elevon and two of them to the left inboard."
Flight: "Okay, I got you."

The Flight Director then continues to discuss indications with other Mission Control Center personnel, including the Guidance, Navigation, and Control officer (GNC).

Flight: "GNC – Flight."
GNC: "Flight – GNC."
Flight: "Everything look good to you, control and rates and everything is nominal, right?"
GNC: "Control's been stable through the rolls that we've done so far, flight. We have good trims. I don't see anything out of the ordinary."
Flight: "Okay. And MMACS, Flight?"
MMACS: "Flight – MMACS."
Flight: "All other indications for your hydraulic system indications are good."
MMACS: "They're all good. We've had good quantities all the way across."
Flight: "And the other temps are normal?"
MMACS: "The other temps are normal, yes sir."
Flight: "And when you say you lost these, are you saying that they went to zero?" [Time: 8:57:59 a.m., EI+830] "Or, off-scale low?"
MMACS: "All four of them are off-scale low. And they were all staggered. They were, like I said, within several seconds of each other."
Flight: "Okay."

At 8:58:00 a.m. (EI+831), Columbia *crossed the New Mexico-Texas state line. Within the minute, a broken call came on the air-to-ground voice loop from* Columbia*'s commander, "And, uh, Hou ..." This was followed by a call from MMACS about failed tire pressure sensors at 8:59:15 a.m. (EI+906).*

MMACS: "Flight – MMACS."
Flight: "Go."
MMACS: "We just lost tire pressure on the left outboard and left inboard, both tires."

[continued on next page]

The Flight Director then told the Capsule Communicator (CAPCOM) to let the crew know that Mission Control saw the messages and that the Flight Control Team was evaluating the indications and did not copy their last transmission.

CAPCOM: "And *Columbia*, Houston, we see your tire pressure messages and we did not copy your last call."
Flight: "Is it instrumentation, MMACS? Gotta be ..."
MMACS: "Flight – MMACS, those are also off-scale low."

At 8:59:32 a.m. (EI+923), Columbia *was approaching Dallas, Texas, at 200,700 feet and Mach 18.1. At the same time, another broken call, the final call from* Columbia*'s commander, came on the air-to-ground voice loop:*

Commander: "Roger, [cut off in mid-word] ..."

This call may have been about the backup flight system tire pressure fault-summary messages annunciated to the crew onboard, and seen in the telemetry by Mission Control personnel. An extended loss of signal began at 08:59:32.136 a.m. (EI+923). This was the last valid data accepted by the Mission Control computer stream, and no further real-time data updates occurred in Mission Control. This coincided with the approximate time when the Flight Control Team would expect a short-duration loss of signal during antenna switching, as the onboard communication system automatically reconfigured from the west Tracking and Data Relay System satellite to either the east satellite or to the ground station at Kennedy Space Center. The following exchange then took place on the Flight Director loop with the Instrumentation and Communication Office (INCO):

INCO: "Flight – INCO."
Flight: "Go."
INCO: "Just taking a few hits here. We're right up on top of the tail. Not too bad."

The Flight Director then resumes discussion with the MMACS officer at 9:00:18 a.m. (EI+969).

Flight: "MMACS – Flight."
MMACS: "Flight – MMACS."
Flight: "And there's no commonality between all these tire pressure instrumentations and the hydraulic return instrumentations."
MMACS: "No sir, there's not. We've also lost the nose gear down talkback and the right main gear down talkback."
Flight: "Nose gear and right main gear down talkbacks?"
MMACS: "Yes sir."

At 9:00:18 a.m. (EI+969), the postflight video and imagery analyses indicate that a catastrophic event occurred. Bright flashes suddenly enveloped the Orbiter, followed by a dramatic change in the trail of superheated air. This is considered the most likely time of the main breakup of Columbia*. Because the loss of signal had occurred 46 seconds earlier, Mission Control had no insight into this event. Mission Control continued to work the loss-of-signal problem to regain communication with* Columbia*:*

INCO: "Flight – INCO, I didn't expect, uh, this bad of a hit on comm [communications]."
Flight: "GC [Ground Control officer] how far are we from UHF? Is that two-minute clock good?"
GC: "Affirmative, Flight."
GNC: "Flight – GNC."
Flight: "Go."
GNC: "If we have any reason to suspect any sort of controllability issue, I would keep the control cards handy on page 4-dash-13."
Flight: "Copy."

At 9:02:21 a.m. (EI+1092, or 18 minutes-plus), the Mission Control Center commentator reported, "Fourteen minutes to touchdown for Columbia *at the Kennedy Space Center. Flight controllers are continuing to stand by to regain communications with the spacecraft."*

Flight: "INCO, we were rolled left last data we had and you were expecting a little bit of ratty comm [communications], but not this long?"
INCO: "That's correct, Flight. I expected it to be a little intermittent. And this is pretty solid right here."
Flight: "No onboard system config [configuration] changes right before we lost data?"
INCO: "That is correct, Flight. All looked good."
Flight: "Still on string two and everything looked good?"
INCO: "String two looking good."

The Ground Control officer then told the Flight Director that the Orbiter was within two minutes of acquiring the Kennedy Space Center ground station for communications, "Two minutes to MILA." The Flight Director told the CAPCOM to try another communications check with Columbia*, including one on the UHF system (via MILA, the Kennedy Space Center tracking station):*

CAPCOM: "*Columbia*, Houston, comm [communications] check."
CAPCOM: "*Columbia*, Houston, UHF comm [communications] check."

At 9:03:45 a.m. (EI+1176, or 19 minutes-plus), the Mission Control Center commentator reported, "CAPCOM Charlie Hobaugh calling Columbia *on a UHF frequency as it approaches the Merritt Island (MILA) tracking station in Florida. Twelve-and-a-half minutes to touchdown, according to clocks in Mission Control."*

MMACS: "Flight – MMACS."
Flight: "MMACS?"
MMACS: "On the tire pressures, we did see them go erratic for a little bit before they went away, so I do believe it's instrumentation."
Flight: "Okay."

The Flight Control Team still had no indications of any serious problems onboard the Orbiter. In Mission Control, there was no way to know the exact cause of the failed sensor measurements, and while there was concern for the extended loss of signal, the recourse was to continue to try to regain communications and in the meantime determine if the other systems, based on the last valid data, continued to appear as expected. The Flight Director told the CAPCOM to continue to try to raise Columbia *via UHF:*

CAPCOM: "*Columbia*, Houston, UHF comm [communications] check."
CAPCOM: "*Columbia*, Houston, UHF comm [communications] check."
GC: "Flight – GC."
Flight: "Go."
GC: "MILA not reporting any RF [radio frequency] at this time."

[continued on next page]

[continued from previous page]

INCO: "Flight – INCO, SPC [stored program command] just should have taken us to STDN low." *[STDN is the Space Tracking and Data Network, or ground station communication mode]*
Flight: "Okay."
Flight: "FDO, when are you expecting tracking? " *[FDO is the Flight Dynamics Officer in the Mission Control Center]*
FDO: "One minute ago, Flight."
GC: "And Flight – GC, no C-band yet."
Flight: "Copy."
CAPCOM: "*Columbia*, Houston, UHF comm [communications] check."
INCO: "Flight – INCO."
Flight: "Go."
INCO: "I could swap strings in the blind."
Flight: "Okay, command us over."
INCO: "In work, Flight."

At 09:08:25 a.m. (EI+1456, or 24 minutes-plus), the Instrumentation and Communications Officer reported, "Flight – INCO, I've commanded string one in the blind," which indicated that the officer had executed a command sequence to Columbia *to force the onboard S-band communications system to the backup string of avionics to try to regain communication, per the Flight Director's direction in the previous call.*

GC: "And Flight – GC."
Flight: "Go."
GC: "MILA's taking one of their antennas off into a search mode [to try to find *Columbia*]."
Flight: "Copy. FDO – Flight?"
FDO: "Go ahead, Flight."
Flight: "Did we get, have we gotten any tracking data?"
FDO: "We got a blip of tracking data, it was a bad data point, Flight. We do not believe that was the Orbiter *[referring to an errant blip on the large front screen in the Mission Control, where Orbiter tracking data is displayed.]* We're entering a search pattern with our C-bands at this time. We do not have any valid data at this time."

By this time, 9:09:29 a.m. (EI+1520), Columbia*'s speed would have dropped to Mach 2.5 for a standard approach to the Kennedy Space Center.*

Flight: "OK. Any other trackers that we can go to?"
FDO: "Let me start talking, Flight, to my navigator."

At 9:12:39 a.m. (E+1710, or 28 minutes-plus), Columbia *should have been banking on the heading alignment cone to line up on Runway 33. At about this time, a member of the Mission Control team received a call on his cell phone from someone who had just seen live television coverage of* Columbia *breaking up during re-entry. The Mission Control team member walked to the Flight Director's console and told him the Orbiter had disintegrated.*

Flight: "GC, – Flight. GC – Flight?"
GC: "Flight – GC."
Flight: "Lock the doors."

Having confirmed the loss of Columbia, *the Entry Flight Director directed the Flight Control Team to begin contingency procedures.*

In order to preserve all material relating to STS-107 as evidence for the accident investigation, NASA officials impounded data, software, hardware, and facilities at NASA and contractor sites in accordance with the pre-existing mishap response plan.

At the Johnson Space Center, the door to Mission Control was locked while personnel at the flight control consoles archived all original mission data. At the Kennedy Space Center, mission facilities and related hardware, including Launch Complex 39-A, were put under guard or stored in secure warehouses. Officials took similar actions at other key Shuttle facilities, including the Marshall Space Flight Center and the Michoud Assembly Facility.

Within minutes of the accident, the NASA Mishap Investigation Team was activated to coordinate debris recovery efforts with local, state, and federal agencies. The team initially operated out of Barksdale Air Force Base in Louisiana and soon after in Lufkin, Texas, and Carswell Field in Fort Worth, Texas.

Debris Search and Recovery

On the morning of February 1, a crackling boom that signaled the breakup of *Columbia* startled residents of East Texas. The long, low-pitched rumble heard just before 8:00 a.m. Central Standard Time (CST) was generated by pieces of debris streaking into the upper atmosphere at nearly 12,000 miles per hour. Within minutes, that debris fell to the ground. Cattle stampeded in Eastern Nacogdoches County. A fisherman on Toledo Bend reservoir saw a piece splash down in the water, while a women driving near Lufkin almost lost control of her car when debris smacked her windshield. As 911 dispatchers across Texas were flooded with calls reporting sonic booms and smoking debris, emergency personnel soon realized that residents were encountering the remnants of the Orbiter that NASA had reported missing minutes before.

The emergency response that began shortly after 8:00 a.m. CST Saturday morning grew into a massive effort to decontaminate and recover debris strewn over an area that in Texas alone exceeded 2,000 square miles (see Figure 2.7-1). Local fire and police departments called in all personnel, who began responding to debris reports that by late afternoon were phoned in at a rate of 18 per minute.

Within hours of the accident, President Bush declared East Texas a federal disaster area, enabling the dispatch of emergency response teams from the Federal Emergency Management Agency and Environmental Protection Agency. As the day wore on, county constables, volunteers on horseback, and local citizens headed into pine forests and bushy thickets in search of debris and crew remains, while National Guard units mobilized to assist local law-enforcement guard debris sites. Researchers from Stephen F. Austin University sent seven teams into the field with Global Positioning System units to mark the exact location of debris. The researchers and later searchers then used this data to update debris distribution on detailed Geographic Information System maps.

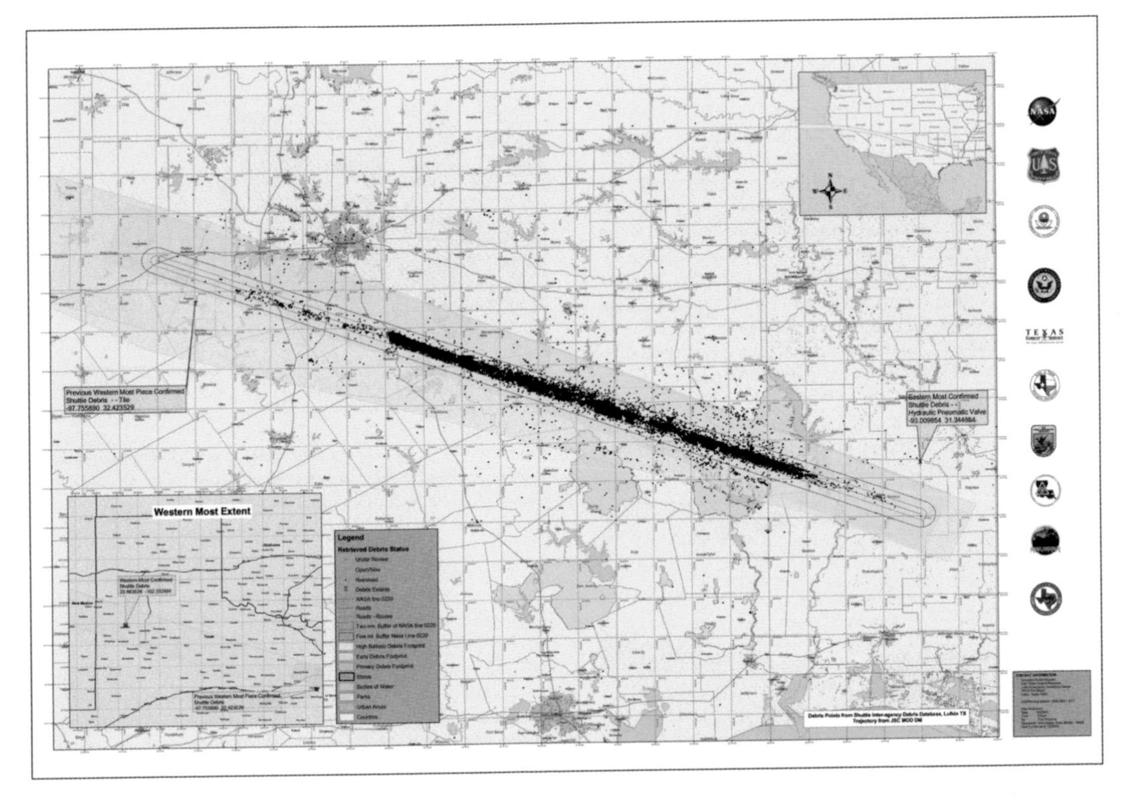

Figure 2.7-1. The debris field in East Texas spread over 2,000 square miles, and eventually over 700,000 acres were searched.

Public Safety Concerns

From the start, NASA officials sought to make the public aware of the hazards posed by certain pieces of debris, as well as the importance of turning over all debris to the authorities. *Columbia* carried highly toxic propellants that maneuvered the Orbiter in space and during early stages of re-entry. These propellants and other gases and liquids were stored in pressurized tanks and cylinders that posed a danger to people who might approach Orbiter debris. The propellants, monomethyl hydrazine and nitrogen tetroxide, as well as concentrated ammonia used in the Orbiter's cooling systems, can severely burn the lungs and exposed skin when encountered in vapor form. Other materials used in the Orbiter, such as beryllium, are also toxic. The Orbiter also contains various pyrotechnic devices that eject or release items such as the Ku-Band antenna, landing gear doors, and hatches in an emergency. These pyrotechnic devices and their triggers, which are designed to withstand high heat and therefore may have survived re-entry, posed a danger to people and livestock. They had to be removed by personnel trained in ordnance disposal.

In light of these and other hazards, NASA officials worked with local media and law enforcement to ensure that no one on the ground would be injured. To determine that Orbiter debris did not threaten air quality or drinking water, the Environmental Protection Agency activated Emergency Response and Removal Service contractors, who surveyed the area.

Land Search

The tremendous efforts mounted by the National Guard, Texas Department of Public Safety, and emergency personnel from local towns and communities were soon overwhelmed by the expanding bounds of the debris field, the densest region of which ran from just south of Fort Worth, Texas, to Fort Polk, Louisiana. Faced with a debris field several orders of magnitude larger than any previous accident site, NASA and Federal Emergency Management Agency officials activated Forest Service wildland firefighters to serve as the primary search teams. As NASA identified the areas to be searched, personnel and equipment were furnished by the Forest Service.

Within two weeks, the number of ground searchers exceeded 3,000. Within a month, more than 4,000 searchers were flown in from around the country to base camps in Corsicana, Palestine, Nacogdoches, and Hemphill, Texas. These searchers, drawn from across the United States and Puerto Rico, worked 12 hours per day on 14-, 21-, or 30-day rotations and were accompanied by Global Positioning System-equipped NASA and Environmental Protection Agency personnel trained to handle and identify debris.

Based on sophisticated mapping of debris trajectories gathered from telemetry, radar, photographs, video, and meteorological data, as well as reports from the general public, teams were dispatched to walk precise grids of East Texas pine brush and thicket (see Figure 2.7-2). In lines 10 feet apart, a distance calculated to provide a 75 percent probability of detecting a six-inch-square object, wildland firefighters scoured snake-infested swamps, mud-filled creek beds, and brush so thick that one team advanced only a few hundred feet in an entire morning. These 20-person ground teams systematically covered an area two miles to either side of the Orbiter's ground track. Initial efforts concentrated on the search for human remains and the debris corridor between Corsicana, Texas, and Fort Polk. Searchers gave highest priority to a list of some 20 "hot items" that potentially contained crucial information, including the Orbiter's General Purpose Computers, film, cameras, and the Modular Auxiliary Data System recorder. Once the wildland firefighters entered the field, recovery rates exceeded 1,000 pieces of debris per day.

Figure 2.7-2. Searching for debris was a laborious task that used thousands of people walking over hundreds of acres of Texas and Louisiana.

After searchers spotted a piece of debris and determined it was not hazardous, its location was recorded with a Global Positioning System unit and photographed. The debris was then tagged and taken to one of four collection centers at Corsicana, Palestine, Nacogdoches, and Hemphill, Texas. There, engineers made a preliminary identification, entered the find into a database, and then shipped the debris to Kennedy Space Center, where it was further analyzed in a hangar dedicated to the debris reconstruction.

Air Search

Air crews used 37 helicopters and seven fixed-wing aircraft to augment ground searchers by searching for debris farther out from the Orbiter's ground track, from two miles from the centerline to five miles on either side. Initially, these crews used advanced remote sensing technologies, including two satellite platforms, hyper-spectral and forward-looking infrared scanners, forest penetration radars, and imagery from Lockheed U-2 reconnaissance aircraft. Because of the density of the East Texas vegetation, the small sizes of the debris, and the inability of sensors to differentiate Orbiter material from other objects, these devices proved of little value. As a result, the detection work fell to spotter teams who visually scanned the terrain. Air search coordinators apportioned grids to allow a 50 percent probability of detection for a one-foot-square object. Civil Air Patrol volunteers and others in powered parachutes, a type of ultralight aircraft, also participated in the search, but were less successful than helicopter and fixed-wing air crews in retrieving debris. During the air search, a Bell 407 helicopter crashed in Angelina National Forest in San Augustine County after a mechanical failure. The accident took the lives of Jules F. "Buzz" Mier Jr., a contract pilot, and Charles Krenek, a Texas Forest Service employee, and injured three others (see Figure 2.7-3).

Figure 2.7-3. Tragically, a helicopter crash during the debris search claimed the lives of Jules "Buzz" Mier (in black coat) and Charles Krenek (yellow coat).

Water Search

The United States Navy Supervisor of Salvage organized eight dive teams to search Lake Nacogdoches and Toledo Bend Reservoir, two bodies of water in dense debris fields. Sonar mapping of more than 31 square miles of lake bottom identified more than 3,100 targets in Toledo Bend and 326 targets in Lake Nacogdoches. Divers explored each target, but in murky water with visibility of only a few inches, underwater forests, and other submerged hazards, they recovered only one object in Toledo Bend and none in Lake Nacogdoches. The 60 divers came from the Navy, Coast Guard, Environmental Protection Agency, Texas Forest Service, Texas Department of Public Safety, Houston and Galveston police and fire departments, and Jasper County Sheriff's Department.

Search Beyond Texas and Louisiana

As thousands of personnel combed the Orbiter's ground track in Texas and Louisiana, other civic and community groups searched areas farther west. Environmental organizations and local law enforcement walked three counties of California coastline where oceanographic data indicated a high probability of debris washing ashore. Prison inmates scoured sections of the Nevada desert. Civil Air Patrol units and other volunteers searched thousands of acres in New Mexico, by air and on foot. Though these searchers failed to find any debris, they provided a valuable service by closing out potential debris sites, including nine areas in Texas, New Mexico, Nevada, and Utah identified by the National Transportation Safety Board as likely to contain debris. NASA's Mishap Investigation Team addressed each of the 1,459 debris reports it received. So eager was the general public to turn in pieces of potential debris that NASA received reports from 37 U.S. states that *Columbia*'s re-entry ground track did not cross, as well as from Canada, Jamaica, and the Bahamas.

Property Damage

No one was injured and little property damage resulted from the tens of thousands of pieces of falling debris (see Chapter 10). A reimbursement program administered by NASA distributed approximately $50,000 to property owners who made claims resulting from falling debris or collateral damage from the search efforts. There were, however, a few close calls that emphasize the importance of selecting the ground track that re-entering Orbiters follow. A 600-pound piece of a main engine dug a six-foot-wide hole in the Fort Polk golf course, while an 800-pound main engine piece, which hit the ground at an estimated 1,400 miles per hour, dug an even larger hole nearby. Disaster was narrowly averted outside Nacogdoches when a piece of debris landed between two highly explosive natural gas tanks set just feet apart.

Debris Amnesty

The response of the public in reporting and turning in debris was outstanding. To reinforce the message that Orbiter debris was government property as well as essential evidence of the accident's cause, NASA and local media officials repeatedly urged local residents to report all debris immediately. For those who might have been keeping debris as souvenirs, NASA offered an amnesty that ran for several days. In the end, only a handful of people were prosecuted for theft of debris.

Final Totals

More than 25,000 people from 270 organizations took part in debris recovery operations. All told, searchers expended over 1.5 million hours covering more than 2.3 million acres, an area approaching the size of Connecticut. Over 700,000 acres were searched by foot, and searchers found over 84,000 individual pieces of Orbiter debris weighing more than 84,900 pounds, representing 38 percent of the Orbiter's dry weight. Though significant evidence from radar returns and video recordings indicate debris shedding across California, Nevada, and New Mexico, the most westerly piece of confirmed debris (at the time this report was published) was the tile found in a field in Littleton, Texas. Heavier objects with higher ballistic coefficients, a measure of how far objects will travel in the air, landed toward the end of the debris trail in western Louisiana. The most easterly debris pieces, including the Space Shuttle Main Engine turbopumps, were found in Fort Polk, Louisiana.

Figure 2.7-4. Recovered debris was returned to the Kennedy Space Center where it was laid out in a large hangar. The tape on the floor helped workers place each piece near where it had been on the Orbiter.

The Federal Emergency Management Agency, which directed the overall effort, expended more than $305 million to fund the search. This cost does not include what NASA spent on aircraft support or the wages of hundreds of civil servants employed at the recovery area and in analysis roles at NASA centers.

The Importance of Debris

The debris collected (see Figure 2.7-4) by searchers aided the investigation in significant ways. Among the most important finds was the Modular Auxiliary Data System recorder that captured data from hundreds of sensors that was not telemetered to Mission Control. Data from these 800 sensors, recorded on 9,400 feet of magnetic tape, provided investigators with millions of data points, including temperature sensor readings from *Columbia*'s left wing leading edge. The data also helped fill a 30-second gap in telemetered data and provided an additional 14 seconds of data after the telemetry loss of signal.

Recovered debris allowed investigators to build a three-dimensional reconstruction of *Columbia*'s left wing leading edge, which was the basis for understanding the order in which the left wing structure came apart, and led investigators to determine that heat first entered the wing in the location where photo analysis indicated the foam had struck.

ENDNOTES FOR CHAPTER 2

The citations that contain a reference to "CAIB document" with CAB or CTF followed by seven to eleven digits, such as CAB001-0010, refer to a document in the Columbia Accident Investigation Board database maintained by the Department of Justice and archived at the National Archives.

[1] The primary source document for this process is NSTS 08117, Requirements and Procedures for Certification and Flight Readiness. CAIB document CTF017-03960413.

[2] Statement of Daniel S. Goldin, Administrator, National Aeronautics and Space Administration, before the Subcommittee on VA-HUD-Independent Agencies, Committee on Appropriations, House of Representatives, March 31, 1998. CAIB document CAB048-04000418.

[3] Roberta L. Gross, Inspector General, NASA, to Daniel S. Goldin, Administrator, NASA, "Assessment of the Triana Mission, G-99-013, Final Report," September 10, 1999. See in particular footnote 3, concerning Triana and the requirements of the Commercial Space Act, and Appendix C, "Accounting for Shuttle Costs." CAIB document CAB048-02680269.

[4] Although there is more volume of liquid hydrogen in the External Tank, liquid hydrogen is very light and its slosh effects are minimal and are generally ignored. At launch, the External Tank contains approximately 1.4 million pounds (140,000 gallons) of liquid oxygen, but only 230,000 pounds (385,000 gallons) of liquid hydrogen.

[5] The Performance Enhancements (PE) flight profile flown by STS-107 is a combination of flight software and trajectory design changes that were introduced in late 1997 for STS-85. These changes to the ascent flight profile allow the Shuttle to carry some 1,600 pounds of additional payload on International Space Station assembly missions. Although developed to meet the Space Station payload lift requirement, a modified PE profile has been used for all Shuttle missions since it was introduced.

CHAPTER 3

Accident Analysis

One of the central purposes of this investigation, like those for other kinds of accidents, was to identify the chain of circumstances that caused the *Columbia* accident. In this case the task was particularly challenging, because the breakup of the Orbiter occurred at hypersonic velocities and extremely high altitudes, and the debris was scattered over a wide area. Moreover, the initiating event preceded the accident by more than two weeks. In pursuit of the sequence of the cause, investigators developed a broad array of information sources. Evidence was derived from film and video of the launch, radar images of *Columbia* on orbit, and amateur video of debris shedding during the in-flight breakup. Data was obtained from sensors onboard the Orbiter – some of this data was downlinked during the flight, and some came from an on-board recorder that was recovered during the debris search. Analysis of the debris was particularly valuable to the investigation. Clues were to be found not only in the condition of the pieces, but also in their location – both where they had been on the Orbiter and where they were found on the ground. The investigation also included extensive computer modeling, impact tests, wind tunnel studies, and other analytical techniques. Each of these avenues of inquiry is described in this chapter.

Because it became evident that the key event in the chain leading to the accident involved both the External Tank and one of the Orbiter's wings, the chapter includes a study of these two structures. The understanding of the accident's physical cause that emerged from this investigation is summarized in the statement at the beginning of the chapter. Included in the chapter are the findings and recommendations of the Columbia Accident Investigation Board that are based on this examination of the physical evidence.

3.1 THE PHYSICAL CAUSE

The physical cause of the loss of *Columbia* and its crew was a breach in the Thermal Protection System on the leading edge of the left wing. The breach was initiated by a piece of insulating foam that separated from the left bipod ramp of the External Tank and struck the wing in the vicinity of the lower half of Reinforced Carbon-Carbon panel 8 at 81.9 seconds after launch. During re-entry, this breach in the Thermal Protection System allowed superheated air to penetrate the leading-edge insulation and progressively melt the aluminum structure of the left wing, resulting in a weakening of the structure until increasing aerodynamic forces caused loss of control, failure of the wing, and breakup of the Orbiter.

Figure 3.1-1. Columbia sitting at Launch Complex 39-A. The upper circle shows the left bipod (–Y) ramp on the forward attach point, while the lower circle is around RCC panel 8-left.

3.2 THE EXTERNAL TANK AND FOAM

The External Tank is the largest element of the Space Shuttle. Because it is the common element to which the Solid Rocket Boosters and the Orbiter are connected, it serves as the main structural component during assembly, launch, and ascent. It also fulfills the role of the low-temperature, or cryogenic, propellant tank for the Space Shuttle Main Engines. It holds 143,351 gallons of liquid oxygen at minus 297 degrees Fahrenheit in its forward (upper) tank and 385,265 gallons of liquid hydrogen at minus 423 degrees Fahrenheit in its aft (lower) tank.[1]

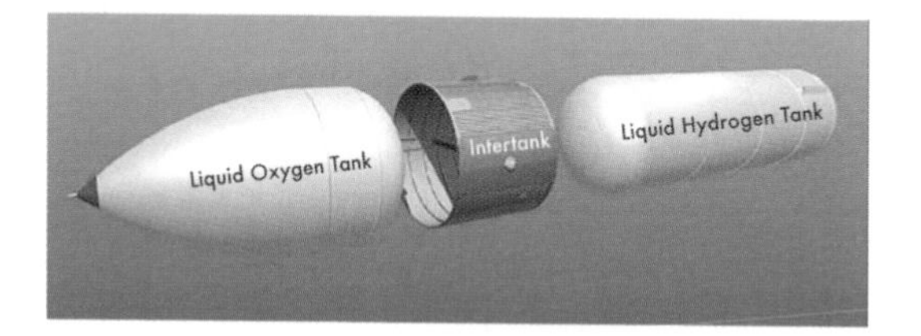

Figure 3.2-1. The major components of the External Tank.

Lockheed Martin builds the External Tank under contract to the NASA Marshall Space Flight Center at the Michoud Assembly Facility in eastern New Orleans, Louisiana.

The External Tank is constructed primarily of aluminum alloys (mainly 2219 aluminum alloy for standard-weight and lightweight tanks, and 2195 Aluminum-Lithium alloy for super-lightweight tanks), with steel and titanium fittings and attach points, and some composite materials in fairings and access panels. The External Tank is 153.8 feet long and 27.6 feet in diameter, and comprises three major sections: the liquid oxygen tank, the liquid hydrogen tank, and the intertank area between them (see Figure 3.2-1). The liquid oxygen and liquid hydrogen tanks are welded assemblies of machined and formed panels, barrel sections, ring frames, and dome and ogive sections. The liquid oxygen tank is pressure-tested with water, and the liquid hydrogen tank with compressed air, before they are incorporated into the External Tank assembly. STS-107 used Lightweight External Tank-93.

The propellant tanks are connected by the intertank, a 22.5-foot-long hollow cylinder made of eight stiffened aluminum alloy panels bolted together along longitudinal joints. Two of these panels, the integrally stiffened thrust panels (so called because they react to the Solid Rocket Booster thrust loads) are located on the sides of the External Tank where the Solid Rocket Boosters are mounted; they consist of single slabs of aluminum alloy machined into panels with solid longitudinal ribs. The thrust panels are joined across the inner diameter by the intertank truss, the major structural element of the External Tank. During propellant loading, nitrogen is used to purge the intertank to prevent condensation and also to prevent liquid oxygen and liquid hydrogen from combining.

The External Tank is attached to the Solid Rocket Boosters by bolts and fittings on the thrust panels and near the aft end of the liquid hydrogen tank. The Orbiter is attached to the External Tank by two umbilical fittings at the bottom (that also contain fluid and electrical connections) and by a "bipod" at the top. The bipod is attached to the External Tank by fittings at the right and left of the External Tank centerline. The bipod fittings, which are titanium forgings bolted to the External Tank, are forward (above) of the intertank-liquid hydrogen flange joint (see Figures 3.2-2 and 3.2-3). Each forging contains a spindle that attaches to one end of a bipod strut and rotates to compensate for External Tank shrinkage during the loading of cryogenic propellants.

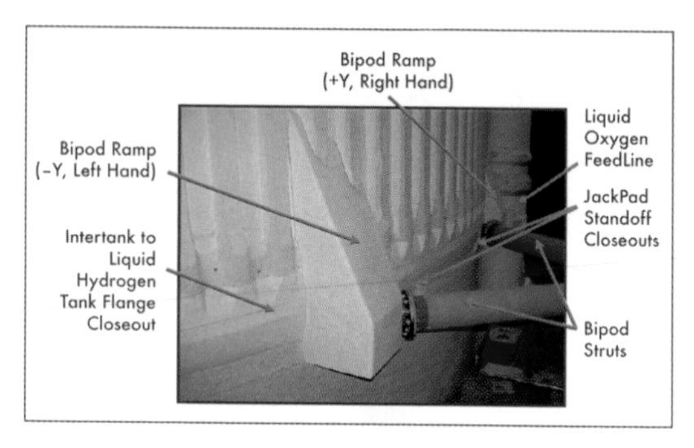

Figure 3.2-2. The exterior of the left bipod attachment area showing the foam ramp that came off during the ascent of STS-107.

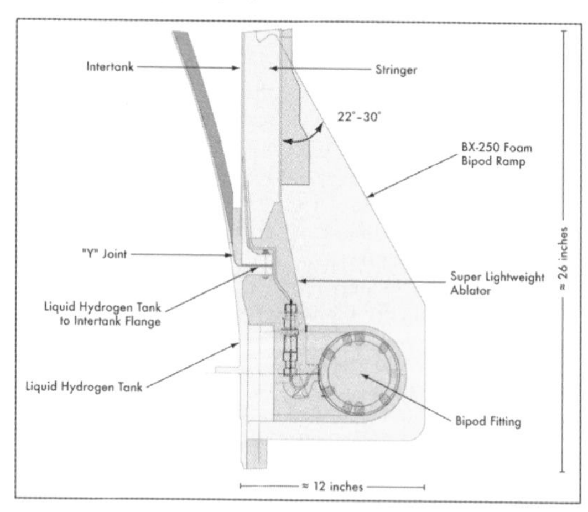

Figure 3.2-3. Cutaway drawing of the bipod ramp and its associated fittings and hardware.

External Tank Thermal Protection System Materials

The External Tank is coated with two materials that serve as the Thermal Protection System: dense composite ablators for dissipating heat, and low density closed-cell foams for high insulation efficiency.[2] (Closed-cell materials consist of small pores filled with air and blowing agents that are separated by thin membranes of the foam's polymeric component.) The External Tank Thermal Protection System is designed to maintain an interior temperature that keeps the

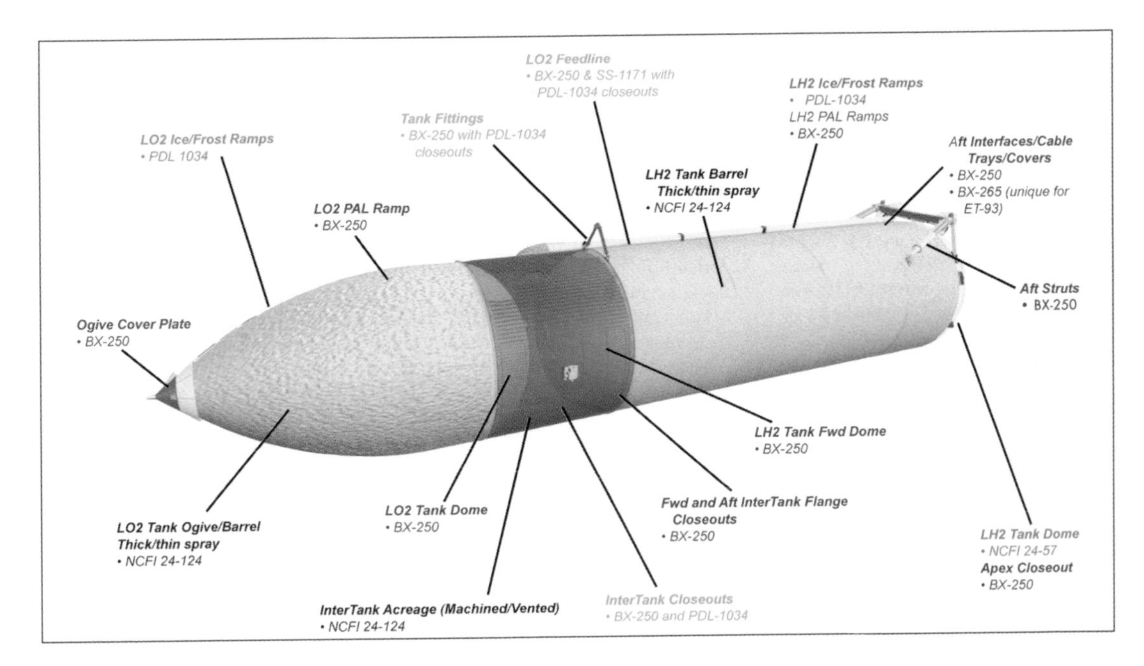

Figure 3.2-4. Locations of the various foam systems as used on ET-93, the External Tank used for STS-107.

oxygen and hydrogen in a liquid state, and to maintain the temperature of external parts high enough to prevent ice and frost from forming on the surface. Figure 3.2-4 summarizes the foam systems used on the External Tank for STS-107.

The adhesion between sprayed-on foam insulation and the External Tank's aluminum substrate is actually quite good, provided that the substrate has been properly cleaned and primed. (Poor surface preparation does not appear to have been a problem in the past.) In addition, large areas of the aluminum substrate are usually heated during foam application to ensure that the foam cures properly and develops the maximum adhesive strength. The interface between the foam and the aluminum substrate experiences stresses due to differences in how much the aluminum and the foam contract when subjected to cryogenic temperatures, and due to the stresses on the External Tank's aluminum structure while it serves as the backbone of the Shuttle stack. While these stresses at the foam-aluminum interface are certainly not trivial, they do not appear to be excessive, since very few of the observed foam loss events indicated that the foam was lost down to the primed aluminum substrate.

Throughout the history of the External Tank, factors unrelated to the insulation process have caused foam chemistry changes (Environmental Protection Agency regulations and material availability, for example). The most recent changes resulted from modifications to governmental regulations of chlorofluorocarbons.

Most of the External Tank is insulated with three types of spray-on foam. NCFI 24-124, a polyisocyanurate foam applied with blowing agent HCFC 141b hydrochlorofluorocarbon, is used on most areas of the liquid oxygen and liquid hydrogen tanks. NCFI 24-57, another polyisocyanurate foam applied with blowing agent HCFC 141b hydrochlorofluorocarbon, is used on the lower liquid hydrogen tank dome. BX-250, a polyurethane foam applied with CFC-11 chlorofluorocarbon, was used on domes, ramps, and areas where the foam is applied by hand. The foam types changed on External Tanks built after External Tank 93, which was used on STS-107, but these changes are beyond the scope of this section.

Metallic sections of the External Tank that will be insulated with foam are first coated with an epoxy primer. In some areas, such as on the bipod hand-sculpted regions, foam is applied directly over ablator materials. Where foam is applied over cured or dried foam, a bonding enhancer called Conathane is first applied to aid the adhesion between the two foam coats.

After foam is applied in the intertank region, the larger areas of foam coverage are machined down to a thickness of about an inch. Since controlling weight is a major concern for the External Tank, this machining serves to reduce foam thickness while still maintaining sufficient insulation.

The insulated region where the bipod struts attach to the External Tank is structurally, geometrically, and materially complex. Because of concerns that foam applied over the fittings would not provide enough protection from the high heating of exposed surfaces during ascent, the bipod fittings are coated with ablators. BX-250 foam is sprayed by hand over the fittings (and ablator materials), allowed to dry, and manually shaved into a ramp shape. The foam is visually inspected at the Michoud Assembly Facility and also at the Kennedy Space Center, but no other non-destructive evaluation is performed.

Since the Shuttle's inaugural flight, the shape of the bipod ramp has changed twice. The bipod foam ramps on External Tanks 1 through 13 originally had a 45-degree ramp angle. On STS-7, foam was lost from the External Tank bipod ramp; subsequent wind tunnel testing showed that shallower angles were aerodynamically preferable. The ramp angle was changed from 45 degrees to between 22 and 30 degrees on External Tank 14 and later tanks. A slight modification to the ramp impingement profile, implemented on External Tank 76 and later, was the last ramp geometry change.

STS-107 Left Bipod Foam Ramp Loss

A combination of factors, rather than a single factor, led to the loss of the left bipod foam ramp during the ascent of STS-107. NASA personnel believe that testing conducted during the investigation, including the dissection of as-built hardware and testing of simulated defects, showed conclusively that pre-existing defects in the foam were a major factor, and in briefings to the Board, these were cited as a necessary condition for foam loss. However, analysis indicated that pre-existing defects alone were not responsible for foam loss.

The basic External Tank was designed more than 30 years ago. The design process then was substantially different than it is today. In the 1970s, engineers often developed particular facets of a design (structural, thermal, and so on) one after another and in relative isolation from other engineers working on different facets. Today, engineers usually work together on all aspects of a design as an integrated team. The bipod fitting was designed first from a structural standpoint, and the application processes for foam (to prevent ice formation) and Super Lightweight Ablator (to protect from high heating) were developed separately. Unfortunately, the structurally optimum fitting design, along with the geometric complexity of its location (near the flange between the intertank and the liquid hydrogen tank), posed many problems in the application of foam and Super Lightweight Ablator that would lead to foam-ramp defects.

Although there is no evidence that substandard methods were used to qualify the bipod ramp design, tests made nearly three decades ago were rudimentary by today's standards and capabilities. Also, testing did not follow the often-used engineering and design philosophy of "Fly what you test and test what you fly." Wind tunnel tests observed the aerodynamics and strength of two geometries of foam bipod enclosures (flat-faced and a 20-degree ramp), but these tests were done on essentially solid foam blocks that were not sprayed onto the complex bipod fitting geometry. Extensive material property tests gauged the strength, insulating potential, and ablative characteristics of foam and Super Lightweight Ablator specimens.

It was – and still is – impossible to conduct a ground-based, simultaneous, full-scale simulation of the combination of loads, airflows, temperatures, pressures, vibration, and acoustics the External Tank experiences during launch and ascent. Therefore, the qualification testing did not truly reflect the combination of factors the bipod would experience during flight. Engineers and designers used the best methods available at the time: test the bipod and foam under as many severe combinations as could be simulated and then interpolate the results. Various analyses determined stresses, thermal gradients, air loads, and other conditions that could not be obtained through testing.

Significant analytical advancements have been made since the External Tank was first conceived, particularly in computational fluid dynamics (see Figure 3.2-5). Computational fluid dynamics comprises a computer-generated model that represents a system or device and uses fluid-flow physics and software to create predictions of flow behavior, and stress or deformation of solid structures. However, analysis must always be verified by test and/or flight data. The External Tank and the bipod ramp were not tested in the complex flight environment, nor were fully instrumented External Tanks ever launched to gather data for verifying analytical tools. The accuracy of the analytical tools used to simulate the External Tank and bipod ramp were verified only by using flight and test data from other Space Shuttle regions.

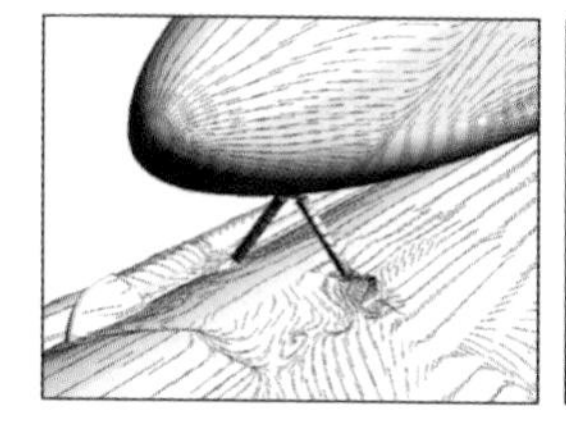

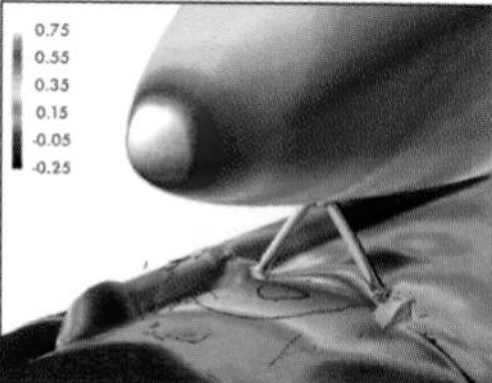

Figure 3.2-5. Computational Fluid Dynamics was used to understand the complex flow fields and pressure coefficients around bipod strut. The flight conditions shown here approximate those present when the left bipod foam ramp was lost from External Tank 93 at Mach 2.46 at a 2.08-degree angle of attack.

Further complicating this problem, foam does not have the same properties in all directions, and there is also variability in the foam itself. Because it consists of small hollow cells, it does not have the same composition at every point. This combination of properties and composition makes foam extremely difficult to model analytically or to characterize physically. The great variability in its properties makes for difficulty in predicting its response in even relatively static conditions, much less during the launch and ascent of the Shuttle. And too little effort went into understanding the origins of this variability and its failure modes.

The way the foam was produced and applied, particularly in the bipod region, also contributed to its variability. Foam consists of two chemical components that must be mixed in an exact ratio and is then sprayed according to strict specifications. Foam is applied to the bipod fitting by hand to make the foam ramp, and this process may be the primary source of foam variability. Board-directed dissection of foam ramps has revealed that defects (voids, pockets, and debris) are likely due to a lack of control of various combinations of parameters in spray-by-hand applications, which is exacerbated by the complexity of the underlying hardware configuration. These defects often occur along "knit lines," the boundaries between each layer that are formed by the repeated application of thin layers – a detail of the spray-by-hand process that contributes to foam variability, suggesting that while foam is sprayed according to approved procedures, these procedures may be questionable if the people who devised them did not have a sufficient understanding of the properties of the foam.

Subsurface defects can be detected only by cutting away the foam to examine the interior. Non-destructive evaluation techniques for determining External Tank foam strength have not been perfected or qualified (although non-destructive testing has been used successfully on the foam on Boeing's new Delta IV booster, a design of much simpler geometry than the External Tank). Therefore, it has been impossible to determine the quality of foam bipod ramps on any External Tank. Furthermore, multiple defects in some cases can combine to weaken the foam along a line or plane.

"Cryopumping" has long been theorized as one of the processes contributing to foam loss from larger areas of coverage. If there are cracks in the foam, and if these cracks lead through the foam to voids at or near the surface of the liquid oxygen and liquid hydrogen tanks, then air, chilled by the extremely low temperatures of the cryogenic tanks, can liquefy in the voids. After launch, as propellant levels fall and aerodynamic heating of the exterior increases, the temperature of the trapped air can increase, leading to boiling and evaporation of the liquid, with concurrent buildup of pressure within the foam. It was believed that the resulting rapid increase in subsurface pressure could cause foam to break away from the External Tank.

"Cryoingestion" follows essentially the same scenario, except it involves gaseous nitrogen seeping out of the intertank and liquefying inside a foam void or collecting in the Super Lightweight Ablator. (The intertank is filled with nitrogen during tanking operations to prevent condensation and also to prevent liquid hydrogen and liquid oxygen from combining.) Liquefying would most likely occur in the circumferential "Y" joint, where the liquid hydrogen tank mates with the intertank, just above the liquid hydrogen-intertank flange. The bipod foam ramps straddle this complex feature. If pooled liquid nitrogen contacts the liquid hydrogen tank, it can solidify, because the freezing temperature of liquid nitrogen (minus 348 degrees Fahrenheit) is higher than the temperature of liquid hydrogen (minus 423 degrees Fahrenheit). As with cryopumping, cryoingested liquid or solid nitrogen could also "flash evaporate" during launch and ascent, causing the foam to crack off. Several paths allow gaseous nitrogen to escape from the intertank, including beneath the flange, between the intertank panels, through the rivet holes that connect stringers to intertank panels, and through vent holes beneath the stringers that prevent over-pressurization of the stringers.

No evidence suggests that defects or cryo-effects alone caused the loss of the left bipod foam ramp from the STS-107 External Tank. Indeed, NASA calculations have suggested that during ascent, the Super Lightweight Ablator remains just slightly above the temperature at which nitrogen liquefies, and that the outer wall of the hydrogen tank near the bipod ramp does not reach the temperature at which nitrogen boils until 150 seconds into the flight,[3] which is too late to explain the only two bipod ramp foam losses whose times during ascent are known. Recent tests at the Marshall Space Flight Center revealed that flight conditions could permit ingestion of nitrogen or air into subsurface foam, but would not permit "flash evaporation" and a sufficient subsurface pressure increase to crack the foam. When conditions are modified to force a flash evaporation, the failure mode in the foam is a crack that provides pressure relief rather than explosive cracking. Therefore, the flight environment itself must also have played a role. Aerodynamic loads, thermal and vacuum effects, vibrations, stress in the External Tank structure, and myriad other conditions may have contributed to the growth of subsurface defects, weakening the foam ramp until it could no longer withstand flight conditions.

Conditions in certain combinations during ascent may also have contributed to the loss of the foam ramp, even if individually they were well within design certification limits. These include a wind shear, associated Solid Rocket Booster and Space Shuttle Main Engine responses, and liquid oxygen sloshing in the External Tank.[4] Each of these conditions, alone, does not appear to have caused the foam loss, but their contribution to the event in combination is unknown.

Negligence on the part of NASA, Lockheed Martin, or United Space Alliance workers does not appear to have been a factor. There is no evidence of sabotage, either during production or pre-launch. Although a Problem Report was written for a small area of crushed foam near the left bipod (a condition on nearly every flight), this affected only a very small region and does not appear to have contributed to the loss of the ramp (see Chapter 4 for a fuller discussion). Nor does the basic quality of the foam appear to be a concern. Many of the basic components are continually and meticulously tested for quality before they are applied. Finally, despite commonly held perceptions, numerous tests show that moisture absorption and ice formation in the foam appears negligible.

Foam loss has occurred on more than 80 percent of the 79 missions for which imagery is available, and foam was lost from the left bipod ramp on nearly 10 percent of missions where the left bipod ramp was visible following External Tank separation. For about 30 percent of all missions, there is no way to determine if foam was lost; these were either night launches, or the External Tank bipod ramp areas were not in view when the images were taken. The External Tank was not designed to be instrumented or recovered after separation, which deprives NASA of physical evidence that could help pinpoint why foam separates from it.

The precise reasons why the left bipod foam ramp was lost from the External Tank during STS-107 may never be known. The specific initiating event may likewise remain a mystery. However, it is evident that a combination of variable and pre-existing factors, such as insufficient testing and analysis in the early design stages, resulted in a highly variable and complex foam material, defects induced by an imperfect

Foam Fracture Under Hydrostatic Pressure

The Board has concluded that the physical cause of the breakup of *Columbia* upon re-entry was the result of damage to the Orbiter's Thermal Protection System, which occurred when a large piece of BX-250 foam insulation fell from the left (–Y) bipod assembly 81.7 seconds after launch and struck the leading edge of the left wing. As the External Tank is covered with insulating foam, it seemed to me essential that we understand the mechanisms that could cause foam to shed.

Many if not most of the systems in the three components of the Shuttle stack (Orbiter, External Tank, and Solid Rocket Boosters) are by themselves complex, and often operate near the limits of their performance. Attempts to understand their complex behavior and failure modes are hampered by their strong interactions with other systems in the stack, through their shared environment. The foam of the Thermal Protection System is no exception. To understand the behavior of systems under such circumstances, one must first understand their behavior in relatively simple limits. Using this understanding as a guide, one is much more likely to determine the mechanisms of complex behavior, such as the shedding of foam from the –Y bipod ramp, than simply creating simulations of the complex behavior itself.

I approached this problem by trying to imagine the fracture mechanism by which fluid pressure built up inside the foam could propagate to the surface. Determining this process is clearly key to understanding foam ejection through the heating of cryogenic fluids trapped in voids beneath the surface of the foam, either through "cryopumping" or "cryoingestion." I started by imagining a fluid under hydrostatic pressure in contact with the surface of such foam. It seemed clear that as the pressure increased, it would cause the weakest cell wall to burst, filling the adjacent cell with the fluid, and exerting the same hydrostatic pressure on all the walls of that cell. What happened next was unclear. It was possible that the next cell wall to burst would not be one of the walls of the newly filled cell, but some other cell that had been on the surface that was initially subjected to the fluid pressure. This seemed like a rather complex process, and I questioned my ability to include all the physics correctly if I tried to model it. Instead, I chose to perform an experiment that seemed straightforward, but which had a result I could not have foreseen.

I glued a 1.25-inch-thick piece of BX-250 foam to a 0.25-inch-thick brass plate. The 3-by-3-inch plate had a 0.25-inch-diameter hole in its center, into which a brass tube was soldered. The tube was filled with a liquid dye, and the air pressure above the dye could be slowly raised, using a battery-operated tire pump to which a pressure regulator was attached until the fluid was forced through the foam to its outer surface. Not knowing what to expect, the first time I tried this experiment with my graduate student, Jim Baumgardner, we did so out on the loading dock of the Stanford Physics Department. If this process were to mimic the cryoejection of foam, we expected a violent explosion when the pressure burst through the surface. To keep from being showered with dye, we put the assembly in a closed cardboard box, and donned white lab coats.

Instead of a loud explosion, we heard nothing. We found, though, that the pressure above the liquid began dropping once the gas pressure reached about 45 pounds per square inch. Releasing the pressure and opening the box, we found a thin crack, about a half-inch long, at the upper surface of the foam. Curious about the path the pressure had taken to reach the surface, I cut the foam off the brass plate, and made two vertical cuts through the foam in line with the crack. When I bent the foam in line with the crack, it separated into two sections along the crack. The dye served as a tracer for where the fluid had traveled in its path through the foam. This path was along a flat plane, and was the shape of a teardrop that intersected perpendicular to the upper surface of the foam. Since the pressure could only exert force in the two directions perpendicular to this fault plane, it could not possibly result in the ejection of foam, because that would require a force perpendicular to the surface of the foam. I repeated this experiment with several pieces of foam and always found the same behavior.

I was curious why the path of the pressure fault was planar, and why it had propagated upward, nearly perpendicular to the outer surface of the foam. For this sample, and most of the samples that NASA had given me, the direction of growth of the foam was vertical, as evidenced by horizontal "knit lines" that result from successive applications of the sprayed foam. The knit lines are perpendicular to the growth direction. I then guessed that the growth of the pressure fault was influenced by the foam's direction of growth. To test this hypothesis, I found a piece of foam for which the growth direction was vertical near the top surface of the foam, but was at an approximately 45-degree angle to the vertical near the bottom. If my hypothesis were correct, the direction of growth of the pressure fault would follow the direction of growth of the foam, and hence would always intersect the knit lines at 90 degrees. Indeed, this was the case.

The reason the pressure fault is planar has to do with the fact that such a geometry can amplify the fluid pressure, creating a much greater stress on the cell walls near the outer edges of the teardrop, for a given hydrostatic pressure, than would exist for a spherical pressure-filled void. A pressure fault follows the direction of foam growth because more cell walls have their surfaces along this direction than along any other. The stiffness of the foam is highest when you apply a force parallel to the cell walls. If you squeeze a cube of foam in various directions, you find that the foam is stiffest along its growth direction. By advancing along the stiff direction, the crack is oriented so that the fluid pressure can more easily force the (nearly) planar walls of the crack apart.

Because the pressure fault intersects perpendicular to the upper surface, hydrostatic pressure will *generally* not lead to foam shedding. There are, however, cases where pressure *can* lead to foam shedding, but this will only occur when the fluid pressure exists over an area whose dimensions are large compared to the thickness of the foam above it, and roughly parallel to the outer surface. This would require a large structural defect within the foam, such as the delamination of the foam from its substrate or the separation of the foam at a knit line. Such large defects are quite different from the small voids that occur when gravity causes uncured foam to "roll over" and trap a small bubble of air.

Experiments like this help us understand how foam shedding does (and doesn't) occur, because they elucidate the properties of "perfect" foam, free from voids and other defects. Thus, this behavior represents the true behavior of the foam, free from defects that may or may not have been present. In addition, these experiments are fast and cheap, since they can be carried out on relatively small pieces of foam in simple environments. Finally, we can understand why the observed behavior occurs from our understanding of the basic physical properties of the foam itself. By contrast, if you wish to mimic left bipod foam loss, keep in mind that such loss could have been detected only 7 times in 72 instances. Thus, not observing foam loss in a particular experiment will not insure that it would never happen under the same conditions at a later time. NASA is now undertaking both kinds of experiments, but it is the simple studies that so far have most contributed to our understanding of foam failure modes.

Douglas Osheroff, Board Member

and variable application, and the results of that imperfect process, as well as severe load, thermal, pressure, vibration, acoustic, and structural launch and ascent conditions.

Findings:

F3.2–1 NASA does not fully understand the mechanisms that cause foam loss on almost all flights from larger areas of foam coverage and from areas that are sculpted by hand.

F3.2–2 There are no qualified non-destructive evaluation techniques for the as-installed foam to determine the characteristics of the foam before flight.

F3.2–3 Foam loss from an External Tank is unrelated to the tank's age and to its total pre-launch exposure to the elements. Therefore, the foam loss on STS-107 is unrelated to either the age or exposure of External Tank 93 before launch.

F3.2–4 The Board found no indications of negligence in the application of the External Tank Thermal Protection System.

F3.2–5 The Board found instances of left bipod ramp shedding on launch that NASA was not aware of, bringing the total known left bipod ramp shedding events to 7 out of 72 missions for which imagery of the launch or External Tank separation is available.

F3.2–6 Subsurface defects were found during the dissection of three bipod foam ramps, suggesting that similar defects were likely present in the left bipod ramp of External Tank 93 used on STS-107.

F3.2–7 Foam loss occurred on more than 80 percent of the 79 missions for which imagery was available to confirm or rule out foam loss.

F3.2–8 Thirty percent of all missions lacked sufficient imagery to determine if foam had been lost.

F3.2–9 Analysis of numerous separate variables indicated that none could be identified as the sole initiating factor of bipod foam loss. The Board therefore concludes that a combination of several factors resulted in bipod foam loss.

Recommendation:

R3.2-1 Initiate an aggressive program to eliminate all External Tank Thermal Protection System debris-shedding at the source with particular emphasis on the region where the bipod struts attach to the External Tank.

3.3 WING LEADING EDGE STRUCTURAL SUBSYSTEM

The components of the Orbiter's wing leading edge provide the aerodynamic load bearing, structural, and thermal control capability for areas that exceed 2,300 degrees Fahrenheit. Key design requirements included flying 100 missions with minimal refurbishment, maintaining the aluminum wing structure at less than 350 degrees Fahrenheit, withstanding a kinetic energy impact of 0.006 foot-pounds, and the ability to withstand 1.4 times the load ever expected in operation.[5] The requirements specifically stated that the wing leading edge would not need to withstand impact from debris or ice, since these objects would not pose a threat during the launch phase.[6]

REINFORCED CARBON-CARBON (RCC)

The basic RCC composite is a laminate of graphite-impregnated rayon fabric, further impregnated with phenolic resin and layered, one ply at a time, in a unique mold for each part, then cured, rough-trimmed, drilled, and inspected. The part is then packed in calcined coke and fired in a furnace to convert it to carbon and is made more dense by three cycles of furfuryl alcohol vacuum impregnation and firing.

To prevent oxidation, the outer layers of the carbon substrate are converted into a 0.02-to-0.04-inch-thick layer of silicon carbide in a chamber filled with argon at temperatures up to 3,000 degrees Fahrenheit. As the silicon carbide cools, "craze cracks" form because the thermal expansion rates of the silicon carbide and the carbon substrate differ. The part is then repeatedly vacuum-impregnated with tetraethyl orthosilicate to fill the pores in the substrate, and the craze cracks are filled with a sealant.

Reinforced Carbon-Carbon

The development of Reinforced Carbon-Carbon (RCC) as part of the Thermal Protection System was key to meeting the wing leading edge design requirements. Developed by Ling-Temco-Vought (now Lockheed Martin Missiles and Fire Control), RCC is used for the Orbiter nose cap, chin panel, forward External Tank attachment point, and wing leading edge panels and T-seals. RCC is a hard structural material, with reasonable strength across its operational temperature range (minus 250 degrees Fahrenheit to 3,000 degrees). Its low thermal expansion coefficient minimizes thermal shock and thermoelastic stress.

Each wing leading edge consists of 22 RCC panels (see Figure 3.3-1), numbered from 1 to 22 moving outward on each wing (the nomenclature is "5-left" or "5-right" to differentiate, for example, the two number 5 panels). Because the shape of the wing changes from inboard to outboard, each panel is unique.

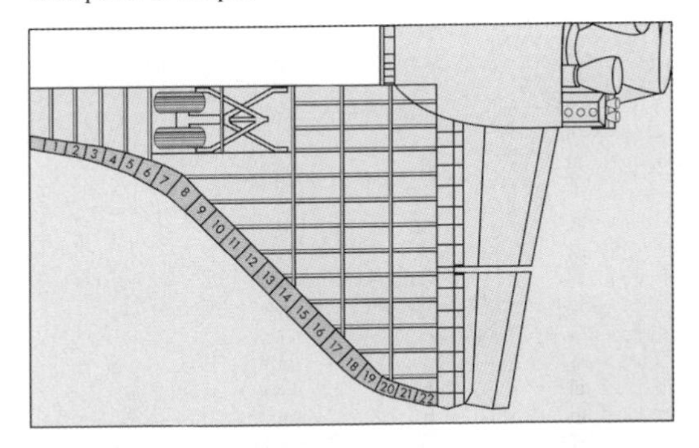

Figure 3.3-1. There are 22 panels of Reinforced Carbon-Carbon on each wing, numbered as shown above.

Wing Leading Edge Damage

The risk of micrometeoroid or debris damage to the RCC panels has been evaluated several times. Hypervelocity impact testing, using nylon, glass, and aluminum projectiles, as well as low-velocity impact testing with ice, aluminum, steel, and lead projectiles, resulted in the addition of a 0.03- to 0.06-inch-thick layer of Nextel-440 fabric between the Inconel foil and Cerachrome insulation. Analysis of the design change predicts that the Orbiter could survive re-entry with a quarter-inch diameter hole in the lower surfaces of RCC panels 8 through 10 or with a one-inch hole in the rest of the RCC panels.

RCC components have been struck by objects throughout their operational life, but none of these components has been completely penetrated. A sampling of 21 post-flight reports noted 43 hypervelocity impacts, the largest being 0.2 inch. The most significant low-velocity impact was to *Atlantis*' panel 10-right during STS-45 in March and April 1992. The damaged area was 1.9 inches by 1.6 inches on the exterior surface and 0.5 inches by 0.1 inches in the interior surface. The substrate was exposed and oxidized, and the panel was scrapped. Analysis concluded that the damage was caused by a strike by a man-made object, possibly during ascent. Figures 3.3-2 and 3.3-3 show the damage to the outer and inner surfaces, respectively.

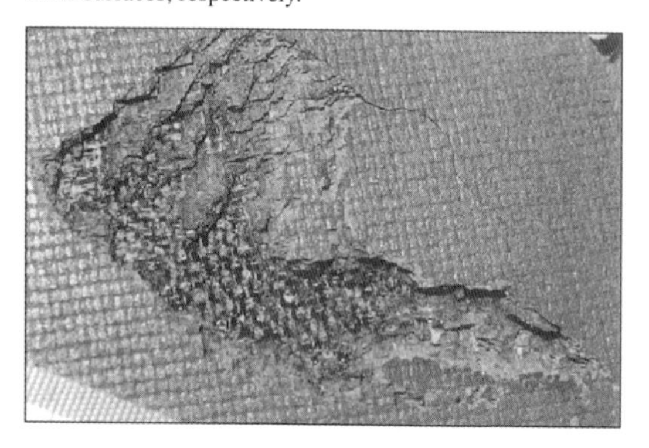

Figure 3.3-2. Damage on the outer surface of RCC panel 10-right from Atlantis after STS-45.

Figure 3.3-3. Damage on the inner surface of RCC panel 10-right from Atlantis after STS-45.

Leading Edge Maintenance

Post-flight RCC component inspections for cracks, chips, scratches, pinholes, and abnormal discoloration are primarily visual, with tactile evaluations (pushing with a finger) of some regions. Boeing personnel at the Kennedy Space Center make minor repairs to the silicon carbide coating and surface defects.

With the goal of a long service life, panels 6 through 17 are refurbished every 18 missions, and panels 18 and 19 every 36 missions. The remaining panels have no specific refurbishment requirement.

At the time of STS-107, most of the RCC panels on *Columbia*'s left wing were original equipment, but panel 10-left, T-seal 10-left, panel 11-left, and T-seal 11-left had been replaced (along with panel 12 on the right wing). Panel 10-left was tested to destruction after 19 flights. Minor surface repairs had been made to panels 5, 7, 10, 11, 12, 13, and 19 and T-seals 3, 11, 12, 13, 14, and 19. Panels and T-seals 6 through 9 and 11 through 17 of the left wing had been refurbished.

Reinforced Carbon-Carbon Mission Life

The rate of oxidation is the most important variable in determining the mission life of RCC components. Oxidation of the carbon substrate results when oxygen penetrates the microscopic pores or fissures of the silicon carbide protective coating. The subsequent loss of mass due to oxidation reduces the load the structure can carry and is the basis for establishing a mission life limit. The oxidation rate is a function of temperature, pressure, time, and the type of heating. Repeated exposure to the Orbiter's normal flight environment degrades the protective coating system and accelerates the loss of mass, which weakens components and reduces mission life capability.

Currently, mass loss of flown RCC components cannot be directly measured. Instead, mass loss and mission life reduction are predicted analytically using a methodology based on mass loss rates experimentally derived in simulated re-entry environments. This approach then uses derived re-entry temperature-time profiles of various portions of RCC components to estimate the actual re-entry mass loss.

For the first five missions of *Columbia*, the RCC components were not coated with Type A sealant, and had shorter mission service lives than the RCC components on the other Orbiters. (*Columbia*'s panel 9 has the shortest mission service life of 50 flights as shown in Figure 3.3-4.) The predicted life for panel/T-seals 7 through 16 range from 54 to 97 flights.[7]

Localized penetration of the protective coating on RCC components (pinholes) were first discovered on *Columbia* in 1992, after STS-50, *Columbia*'s 12th flight. Pinholes were later found in all Orbiters, and their quantity and size have increased as flights continue. Tests showed that pinholes were caused by zinc oxide contamination from a primer used on the launch pad.

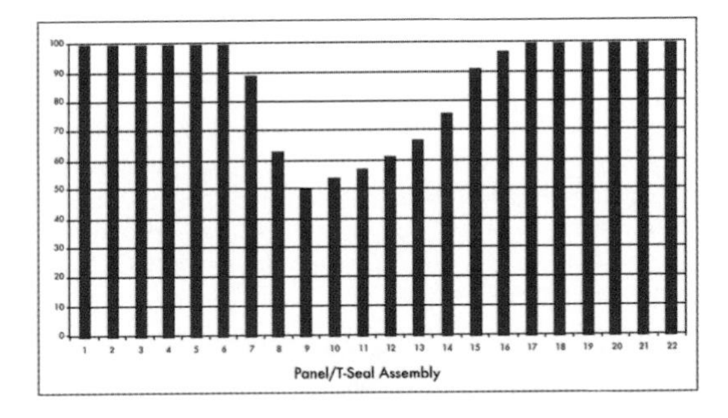

Figure 3.3-4. The expected mission life for each of the wing leading edge RCC panels on Columbia. Note that panel 9 has the shortest life expectancy.

In October 1993, panel 12-right was removed from *Columbia* after its 15th flight for destructive evaluation. Optical and scanning electron microscope examinations of 15 pinholes revealed that a majority occurred along craze cracks in the thick regions of the silicon carbide layer. Pinhole glass chemistry revealed the presence of zinc, silicon, oxygen, and aluminum. There is no zinc in the leading edge support system, but the launch pad corrosion protection system uses an inorganic zinc primer under a coat of paint, and this coat of paint is not always refurbished after a launch. Rain samples from the Rotating Support Structure at Launch Complex 39-A in July 1994 confirmed that rain washed the unprotected primer off the service structure and deposited it on RCC panels while the Orbiter sat on the launch pad. At the request of the Columbia Accident Investigation Board, rain samples were again collected in May 2003. The zinc

Left Wing and Wing Leading Edge

The Orbiter wing leading edge structural subsystem consists of the RCC panels, the upper and lower access panels (also called carrier panels), and the associated attachment hardware for each of these components.

On *Columbia*, two upper and lower A-286 stainless steel spar attachment fittings connected each RCC panel to the aluminum wing leading edge spar. On later Orbiters, each upper and lower spar attachment fitting is a one-piece assembly.

The space between each RCC panel is covered by a gap seal, also known as a T-seal. Each T-seal, also manufactured from RCC, is attached to its associated RCC panel by two Inconel 718 attachment clevises. The upper and lower carrier panels, which allow access behind each RCC panel, are attached to the spar attachment fittings after the RCC panels and T-seals are installed. The lower carrier panel prevents superheated air from entering the RCC panel cavity. A small space between the upper carrier panel and the RCC panel allows air pressure to equalize behind the RCC panels during ascent and re-entry.

The mid-wing area on the left wing, behind where the breach occurred, is supported by a series of trusses, as shown in red in the figure below. The mid-wing area is bounded in the front and back by the Xo1040 and Xo1191 cross spars, respectively. The numerical designation of each spar comes from its location along the Orbiter's X-axis; for example, the Xo1040 spar is 1,040 inches from the zero point on the X-axis. The cross spars provide the wing's structural integrity. Three major cross spars behind the Xo1191 spar provide the primary structural strength for the aft portion of the wing. The inboard portion of the mid-wing is the outer wall of the left wheel-well, and the outboard portion of the mid-wing is the wing leading edge spar, where the RCC panels attach.

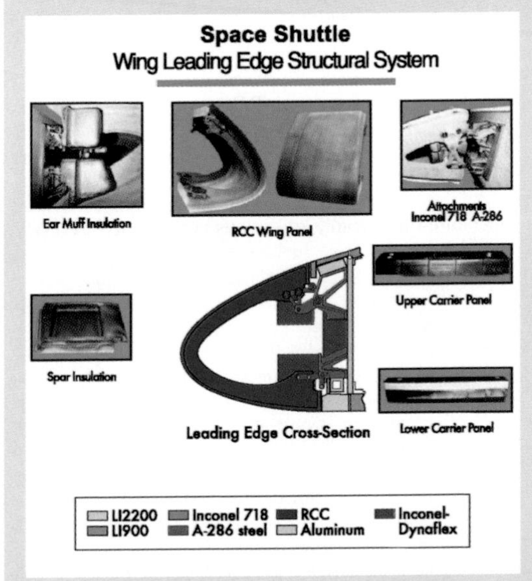

The Wing Leading Edge Structural System on Columbia.

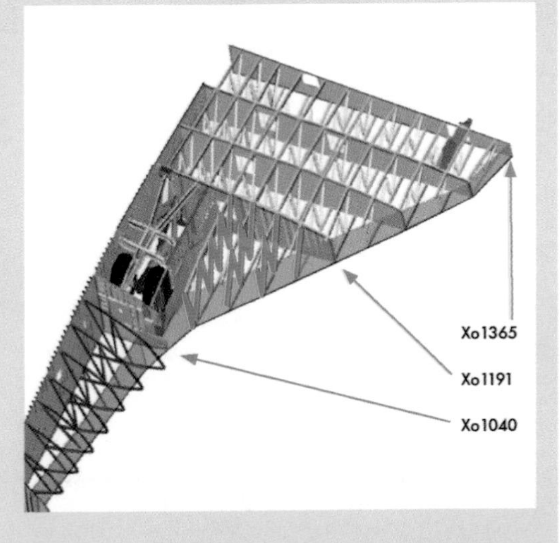

The major internal support structures in the mid-wing are constructed from aluminum alloy. Since aluminum melts at 1,200 degrees Fahrenheit, it is likely these truss tubes in the mid-wing were destroyed and wing structural integrity was lost.

fallout rate was generally less than previously recorded except for one location, which had the highest rate of zinc fallout of all the samples from both evaluations. Chemical analysis of the most recent rainwater samples determined the percentage of zinc to be consistently around nine percent, with that one exception.

Specimens with pinholes were fabricated from RCC panel 12-right and arc-jet-tested, but the arc-jet testing did not substantially change the pinhole dimensions or substrate oxidation. (Arc jet testing is done in a wind tunnel with an electrical arc that provides an airflow of up to 2,800 degrees Fahrenheit.) As a result of the pinhole investigation, the sealant refurbishment process was revised to include cleaning the part in a vacuum at 2,000 degrees Fahrenheit to bake out contaminants like zinc oxide and salt, and forcing sealant into pinholes.

Post-flight analysis of RCC components confirms that sealant is ablated during each mission, which increases subsurface oxidation and reduces component strength and mission life. Based on the destructive evaluation of *Columbia*'s panel 12-right and various arc-jet tests, refurbishment intervals were established to achieve the desired service life.

In November 2001, white residue was discovered on about half the RCC panels on *Columbia*, *Atlantis*, and *Endeavour*. Investigations revealed that the deposits were sodium carbonate that resulted from the exposure of sealant to rainwater, with three possible outcomes: (1) the deposits are washed off, which decreases sealant effectiveness; (2) the deposits remain on the part's surface, melt on re-entry, and combine with the glass, restoring the sealant composition; or (3) the deposits remain on the part's surface, melt on re-entry, and flow onto metal parts.

The root cause of the white deposits on the surface of RCC parts was the breakdown of the sealant. This does not damage RCC material.

Non-Destructive Evaluations of Reinforced Carbon-Carbon Components

Over the 20 years of Space Shuttle operations, RCC has performed extremely well in the harsh environment it is exposed to during a mission. Within the last several years, a few instances of damage to RCC material have resulted in a re-examination of the current visual inspection process. Concerns about potential oxidation between the silicon carbide layer and the substrate and within the substrate has resulted in further efforts to develop improved Non-Destructive Evaluation methods and a better understanding of subsurface oxidation.

Since 1997, inspections have revealed five instances of RCC silicon carbide layer loss with exposed substrate. In November 1997, *Columbia* returned from STS-87 with three damaged RCC parts with carbon substrate exposed. Panel 19-right had a 0.04 inch-diameter by 0.035 inch-deep circular dimple, panel 17-right had a 0.1 inch-wide by 0.2 inch-long by 0.025-inch-deep dimple, and the Orbiter forward External Tank attachment point had a 0.2-inch by 0.15-inch by 0.026-inch-deep dimple. In January 2000, after STS-103, *Discovery*'s panel 8-left was scrapped because of similar damage (see Figure 3.3-5).

In April 2001, after STS-102, *Columbia*'s panel 10-left had a 0.2-inch by 0.3-inch wide by 0.018-inch-deep dimple in the panel corner next to the T-seal. The dimple was repaired and the panel flew one more mission, then was scrapped because of damage found in the repair.

Figure 3.3-5. RCC panel 8-left from Discovery had to be scrapped after STS-103 because of the damage shown here.

Findings:

F3.3-1 The original design specifications required the RCC components to have essentially no impact resistance.

F3.3-2 Current inspection techniques are not adequate to assess structural integrity of the RCC components.

F3.3-3 After manufacturer's acceptance non-destructive evaluation, only periodic visual and touch tests are conducted.

F3.3-4 RCC components are weakened by mass loss caused by oxidation within the substrate, which accumulates with age. The extent of oxidation is not directly measurable, and the resulting mission life reduction is developed analytically.

F3.3-5 To date, only two flown RCC panels, having achieved 15 and 19 missions, have been destructively tested to determine actual loss of strength due to oxidation.

F3.3-6 Contamination from zinc leaching from a primer under the paint topcoat on the launch pad structure increases the opportunities for localized oxidation.

Recommendations:

- R3.3-1 Develop and implement a comprehensive inspection plan to determine the structural integrity of all Reinforced Carbon-Carbon system components. This inspection plan should take advantage of advanced non-destructive inspection technology.
- R3.3-2 Initiate a program designed to increase the Orbiter's ability to sustain minor debris damage by measures such as improved impact-resistant Reinforced Carbon-Carbon and acreage tiles. This program should determine the actual impact resistance of current materials and the effect of likely debris strikes.
- R3.3-3 To the extent possible, increase the Orbiter's ability to successfully re-enter the Earth's atmosphere with minor leading edge structural sub-system damage.
- R3.3-4 In order to understand the true material characteristics of Reinforced Carbon-Carbon components, develop a comprehensive database of flown Reinforced Carbon-Carbon material characteristics by destructive testing and evaluation.
- R3.3-5 Improve the maintenance of launch pad structures to minimize the leaching of zinc primer onto Reinforced Carbon-Carbon components.

3.4 Image and Transport Analyses

At 81.9 seconds after launch of STS-107, a sizable piece of foam struck the leading edge of *Columbia*'s left wing. Visual evidence established the source of the foam as the left bipod ramp area of the External Tank. The widely accepted implausibility of foam causing significant damage to the wing leading edge system led the Board to conduct independent tests to characterize the impact. While it was impossible to determine the precise impact parameters because of uncertainties about the foam's density, dimensions, shape, and initial velocity, intensive work by the Board, NASA, and contractors provided credible ranges for these elements. The Board used a combination of tests and analyses to conclude that the foam strike observed during the flight of STS-107 was the direct, physical cause of the accident.

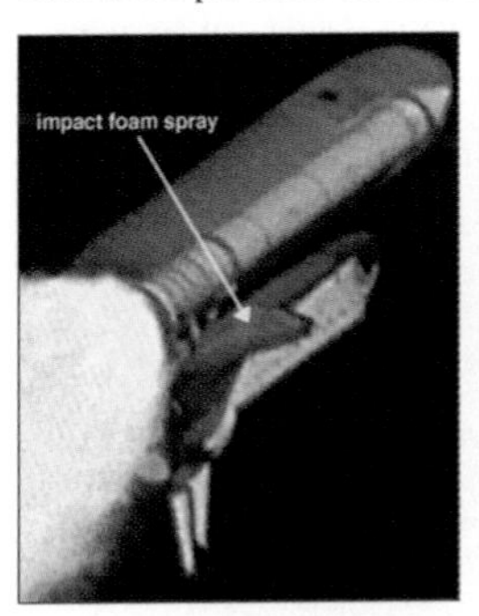

Figure 3.4-1 (color enhanced and "de-blurred" by Lockheed Martin Gaithersburg) and Figure 3.4-2 (processed by the National Imagery and Mapping Agency) are samples of the type of visual data used to establish the time of the impact (81.9 seconds), the altitude at which it occurred (65,860 feet), and the object's relative velocity at impact (about 545 mph relative to the Orbiter).

Image Analysis: Establishing Size, Velocity, Origin, and Impact Area

The investigation image analysis team included members from Johnson Space Center Image Analysis, Johnson Space Center Engineering, Kennedy Space Center Photo Analysis, Marshall Space Flight Center Photo Analysis, Lockheed Martin Management and Data Systems, the National Imagery and Mapping Agency, Boeing Systems Integration, and Langley Research Center. Each member of the image analysis team performed independent analyses using tools and methods of their own choosing. Representatives of the Board participated regularly in the meetings and deliberations of the image analysis team.

A 35-mm film camera, E212, which recorded the foam strike from 17 miles away, and video camera E208, which recorded it from 26 miles away, provided the best of the available evidence. Analysis of this visual evidence (see Figures 3.4-1 and 3.4-2) along with computer-aided design analysis, refined the potential impact area to less than 20 square feet in RCC panels 6 through 9 (see Figure 3.4-3), including a portion of the corresponding carrier panels and adjacent tiles. The investigation image analysis team found no conclusive visual evidence of post-impact debris flowing over the top of the wing.

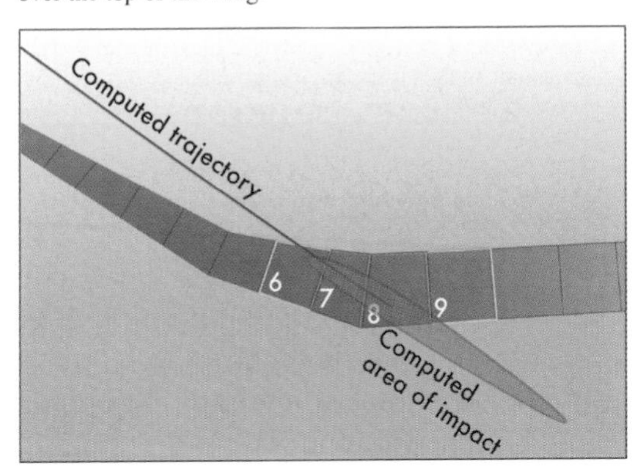

Figure 3.4-3: The best estimate of the site of impact by the center of the foam.

The image analysis team established impact velocities from 625 to 840 feet per second (about 400 to 600 mph) relative to the Orbiter, and foam dimensions from 21 to 27 inches long by 12 to 18 inches wide.[8] The wide range for these measurements is due primarily to the cameras' relatively slow frame rate and poor resolution. For example, a 20-inch change in the position of the foam near the impact point would change the estimated relative impact speed from 675 feet per second to 825 feet per second. The visual evidence could not reveal the foam's shape, but the team was able to describe it as flat and relatively thin. The mass and hence the volume of the foam was determined from the velocity estimates and their ballistic coefficients.

Image analysis determined that the foam was moving almost parallel to the Orbiter's fuselage at impact, with about a five-degree angle upward toward the bottom of the wing and slight motion in the outboard direction. If the foam had hit the tiles adjacent to the leading edge, the angle of incidence would have been about five degrees (the angle of incidence is the angle between the relative velocity of the projectile and the plane of the impacted surface). Because the wing leading edge curves, the angle of incidence increases as the point of impact approaches the apex of an RCC panel. Image and transport analyses estimated that for impact on RCC panel 8, the angle of incidence was between 10 and 20 degrees (see Figure 3.4-4).[9] Because the total force delivered by the impact depends on the angle of incidence, a foam strike near the apex of an RCC panel could have delivered about twice the force as an impact close to the base of the panel.

Despite the uncertainties and potential errors in the data, the Board concurred with conclusions made unanimously by the post-flight image analysis team and concludes the information available about the foam impact during the mission was adequate to determine its effect on both the thermal tiles and RCC. Those conclusions made during the mission follow:

- The bipod ramp was the source of the foam.
- Multiple pieces of foam were generated, but there was no evidence of more than one strike to the Orbiter.
- The center of the foam struck the leading edge structural subsystem of the left wing between panels 6 to 9. The potential impact location included the corresponding carrier panels, T-seals, and adjacent tiles. (Based on further image analysis performed by the National Imagery and Mapping Agency, the transport analysis that follows, and forensic evidence, the Board concluded that a smaller estimated impact area in the immediate vicinity of panel 8 was credible.)
- Estimates of the impact location and velocities rely on timing of camera images and foam position measurements.
- The relative velocity of the foam at impact was 625 to 840 feet per second. (The Board agreed on a narrower speed range based on a transport analysis that follows.)
- The trajectory of the foam at impact was essentially parallel to the Orbiter's fuselage.
- The foam was making about 18 revolutions per second as it fell.
- The orientation at impact could not be determined.
- The foam that struck the wing was 24 (plus or minus 3) inches by 15 (plus or minus 3) inches. The foam shape could only be described as flat. (A subsequent transport analysis estimated a thickness.)
- Ice was not present on the external surface of the bipod ramp during the last Ice Team camera scan prior to launch (at approximately T–5 minutes).
- There was no visual evidence of the presence of other materials inside the bipod ramp.
- The foam impact generated a cloud of pulverized debris with very little component of velocity away from the wing.

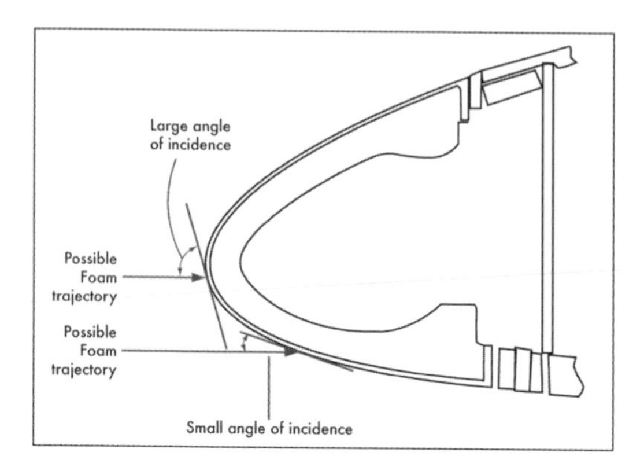

Figure 3.4-4. This drawing shows the curve of the wing leading edge and illustrates the difference the angle of incidence has on the effect of the foam strike.

- In addition, the visual evidence showed two sizable, traceable post-strike debris pieces with a significant component of velocity away from the wing.

Although the investigation image analysis team found no evidence of post-strike debris going over the top of the wing before or after impact, a colorimetric analysis by the National Imagery and Mapping Agency indicated the potential presence of debris material over the top of the left wing immediately following the foam strike. This analysis suggests that some of the foam may have struck closer to the apex of the wing than what occurred during the impact tests described below.

Imaging Issues

The image analysis was hampered by the lack of high resolution and high speed ground-based cameras. The existing camera locations are a legacy of earlier NASA programs, and are not optimum for the high-inclination Space Shuttle missions to the International Space Station and oftentimes

The Orbiter "Ran Into" the Foam

"How could a lightweight piece of foam travel so fast and hit the wing at 545 miles per hour?"

Just prior to separating from the External Tank, the foam was traveling with the Shuttle stack at about 1,568 mph (2,300 feet per second). Visual evidence shows that the foam debris impacted the wing approximately 0.161 seconds after separating from the External Tank. In that time, the velocity of the foam debris slowed from 1,568 mph to about 1,022 mph (1,500 feet per second). Therefore, the Orbiter hit the foam with a relative velocity of about 545 mph (800 feet per second). In essence, the foam debris slowed down and the Orbiter did not, so the Orbiter ran into the foam. The foam slowed down rapidly because such low-density objects have low ballistic coefficients, which means their speed rapidly decreases when they lose their means of propulsion.

	Minimum Impact Speed (mph)	Maximum Impact Speed (mph)	Best Estimated Impact Speed (mph)	Minimum Volume (cubic inches)	Maximum Volume (cubic inches)	Best Estimated Volume (cubic inches)
During STS-107	375	654	477	400	1,920	1,200
After STS-107	528	559	528	1,026	1,239	1,200

Figure 3.4-5. The best estimates of velocities and volumes calculated during the mission and after the accident based on visual evidence and computer analyses. Information available during the mission was adequate to determine the foam's effect on both thermal tiles and RCC.

cameras are not operating or, as in the case of STS-107, out of focus. Launch Commit Criteria should include that sufficient cameras are operating to track the Shuttle from liftoff to Solid Rocket Booster separation.

Similarly, a developmental vehicle like the Shuttle should be equipped with high resolution cameras that monitor potential hazard areas. The wing leading edge system, the area around the landing gear doors, and other critical Thermal Protection System elements need to be imaged to check for damage. Debris sources, such as the External Tank, also need to be monitored. Such critical images need to be downlinked so that potential problems are identified as soon as possible.

Transport Analysis: Establishing Foam Path by Computational Fluid Dynamics

Transport analysis is the process of determining the path of the foam. To refine the Board's understanding of the foam strike, a transport analysis team, consisting of members from Johnson Space Center, Ames Research Center, and Boeing, augmented the image analysis team's research.

A variety of computer models were used to estimate the volume of the foam, as well as to refine the estimates of its velocity, its other dimensions, and the impact location. Figure 3.4-5 lists the velocity and foam size estimates produced during the mission and at the conclusion of the investigation.

The results listed in Figure 3.4-5 demonstrate that reasonably accurate estimates of the foam size and impact velocity were available during the mission. Despite the lack of high-quality visual evidence, the input data available to assess the impact damage during the mission was adequate.

The input data to the transport analysis consisted of the computed airflow around the Shuttle stack when the foam was shed, the estimated aerodynamic characteristics of the foam, the image analysis team's trajectory estimates, and the size and shape of the bipod ramp.

The transport analysis team screened several of the image analysis team's location estimates, based on the feasible aerodynamic characteristics of the foam and the laws of physics. Optical distortions caused by the atmospheric density gradients associated with the shock waves off the Orbiter's nose, External Tank, and Solid Rocket Boosters may have compromised the image analysis team's three position estimates closest to the bipod ramp. In addition, the image analysis team's position estimates closest to the wing were compromised by the lack of two camera views and the shock region ahead of the wing, making triangulation impossible and requiring extrapolation. However, the transport analysis confirmed that the image analysis team's estimates for the central portion of the foam trajectory were well within the computed flow field and the estimated range of aerodynamic characteristics of the foam.

The team identified a relatively narrow range of foam impact velocities and ballistic coefficients. The ballistic coefficient of an object expresses the relative influence of weight and atmospheric drag on it, and is the primary aerodynamic characteristic of an object that does not produce lift. An object with a large ballistic coefficient, such as a cannon ball, has a trajectory that can be computed fairly accurately without accounting for drag. In contrast, the foam that struck the wing had a relatively small ballistic coefficient with a large drag force relative to its weight, which explains why it slowed down quickly after separating from the External Tank. Just prior to separation, the speed of the foam was equal to the speed of the Shuttle, about 1,568 mph (2,300 feet per second). Because of a large drag force, the foam slowed to about 1,022 mph (1,500 feet per second) in about 0.2 seconds, and the Shuttle struck the foam at a relative

Figure 3.4-6. These are the results of a trajectory analysis that used a computational fluid dynamics approach in a program called CART-3D, a comprehensive (six-degree-of-freedom) computer simulation based on the laws of physics. This analysis used the aerodynamic and mass properties of bipod ramp foam, coupled with the complex flow field during ascent, to determine the likely position and velocity histories of the foam.

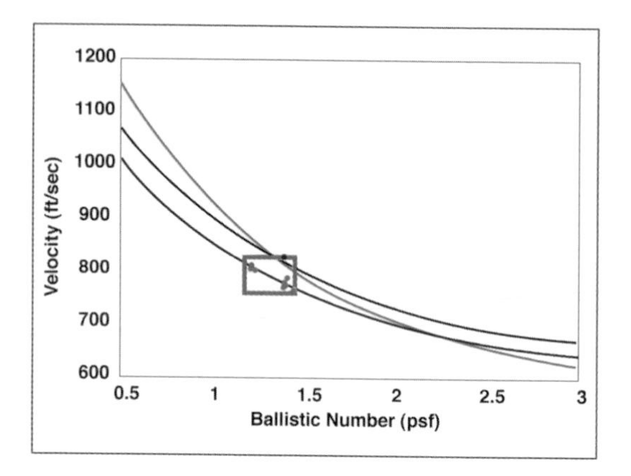

Figure 3.4-7. The results of numerous possible trajectories based on various assumed sizes, shapes, and densities of the foam. Either the foam had a slightly higher ballistic coefficient and the Orbiter struck the foam at a lower speed relative to the Orbiter, or the foam was more compact and the wing struck the foam at a higher speed. The "best fit" box represents the overlay of the data from the image analysis with the transport analysis computations. This data enabled a final selection of projectile characteristics for impact testing.

speed of about 545 mph (800 feet per second). (See Appendix D.8.)

The undetermined and yet certainly irregular shape of the foam introduced substantial uncertainty about its estimated aerodynamic characteristics. Appendix D.8 contains an independent analysis conducted by the Board to confirm that the estimated range of ballistic coefficients of the foam in Figure 3.4-6 was credible, given the foam dimension results from the image analyses and the expected range of the foam density. Based on the results in Figure 3.4-7, the physical dimensions of the bipod ramp, and the sizes and shapes of the available barrels for the compressed-gas gun used in the impact test program described later in this chapter, the Board and the NASA Accident Investigation Team decided that a foam projectile 19 inches by 11.5 inches by 5.5 inches, weighing 1.67 pounds, and with a weight density of 2.4 pounds per cubic foot, would best represent the piece of foam that separated from the External Tank bipod ramp and was hit by the Orbiter's left wing. See Section 3.8 for a full discussion of the foam impact testing.

Findings:

F3.4-1 Photographic evidence during ascent indicates the projectile that struck the Orbiter was the left bipod ramp foam.

F3.4-2 The same photographic evidence, confirmed by independent analysis, indicates the projectile struck the underside of the leading edge of the left wing in the vicinity of RCC panels 6 through 9 or the tiles directly behind, with a velocity of approximately 775 feet per second.

F3.4-3 There is a requirement to obtain and downlink on-board engineering quality imaging from the Shuttle during launch and ascent.

F3.4-4 The current long-range camera assets on the Kennedy Space Center and Eastern Range do not provide best possible engineering data during Space Shuttle ascents.

F3.4-5 Evaluation of STS-107 debris impact was hampered by lack of high resolution, high speed cameras (temporal and spatial imagery data).

F3.4-6 Despite the lack of high quality visual evidence, the information available about the foam impact during the mission was adequate to determine its effect on both the thermal tiles and RCC.

Recommendations:

R3.4-1 Upgrade the imaging system to be capable of providing a minimum of three useful views of the Space Shuttle from liftoff to at least Solid Rocket Booster separation, along any expected ascent azimuth. The operational status of these assets should be included in the Launch Commit Criteria for future launches. Consider using ships or aircraft to provide additional views of the Shuttle during ascent.

R3.4-2 Provide a capability to obtain and downlink high-resolution images of the External Tank after it separates.

R3.4-3 Provide a capability to obtain and downlink high-resolution images of the underside of the Orbiter wing leading edge and forward section of both wings' Thermal Protection System.

3.5 On-Orbit Debris Separation – The "Flight Day 2" Object

Immediately after the accident, Air Force Space Command began an in-depth review of its Space Surveillance Network data to determine if there were any detectable anomalies during the STS-107 mission. A review of the data resulted in no information regarding damage to the Orbiter. However, Air Force processing of Space Surveillance Network data yielded 3,180 separate radar or optical observations of the Orbiter from radar sites at Eglin, Beale, and Kirtland Air Force Bases, Cape Cod Air Force Station, the Air Force Space Command's Maui Space Surveillance System in Hawaii, and the Navy Space Surveillance System. These observations, examined after the accident, showed a small object in orbit with *Columbia*. In accordance with the International Designator system, the object was named 2003-003B (*Columbia* was designated 2003-003A). The timeline of significant events includes:

1. January 17, 2003, 9:42 a.m. Eastern Standard Time: Orbiter moves from tail-first to right-wing-first orientation
2. January 17, 10:17 a.m.: Orbiter returns to tail-first orientation
3. January 17, 3:57 p.m.: First confirmed sensor track of object 2003-003B
4. January 17, 4:46 p.m.: Last confirmed sensor track for this date

5. January 18: Object reacquired and tracked by Cape Cod Air Force Station PAVE PAWS
6. January 19: Object reacquired and tracked by Space Surveillance Network
7. January 20, 8:45 – 11:45 p.m.: 2003-003B orbit decays. Last track by Navy Space Surveillance System

Events around the estimated separation time of the object were reviewed in great detail. Extensive on-board sensor data indicates that no unusual crew activities, telemetry data, or accelerations in Orbiter or payload can account for the release of an object. No external mechanical systems were active, nor were any translational (forward, backward, or sideways, as opposed to rotational) maneuvers attempted in this period. However, two attitude maneuvers were made: a 48-degree yaw maneuver to a left-wing-forward and payload-bay-to-Earth attitude from 9:42 to 9:46 a.m. EST), and a maneuver back to the bay-to-Earth, tail-forward attitude from 10:17 to 10:21 a.m. It is possible that this maneuver imparted the initial departure velocity to the object.

ON-ORBIT COLLISION AVOIDANCE

The Space Control Center, operated by the 21st Space Wing's 1st Space Control Squadron (a unit of Air Force Space Command), maintains an orbital data catalog on some 9,000 Earth-orbiting objects, from active satellites to space debris, some of which may be as small as four inches. The Space Control Center ensures that no known orbiting objects will transit an Orbiter "safety zone" measuring 6 miles deep by 25 miles wide and long (Figure A) during a Shuttle mission by projecting the Orbiter's flight path for the next 72 hours (Figure B) and comparing it to the flight paths of all known orbiting or re-entering objects, which generally travel at 17,500 miles per hour. Whenever possible, the Orbiter moves tail-first while on orbit to minimize the chances of orbital debris or micrometeoroids impacting the cabin windscreen or the Orbiter's wing leading edge.

If an object is determined to be within 36-72 hours of colliding with the Orbiter, the Space Control Center notifies NASA, and the agency then determines a maneuver to avoid a collision. There were no close approaches to *Columbia* detected during STS-107.

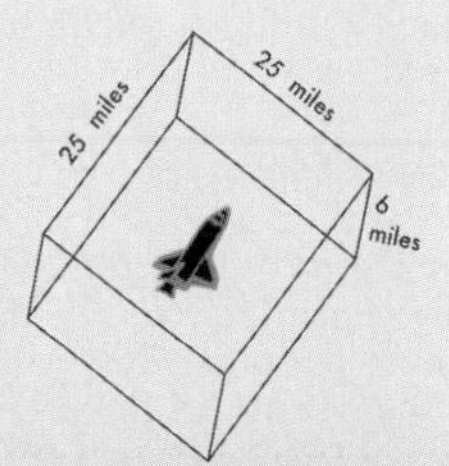

Figure A. Orbiter Safety Zone

Figure B. Protecting the Orbiter's flight path

Although various Space Surveillance Network radars tracked the object, the only reliable physical information includes the object's ballistic coefficient in kilograms per square meter and its radar cross-section in decibels per square meter. An object's radar cross-section relates how much radar energy the object scatters. Since radar cross-section depends on the object's material properties, shape, and orientation relative to the radar, the Space Surveillance Network could not independently estimate the object's size or shape. By radar observation, the object's Ultra-High Frequency (UHF) radar cross-section varied between 0.0 and minus 18.0 decibels per square meter (plus or minus 1.3 decibels), and its ballistic coefficient was known to be 0.1 kilogram per meter squared (plus or minus 15 percent). These two quantities were used to test and ultimately eliminate various objects.

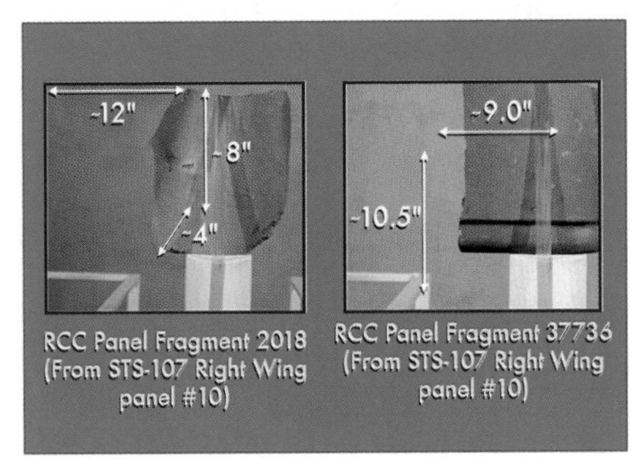

Figure 3.5-1. These representative RCC acreage pieces matched the radar cross-section of the Flight Day 2 object.

In the Advanced Compact Range at the Air Force Research Laboratory in Dayton, Ohio, analysts tested 31 materials from the Orbiter's exterior and payload bay. Additional supercomputer radar cross-section predictions were made for Reinforced Carbon-Carbon T-seals. After exhaustive radar cross-section analysis and testing, coupled with ballistic analysis of the object's orbital decay, only a fragment of RCC panel would match the UHF radar cross-section and ballistic coefficients observed by the Space Surveillance network. Such an RCC panel fragment must be approximately 140 square inches or greater in area to meet the observed radar cross-section characteristics. Figure 3.5-1 shows RCC panel fragments from *Columbia*'s right wing that represent those meeting the observed characteristics of object 2003-003B.[10]

Note that the Southwest Research Institute foam impact test on panel 8 (see Section 3.8) created RCC fragments that fell into the wing cavity. These pieces are consistent in size with the RCC panel fragments that exhibited the required physical characteristics consistent with the Flight Day 2 object.

Findings:

F3.5-1 The object seen on orbit with *Columbia* on Flight Day 2 through 4 matches the radar cross-section and area-to-mass measurements of an RCC panel fragment.

F3.5-2 Though the Board could not positively identify the Flight Day 2 object, the U.S. Air Force exclusionary test and analysis processes reduced the potential Flight Day 2 candidates to an RCC panel fragment.

Recommendations:

- None

3.6 DE-ORBIT/RE-ENTRY

As *Columbia* re-entered Earth's atmosphere, sensors in the Orbiter relayed streams of data both to entry controllers on the ground at Johnson Space Center and to the Modular Auxiliary Data System recorder, which survived the breakup of the Orbiter and was recovered by ground search teams. This data – temperatures, pressures, and stresses – came from sensors located throughout the Orbiter. Entry controllers were unaware of any problems with re-entry until telemetry data indicated errant readings. During the investigation data from these two sources was used to make aerodynamic, aerothermal, and mechanical reconstructions of re-entry that showed how these stresses affected the Orbiter.

The re-entry analysis and testing focused on eight areas:

1. Analysis of the Modular Auxiliary Data System recorder information and the pattern of wire runs and sensor failures throughout the Orbiter.
2. Physical and chemical analysis of the recovered debris to determine where the breach in the RCC panels likely occurred.
3. Analysis of videos and photography provided by the general public.
4. Abnormal heating on the outside of the Orbiter body. Sensors showed lower heating and then higher heating than is usually seen on the left Orbital Maneuvering System pod and the left side of the fuselage.
5. Early heating inside the wing leading edge. Initially, heating occurred inside the left wing RCC panels before the wing leading edge spar was breached.
6. Later heating inside the left wing structure. This analysis focused on the inside of the left wing after the wing leading edge spar had been breached.
7. Early changes in aerodynamic performance. The Orbiter began reacting to increasing left yaw and left roll, consistent with developing drag and loss of lift on the left wing.
8. Later changes in aerodynamic performance. Almost 600 seconds after Entry Interface, the left-rolling tendency of the Orbiter changes to a right roll, indicating an increase in lift on the left wing. The left yaw also increased, showing increasing drag on the left wing.

For a complete compilation of all re-entry data, see the CAIB/NAIT Working Scenario (Appendix D.7), Qualification and Interpretation of Sensor Data from STS-107 (Appendix D.19) and the Re-entry Timeline (Appendix D.9). The extensive aerothermal calculations and wind tunnel tests performed to investigate the observed re-entry phenomenon are documented in NASA report NSTS-37398.

Re-Entry Environment

In the demanding environment of re-entry, the Orbiter must withstand the high temperatures generated by its movement through the increasingly dense atmosphere as it decelerates from orbital speeds to land safely. At these velocities, shock waves form at the nose and along the leading edges of the wing, intersecting near RCC panel 9. The interaction between these two shock waves generates extremely high temperatures, especially around RCC panel 9, which experiences the highest surface temperatures of all the RCC panels. The flow behind these shock waves is at such a high temperature that air molecules are torn apart, or "dissociated." The air immediately around the leading edge surface can reach 10,000 degrees Fahrenheit; however, the boundary layer shields the Orbiter so that the actual temperature is only approximately 3,000 degrees Fahrenheit at the leading edge. The RCC panels and internal insulation protect the aluminum wing leading edge spar. A breach in one of the leading-edge RCC panels would expose the internal wing structure to temperatures well above 3,000 degrees Fahrenheit.

In contrast to the aerothermal environment, the aerodynamic environment during *Columbia*'s re-entry was relatively benign, especially early in re-entry. The re-entry dynamic pressure ranged from zero at Entry Interface to 80 pounds per square foot when the Orbiter went out of control, compared with a dynamic pressure during launch and ascent of nearly 700 pounds per square foot. However, the aerodynamic forces were increasing quickly during the final minutes of *Columbia*'s flight, and played an important role in the loss of control.

Orbiter Sensors

The Operational Flight Instrumentation monitors physical sensors and logic signals that report the status of various Orbiter functions. These sensor readings and signals are telemetered via a 128 kilobit-per-second data stream to the Mission Control Center, where engineers ascertain the real-time health of key Orbiter systems. An extensive review of this data has been key to understanding what happened to STS-107 during ascent, orbit, and re-entry.

The Modular Auxiliary Data System is a supplemental instrumentation system that gathers Orbiter data for processing after the mission is completed. Inputs are almost exclusively physical sensor readings of temperatures, pressures, mechanical strains, accelerations, and vibrations. The Modular Auxiliary Data System usually records only the mission's first and last two hours (see Figure 3.6-1).

The Orbiter Experiment instrumentation is an expanded suite of sensors for the Modular Auxiliary Data System that was installed on *Columbia* for engineering development purposes. Because *Columbia* was the first Orbiter launched,

Figure 3.6-1. The Modular Auxiliary Data System recorder, found near Hemphill, Texas. While not designed to withstand impact damage, the recorder was in near-perfect condition when recovered on March 19, 2003.

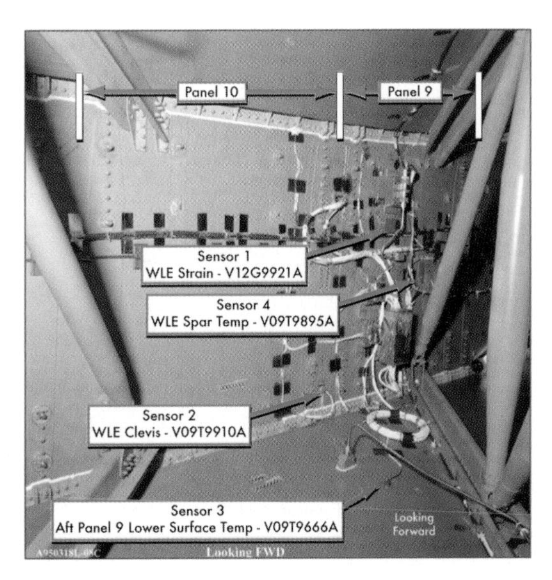

Figure 3.6-2. Location of sensors on the back of the left wing leading edge spar (vertical aluminum structure in picture). Also shown are the round truss tubes and ribs that provided the structural support for the mid-wing in this area.

engineering teams needed a means to gather more detailed flight data to validate their calculations of conditions the vehicle would experience during critical flight phases. The instrumentation remained on *Columbia* as a legacy of the development process, and was still providing valuable flight data from ascent, de-orbit, and re-entry for ongoing flight analysis and vehicle engineering. Nearly all of *Columbia*'s sensors were specified to have only a 10-year shelf life, and in some cases an even shorter service life.

At 22 years old, the majority of the Orbiter Experiment instrumentation had been in service twice as long as its specified service life, and in fact, many sensors were already failing. Engineers planned to stop collecting and analyzing data once most of the sensors had failed, so failed sensors and wiring were not repaired. For instance, of the 181 sensors in *Columbia*'s wings, 55 had already failed or were producing questionable readings before STS-107 was launched.

Re-Entry Timeline

Times in the following section are noted in seconds elapsed from the time *Columbia* crossed Entry Interface (EI) over the Pacific Ocean at 8:44:09 a.m. EST. *Columbia*'s destruction occurred in the period from Entry Interface at 400,000 feet (EI+000) to about 200,000 feet (EI+970) over Texas. The Modular Auxiliary Data System recorded the first indications of problems at EI plus 270 seconds (EI+270). Because data from this system is retained onboard, Mission Control did not notice any troubling indications from telemetry data until 8:54:24 a.m. (EI+613), some 10 minutes after Entry Interface.

Left Wing Leading Edge Spar Breach (EI+270 through EI+515)

At EI+270, the Modular Auxiliary Data System recorded the first unusual condition while the Orbiter was still over the Pacific Ocean. Four sensors, which were all either inside or outside the wing leading edge spar near Reinforced Carbon-Carbon (RCC) panel 9-left, helped tell the story of what happened on the left wing of the Orbiter early in the re-entry. These four sensors were: strain gauge V12G9921A (Sensor 1), resistance temperature detector V09T9910A on the RCC clevis between panel 9 and 10 (Sensor 2), thermocouple V07T9666A, within a Thermal Protection System tile (Sensor 3), and resistance temperature detector V09T9895A (Sensor 4), located on the back side of the wing leading edge spar behind RCC panels 8 and 9 (see Figure 3.6-2).

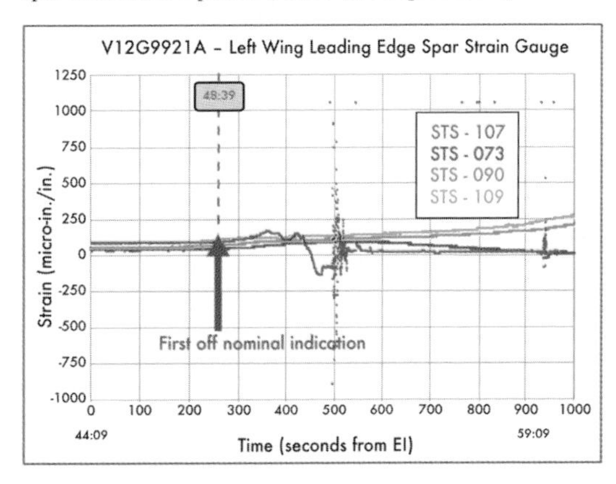

Figure 3.6-3. The strain gauge (Sensor 1) on the back of the left wing leading edge spar was the first sensor to show an anomalous reading. In this chart, and the others that follow, the red line indicates data from STS-107. Data from other Columbia re-entries, similar to the STS-107 re-entry profile, are shown in the other colors.

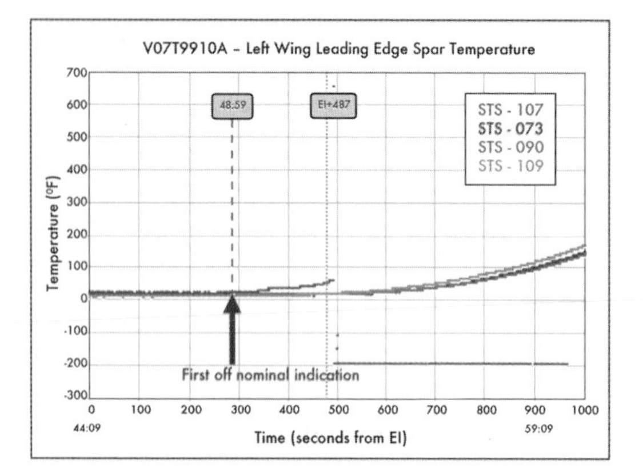

Figure 3.6-4. This temperature thermocouple (Sensor 2) was mounted on the outside of the wing leading edge spar behind the insulation that protects the spar from radiated heat from the RCC panels. It clearly showed an off-nominal trend early in the re-entry sequence and began to show an increase in temperature much earlier than the temperature sensor behind the spar.

Sensor 1 provided the first anomalous reading (see Figure 3.6-3). From EI+270 to EI+360, the strain is higher than that on previous *Columbia* flights. At EI+450, the strain reverses, and then peaks again in a negative direction at EI+475. The strain then drops slightly, and remains constant and negative until EI+495, when the sensor pattern becomes unreliable, probably due to a propagating soft short, or "burn-through" of the insulation between cable conductors caused by heating or combustion. This strain likely indicates significant damage to the aluminum honeycomb spar. In particular, strain reversals, which are unusual, likely mean there was significant high-temperature damage to the spar during this time.

At EI+290, 20 seconds after Sensor 1 gave its first anomalous reading, Sensor 2, the only sensor in the front of the left wing leading edge spar, recorded the beginning of a gradual and abnormal rise in temperature from an expected 30 degrees Fahrenheit to 65 degrees at EI+493, when it then dropped to "off-scale low," a reading that drops off the scale at the low end of the sensor's range (see Figure 3.6-4). Sensor 2, one of the first to fail, did so abruptly. It had indicated only a mild warming of the RCC attachment clevis before the signal was lost.

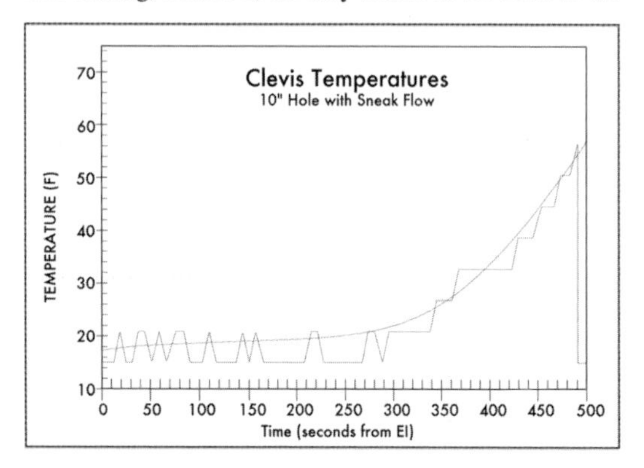

Figure 3.6-5. The analysis of the effect of a 10-inch hole in RCC panel 8 on Sensor 2 from EI to EI+500 seconds. The jagged line shows the actual flight data readings and the smooth line the calculated result for a 10-inch hole with some sneak flow of superheated air behind the spar insulation.

A series of thermal analyses were performed for different sized holes in RCC panel 8 to compute the time required to heat Sensor 2 to the temperature recorded by the Modular Auxiliary Data System. To heat the clevis, various insulators would have to be bypassed with a small amount of leakage, or "sneak flow." Figure 3.6-5 shows the results of these calculations for, as an example, a 10-inch hole, and demonstrates that with sneak flow around the insulation, the temperature profile of the clevis sensor was closely matched by the engineering calculations. This is consistent with the same sneak flow required to match a similar but abnormal ascent temperature rise of the same sensor, which further supports the premise that the breach in the leading edge of the wing occurred during ascent. While the exact size of the breach will never be known, and may have been smaller or larger than 10 inches, these analyses do provide a plausible explanation for the observed rises in temperature sensor data during re-entry.

Investigators initially theorized that the foam might have broken a T-seal and allowed superheated air to enter the wing between the RCC panels. However, the amount of T-seal debris from this area and subsequent aerothermal analysis showing this type of breach did not match the observed damage to the wing, led investigators to eliminate a missing T-seal as the source of the breach.

Although abnormal, the re-entry temperature rise was slow and small compared to what would be expected if Sensor 2 were exposed to a blast of superheated air from an assumed breach in the RCC panels. The slow temperature rise is at-

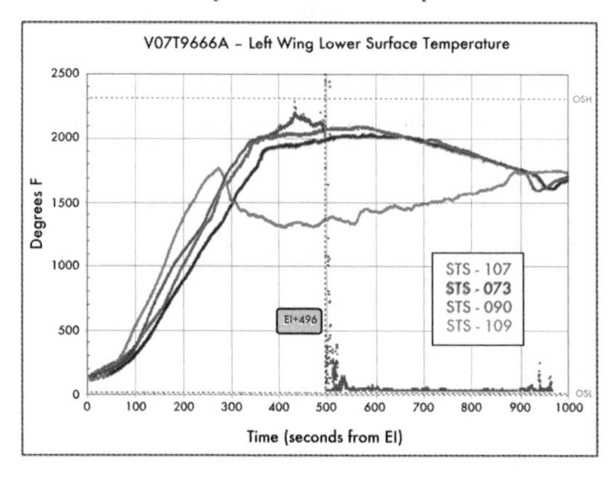

Figure 3.6-6. As early as EI+370, Sensor 3 began reading significantly higher than on previous flights. Since this sensor was located in a thermal tile on the lower surface of the left wing, its temperatures are much higher than those for the other sensors.

tributed to the presence of a relatively modest breach in the RCC, the thick insulation that surrounds the sensor, and the distance from the site of the breach in RCC panel 8 to the clevis sensor.

The readings of Sensor 3, which was in a thermal tile, began rising abnormally high and somewhat erratically as early as EI+370, with several brief spikes to 2,500 degrees Fahrenheit, significantly higher than the 2,000-degree peak temperature on a normal re-entry (Figure 3.6-6). At EI+496, this reading became unreliable, indicating a failure of the wire or the sensor. Because this thermocouple was on the wing lower surface, directly behind the junction of RCC panel 9 and 10, the high temperatures it initially recorded were almost certainly a result of air jetting through the damaged area of RCC panel 8, or of the normal airflow being disturbed by the damage. Note that Sensor 3 provided an external temperature measurement, while Sensors 2 and 4 provided internal temperature measurements.

Sensor 4 also recorded a rise in temperature that ended in an abrupt fall to off-scale low. Figure 3.6-7 shows that an abnormal temperature rise began at EI+425 and abruptly fell at EI+525. Unlike Sensor 2, this temperature rise was extreme, from an expected 20 degrees Fahrenheit at EI+425 to 40 degrees at EI+485, and then rising much faster to 120 degrees at EI+515, then to an off-scale high (a reading that climbs off the scale at the high end of the range) of 450 degrees at EI+522. The failure pattern of this sensor likely indicates destruction by extreme heat.

The timing of the failures of these four sensors and the path of their cable routing enables a determination of both the timing and location of the breach of the leading edge spar, and indirectly, the breach of the RCC panels. All the cables from these sensors, and many others, were routed into wiring harnesses that ran forward along the back side of the leading edge spar up to a cross spar (see Figure 3.6-8), where they passed through the service opening in the cross spar and then ran in front of the left wheel well before reaching interconnect panel 65P, where they entered the fuselage. All sensors with wiring in this set of harnesses failed between EI+487 to EI+497, except Sensor 4, which survived until EI+522. The diversity of sensor types (temperature, pressure, and strains) and their locations in the left wing indicates that they failed because their wiring was destroyed at spar burn-through, as opposed to destruction of each individual sensor by direct heating.

Examination of wiring installation closeout photographs (pictures that document the state of the area that are normally taken just before access is closed) and engineering drawings show five main wiring harness bundles running forward along the spar, labeled top to bottom as A through E (see Figure 3.6-8). The top four, A through D, are spaced 3 inches apart, while the fifth, E, is 6 inches beneath them. The separation between bundle E and the other four is consistent with the later failure time of Sensor 4 by 25 to 29 seconds, and indicates that the breach was in the upper two-thirds of the spar, causing all but one of the cables in this area to fail between EI+487 to EI+497. The breach then expanded vertically, toward the underside of the wing, causing Sensor 4 to fail 25 seconds

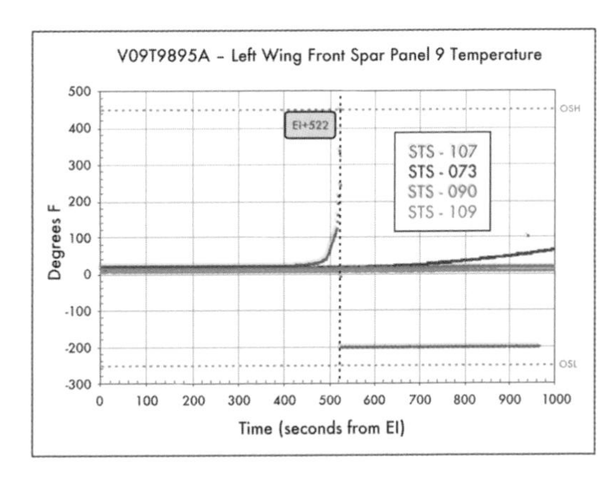

Figure 3.6-7. Sensor 4 also began reading significantly higher than previous flights before it fell off-scale low. The relatively late reaction of this sensor compared to Sensor 2, clearly indicated that superheated air started on the outside of the wing leading edge spar and then moved into the mid-wing after the spar was burned through. Note that immediately before the sensor (or the wire) fails, the temperature is at 450 degrees Fahrenheit and climbing rapidly. It was the only temperature sensor that showed this pattern.

later. Because the distance between bundle A and bundle E is 9 inches, the failure of all these wires indicates that the breach in the wing leading edge spar was at least 9 inches from top to bottom by EI+522 seconds.

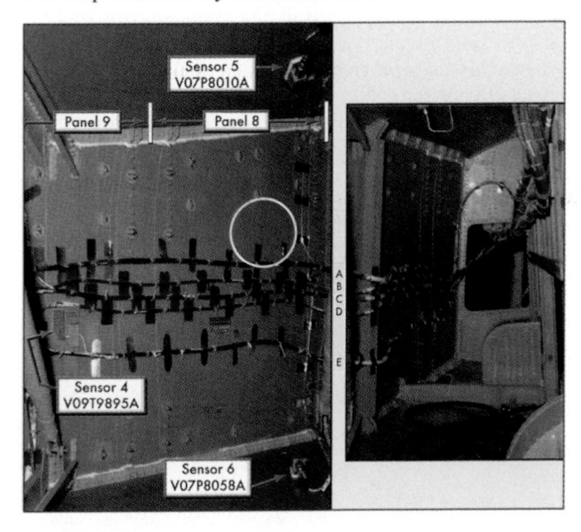

Figure 3.6-8. The left photo above shows the wiring runs on the backside of the wing leading edge behind RCC panel 8 – the circle marks the most likely area where the burn through of the wing leading edge spar initially occurred at EI+487 seconds. The right photo shows the wire bundles as they continue forward behind RCC panels 7 and 6. The major cable bundles in the upper right of the right photo carried the majority of the sensor data inside the wing. As these bundles were burned, controllers on the ground began seeing off-nominal sensor indications.

Also directly behind RCC panel 8 were pressure sensors V07P8010A (Sensor 5), on the upper interior surface of the wing, and V07P8058A (Sensor 6), on the lower interior surface of the wing. Sensor 5 failed abruptly at EI+497. Sensor 6, which was slightly more protected, began falling at EI+495, and failed completely at EI+505. Closeout photographs show that the wiring from Sensor 5 travels down from the top of the wing to join the uppermost harness, A, which then travels along the leading edge spar. Similarly, wiring from Sensor 6 travels up from the bottom of the wing, joins harness A, and continues along the spar. It appears that Sensor 5's wiring, on the upper wing surface, was damaged at EI+497, right after Sensor 1 failed. Noting the times of the sensor failures, and the locations of Sensors 5 and 6 forward of Sensors 1 through 4, spar burn-through must have occurred near where these wires came together.

Two of the 45 left wing strain gauges also recorded an anomaly around EI+500 to EI+580, but their readings were not erratic or off-scale until late in the re-entry, at EI+930. Strain gauge V12G9048A was far forward on a cross spar in the front of the wheel well on the lower spar cap, and strain gauge V12G9049A was on the upper spar cap. Their responses appear to be the actual strain at that location until their failure at EI+935. The exposed wiring for most of the left wing sensors runs along the front of the spar that crosses in front of the left wheel well. The very late failure times of these two sensors indicate that the damage did not spread into the wing cavity forward of the wheel well until at least EI+935, which implies that the breach was aft of the cross spar. Because the cross spar attaches to the transition spar behind RCC panel 6, the breach must have been aft (outboard) of panel 6. The superheated air likely burned through the outboard wall of the wheel well, rather than snaking forward and then back through the vent at the front of the wheel well. Had the gases flowed through the access opening in the cross spar and then through the vent into the wheel well, it is unlikely that the lower strain gauge wiring would have survived.

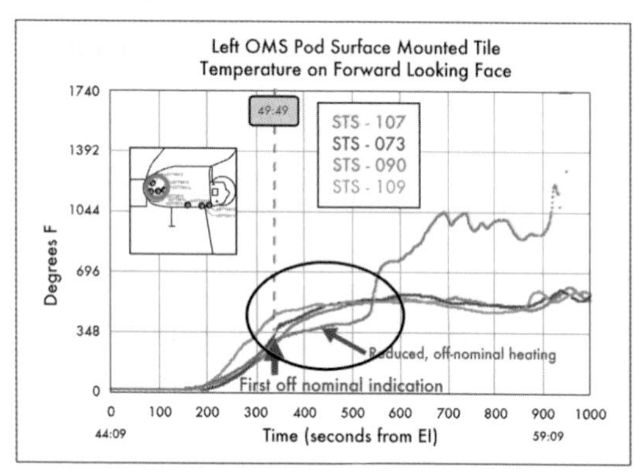

Figure 3.6-9. Orbital Maneuvering System (OMS) pod heating was initially significantly lower than that seen on previous Columbia missions. As wing leading edge damage later increased, the OMS pod heating increased dramatically. Debris recovered from this area of the OMS pod showed substantial pre-breakup heat damage and imbedded drops of once-molten metal from the wing leading edge in the OMS pod thermal tiles.

Finally, the rapid rise in Sensor 4 at EI+425, before the other sensors began to fail, indicates that high temperatures were responsible. Comparisons of sensors on the outside of the wing leading edge spar, those inside of the spar, and those in the wing and left wheel well indicate that abnormal heating first began on the outside of the spar behind the RCC panels and worked through the spar. Since the aluminum spar must have burned through before any cable harnesses attached to it failed, the breach through the wing leading edge spar must have occurred at or before EI+487.

Other abnormalities also occurred during re-entry. Early in re-entry, the heating normally seen on the left Orbital Maneuvering System pod was much lower than usual for this point in the flight (see Figure 3.6-9). Wind tunnel testing demonstrated that airflow into a breach in an RCC panel would then escape through the wing leading edge vents behind the upper part of the panel and interrupt the weak aerodynamic flow field on top of the wing. During re-entry, air normally flows into these vents to equalize air pressure across the RCC panels. The interruption in the flow field behind the wing caused a displacement of the vortices that normally hit the leading edge of the left pod, and resulted in a slowing of pod heating. Heating of the side fuselage slowed, which wind tunnel testing also predicted.

To match this scenario, investigators had to postulate damage to the tiles on the upper carrier panel 9, in order to allow sufficient mass flow through the vent to cause the observed decrease in sidewall heating. No upper carrier panels were found from panels 9, 10, and 11, which supports this hypothesis. Although this can account for the abnormal temperatures on the body of the Orbiter and at the Orbital Maneuvering System pod, flight data and wind tunnel tests confirmed that this venting was not strong enough to alter the aerodynamic force on the Orbiter, and the aerodynamic analysis of mission data showed no change in Orbiter flight control parameters during this time.

During re-entry, a change was noted in the rate of the temperature rise around the RCC chin panel clevis temperature sensor and two water supply nozzles on the left side of the fuselage, just aft of the main bulkhead that divides the crew cabin from the payload bay. Because these sensors were well forward of the damage in the left wing leading edge, it is still unclear how their indications fit into the failure scenario.

Sensor Loss and the Onset of Unusual Aerodynamic Effects (EI+500 through EI+611)

Fourteen seconds after the loss of the first sensor wire on the wing leading edge spar at EI+487, a sensor wire in a bundle of some 150 wires that ran along the upper outside corner of the left wheel well showed a burn-through. In the next 50 seconds, more than 70 percent of the sensor wires in three cables in this area also burned through (see Figure 3.6-10). Investigators plotted the wiring run for every left-wing sensor, looking for a relationship between their location and time of failure.

Only two sensor wires of 169 remained intact when the Modular Auxiliary Data System recorder stopped, indicat-

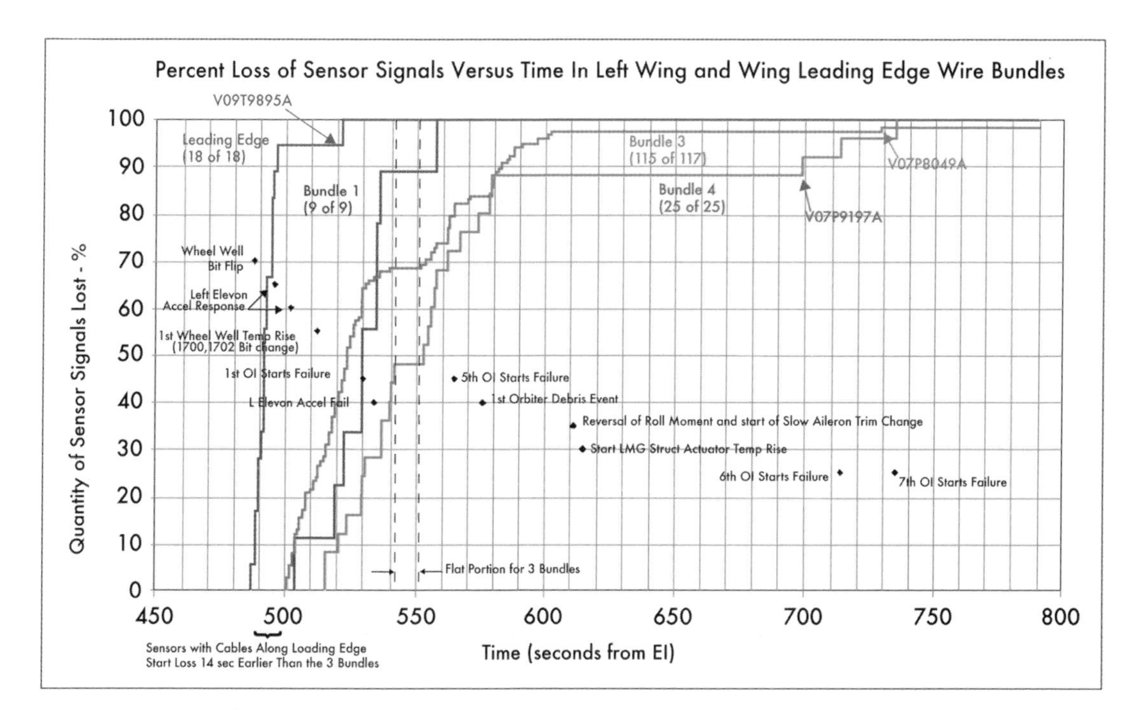

Figure 3.6-10. This chart shows how rapidly the wire bundles in the left wing were destroyed. Over 70 percent of the sensor wires in the wiring bundles burned through in under a minute. The black diamonds show the times of significant timeline sensor events.

ing that the burn-throughs had to occur in an area that nearly every wire ran through. To sustain this type of damage, the wires had to be close enough to the breach for the gas plume to hit them. Arc jet testing (in a wind tunnel with an electrical arc that provides up to a 2,800-degree Fahrenheit airflow) on a simulated wing leading edge spar and simulated wire bundles showed how the leading edge spar would burn through in a few seconds. It also showed that wire bundles would burn through in a timeframe consistent with those seen in the Modular Auxiliary Data System information and the telemetered data.

Later computational fluid dynamics analysis of the mid-wing area behind the spar showed that superheated air flowing into a breached RCC panel 8 and then interacting with the internal structure behind the RCC cavity (RCC ribs and spar insulation) would have continued through the wing leading edge spar as a jet, and would have easily allowed superheated air to traverse the 56.5 inches from the spar to the outside of the wheel well and destroy the cables (Figure 3.6-11). Controllers on the ground saw these first anomalies in the telemetry data at EI+613, when four hydraulic sensor cables that ran from the aft part of the left wing through the wiring bundles outside the wheel well failed.

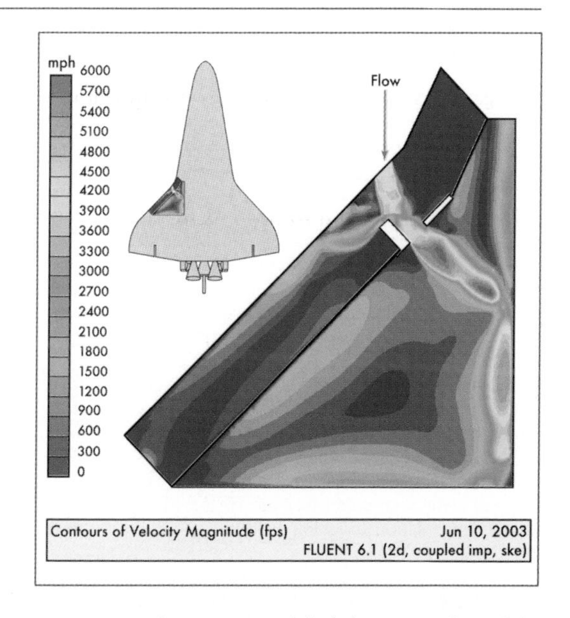

Figure 3.6-11. The computational fluid dynamics analysis of the speed of the superheated air as it entered the breach in RCC panel 8 and then traveled through the wing leading edge spar. The darkest red color indicates speeds of over 4,000 miles per hour. Temperatures in this area likely exceeded 5,000 degrees Fahrenheit. The area of detail is looking down at the top of the left wing.

Aerodynamic roll and yaw forces began to differ from those on previous flights at about EI+500 (see Figure 3.6-12). Investigators used flight data to reconstruct the aerodynamic forces acting on the Orbiter. This reconstructed data was then compared to forces seen on other similar flights of *Columbia*

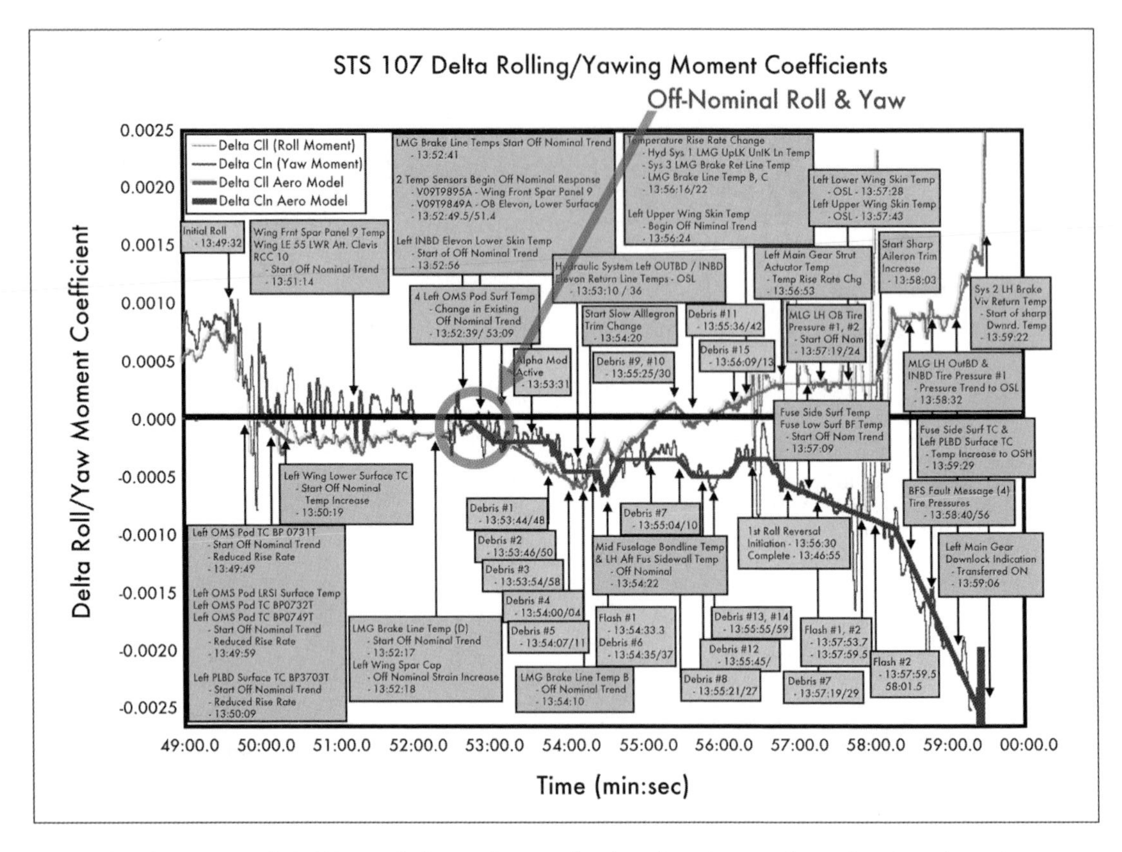

Figure 3.6-12. At approximately EI+500 seconds, the aerodynamic roll and yaw forces began to diverge from those observed on previous flights. The blue line shows the Orbiter's tendency to yaw while the red line shows its tendency to roll. Nominal values would parallel the solid black line. Above the black line, the direction of the force is to the right, while below the black line, the force is to the left.

and to the forces predicted for STS-107. In the early phase of fight, these abnormal aerodynamic forces indicated that *Columbia*'s flight control system was reacting to a change in the external shape of the wing, which was caused by progressive RCC damage that caused a continuing decrease in lift and a continuing increase in drag on the left wing.

Between EI+530 and EI+562, four sensors on the left inboard elevon failed. These sensor readings were part of the data telemetered to the ground. Noting the system failures, the Maintenance, Mechanical, and Crew Systems officer notified the Flight Director of the failures. (See sidebar in Chapter 2 for a complete version of the Mission Control Center conversation about this data.)

At EI+555, *Columbia* crossed the California coast. People on the ground now saw the damage developing on the Orbiter in the form of debris being shed, and documented this with video cameras. In the next 15 seconds, temperatures on the fuselage sidewall and the left Orbital Maneuvering System pod began to rise. Hypersonic wind tunnel tests indicated that the increased heating on the Orbital Maneuvering System pod and the roll and yaw changes were caused by substantial leading edge damage around RCC panel 9. Data on Orbiter temperature distribution as well as aerodynamic forces for various damage scenarios were obtained from wind tunnel testing.

Figure 3.6-13 shows the comparison of surface temperature distribution with an undamaged Orbiter and one with an entire panel 9 removed. With panel 9 removed, a strong vortex flow structure is positioned to increase the temperature on the leading edge of the Orbital Maneuvering System pod. The aim is not to demonstrate that all of panel 9 was missing at this point, but rather to indicate that major damage to panels near panel 9 can shift the strong vortex flow pattern and change the Orbiter's temperature distribution to match the Modular Auxiliary Data System information. Wind tunnel tests also demonstrated that increasing damage to leading edge RCC panels would result in increasing drag and decreasing lift on the left wing.

Recovered debris showed that Inconel 718, which is only found in wing leading edge spanner beams and attachment fittings, was deposited on the left Orbital Maneuvering System pod, verifying that airflow through the breach and out of the upper slot carried molten wing leading edge material back to the pod. Temperatures far exceeded those seen on previous re-entries and further confirmed that the wing leading-edge damage was increasing.

By this time, superheated air had been entering the wing since EI+487, and significant internal damage had probably occurred. The major internal support structure in the mid-wing consists of aluminum trusses with a melting point of 1,200 degrees Fahrenheit. Because the ingested air may have been as hot as 8,000 degrees near the breach, it is likely that the internal support structure that maintains the shape of the wing was severely compromised.

As the Orbiter flew east, people on the ground continued to record the major shedding of debris. Investigators later scrutinized these videos to compare *Columbia*'s re-entry with recordings of other re-entries and to identify the debris. The video analysis was also used to determine additional search areas on the ground and to estimate the size of various pieces of debris as they fell from the Orbiter.

Temperatures in the wheel well began to rise rapidly at EI+601, which indicated that the superheated air coming through the wing leading edge spar had breached the wheel well wall. At the same time, observers on the ground noted additional significant shedding of debris. Analysis of one of these "debris events" showed that the photographed object could have weighed nearly 190 pounds, which would have significantly altered *Columbia*'s physical condition.

At EI+602, the tendency of the Orbiter to roll to the left in response to a loss of lift on the left wing transitioned to a right-rolling tendency, now in response to increased lift on the left wing. Observers on the ground noted additional significant shedding of debris in the next 30 seconds. Left yaw continued to increase, consistent with increasing drag on the left wing. Further damage to the RCC panels explains the increased drag on the left wing, but it does not explain the sudden increase in lift, which can be explained only by some other type of wing damage.

Investigators ran multiple analyses and wind tunnel tests to understand this significant aerodynamic event. Analysis showed that by EI+850, the temperatures inside the wing were high enough to substantially damage the wing skins, wing leading edge spar, and the wheel well wall, and melt the wing's support struts. Once structural support was lost, the wing likely deformed, effectively changing shape and resulting in increased lift and a corresponding increase in drag on the left wing. The increased drag on the left wing further increased the Orbiter's tendency to yaw left.

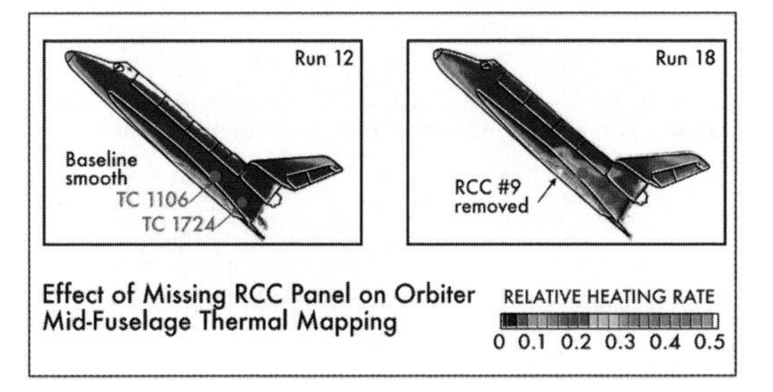

Figure 3.6-13. The effects of removing RCC panel 9 are shown in this figure. Note the brighter colors on the front of the OMS pod show increased heating, a phenomenon supported by both the OMS pod temperature sensors and the debris analysis.

The Kirtland Image

As *Columbia* passed over Albuquerque, New Mexico, during re-entry (around EI+795), scientists at the Air Force Starfire Optical Range at Kirtland Air Force Base acquired images of the Orbiter. This imaging had not been officially assigned, and the photograph was taken using commercial equipment located at the site, not with the advanced Starfire adaptive-optics telescope.

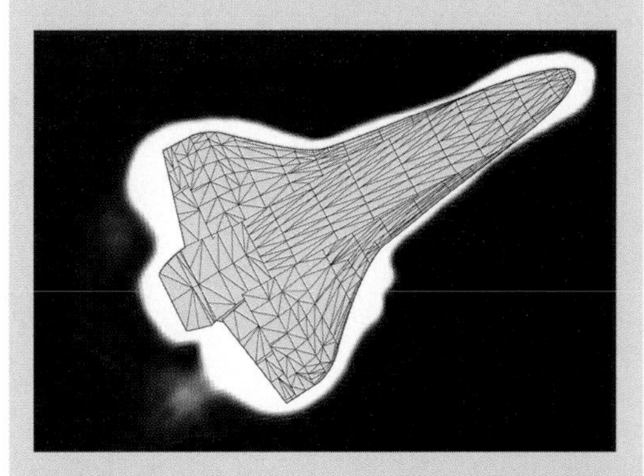

The image shows an unusual condition on the left wing, a leading-edge disturbance that might indicate damage. Several analysts concluded that the distortion evident in the image likely came from the modification and interaction of shock waves due to the damaged leading edge. The overall appearance of the leading-edge damage at this point on the trajectory is consistent with the scenario.

Loss of Vehicle Control (EI+612 through EI+970)

A rise in hydraulic line temperatures inside the left wheel well indicated that superheated air had penetrated the wheel well wall by EI+727. This temperature rise, telemetered to Mission Control, was noted by the Maintenance, Mechanical, and Crew Systems officer. The Orbiter initiated and completed its roll reversal by EI+766 and was positioned left-wing-down for this portion of re-entry. The Guidance and Flight Control Systems performed normally, although the aero-control surfaces (aileron trim) continued to counteract the additional drag and lift from the left wing.

At EI+790, two left main gear outboard tire pressure sensors began trending slightly upward, followed very shortly by going off-scale low, which indicated extreme heating of both the left inboard and outboard tires. The tires, with their large mass, would require substantial heating to produce the sensors' slight temperature rise. Another sharp change in the rolling tendency of the Orbiter occurred at EI+834, along

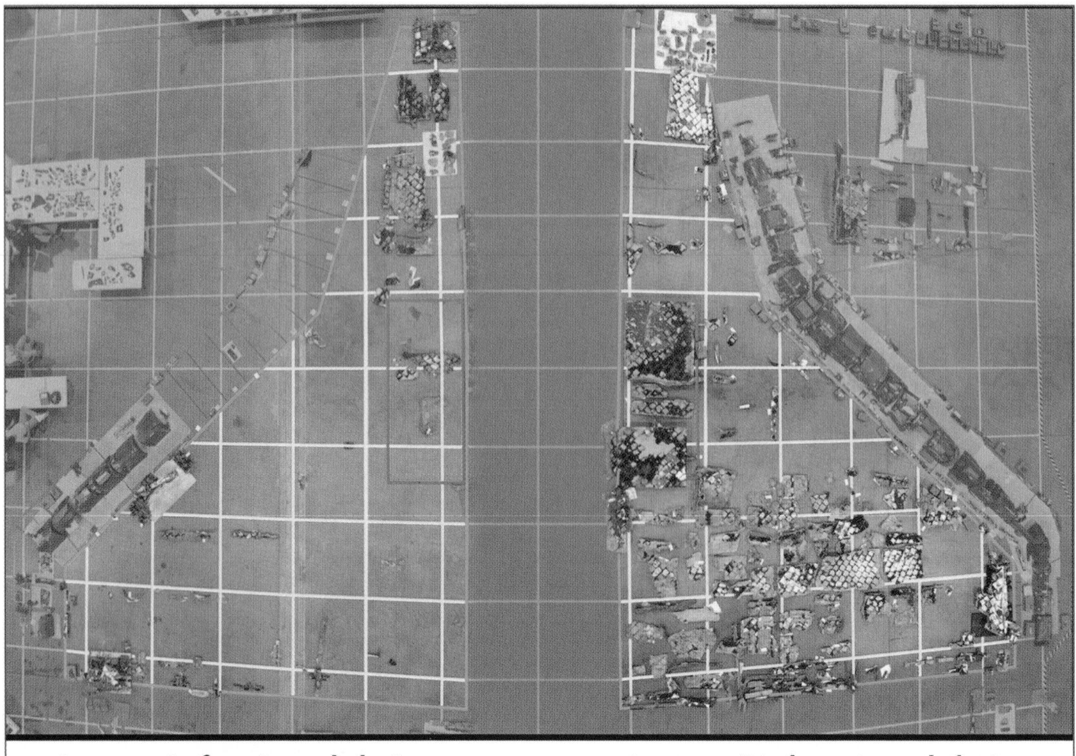

Figure 3.7-1. Comparison of amount of debris recovered from the left and right wings of Columbia. Note the amount of debris recovered from areas in front of the wheel well (the red boxes on each wing) were similar, but there were dramatic differences in the amount of debris recovered aft of each wheel well.

with additional shedding of debris. In an attempt to maintain attitude control, the Orbiter responded with a sharp change in aileron trim, which indicated there was another significant change to the left wing configuration, likely due to wing deformation. By EI+887, all left main gear inboard and outboard tire pressure and wheel temperature measurements were lost, indicating burning wires and a rapid progression of damage in the wheel well.

At EI+897, the left main landing gear downlock position indicator reported that the gear was now down and locked. At the same time, a sensor indicated the landing gear door was still closed, while another sensor indicated that the main landing gear was still locked in the up position. Wire burn-through testing showed that a burn-induced short in the downlock sensor wiring could produce these same contradictions in gear status indication. Several measurements on the strut produced valid data until the final loss of telemetry data. This suggests that the gear-down-and locked indication was the result of a wire burn-through, not a result of the landing gear actually deploying. All four corresponding proximity switch sensors for the right main landing gear remained normal throughout re-entry until telemetry was lost.

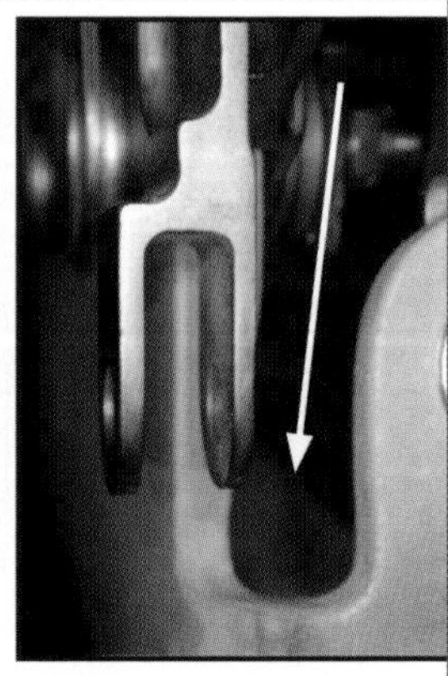

Figure 3.7-2. Each RCC panel has a U-shaped slot (see arrow) in the back of the panel. Once superheated air entered the breach in RCC panel 8, some of that superheated air went through this slot and caused substantial damage to the Thermal Protection System tiles behind this area.

Post-accident analysis of flight data that was generated after telemetry information was lost showed another abrupt change in the Orbiter's aerodynamics caused by a continued progression of left wing damage at EI+917. The data showed a significant increase in positive roll and negative yaw, again indicating another increase in drag on and lift from the damaged left wing. *Columbia*'s flight control system attempted to compensate for this increased left yaw by firing all four right yaw jets. Even with all thrusters firing, combined with a maximum rate of change of aileron trim, the flight control system was unable to control the left yaw, and control of the Orbiter was lost at EI+970 seconds. Mission Control lost all telemetry data from the Orbiter at EI+923 (8:59:32 a.m.). Civilian and military video cameras on the ground documented the final breakup. The Modular Auxiliary Data System stopped recording at EI+970 seconds.

Findings:

F3.6–1 The de-orbit burn and re-entry flight path were normal until just before Loss of Signal.

F3.6–2 *Columbia* re-entered the atmosphere with a pre-existing breach in the left wing.

F3.6–3 Data from the Modular Auxiliary Data System recorder indicates the location of the breach was in the RCC panels on the left wing leading edge.

F3.6–4 Abnormal heating events preceded abnormal aerodynamic events by several minutes.

F3.6–5 By the time data indicating problems was telemetered to Mission Control Center, the Orbiter had already suffered damage from which it could not recover.

Recommendations:

R3.6-1 The Modular Auxiliary Data System instrumentation and sensor suite on each Orbiter should be maintained and updated to include current sensor and data acquisition technologies.

R3.6-2 The Modular Auxiliary Data System should be redesigned to include engineering performance and vehicle health information, and have the ability to be reconfigured during flight in order to allow certain data to be recorded, telemetered, or both, as needs change.

3.7 DEBRIS ANALYSIS

The Board performed a detailed and exhaustive investigation of the debris that was recovered. While sensor data from the Orbiter pointed to early problems on the left wing, it could only isolate the breach to the general area of the left wing RCC panels. Forensics analysis independently determined that RCC panel 8 was the most likely site of the breach, and this was subsequently corroborated by other analyses. (See Appendix D.11.)

Pre-Breakup and Post-Breakup Damage Determination

Differentiating between pre-breakup and post-breakup damage proved a challenge. When *Columbia*'s main body breakup occurred, the Orbiter was at an altitude of about 200,000 feet and traveling at Mach 19, well within the peak-heating region calculated for its re-entry profile. Consequently, as individual pieces of the Orbiter were exposed to the atmosphere at breakup, they experienced temperatures high enough to damage them. If a part had been damaged by heat prior to breakup, high post-breakup temperatures could easily conceal the pre-breakup evidence. In some cases, there was no clear way to determine what happened when. In other cases, heat erosion occurred over fracture surfaces, indicating the piece had first broken and had then experienced high temperatures. Investigators concluded that pre- and post-breakup damage had to be determined on a part-by-part basis; it was impossible to make broad generalizations based on the gross physical evidence.

Amount of Right Wing Debris versus Left Wing Debris

Detailed analysis of the debris revealed unique features and convincing evidence that the damage to the left wing differed significantly from damage to the right, and that significant differences existed in pieces from various areas of the left wing. While a substantial amount of upper and lower right wing structure was recovered, comparatively little of the upper and lower left wing structure was recovered (see Figure 3.7-1).

The difference in recovered debris from the Orbiter's wings clearly indicates that after the breakup, most of the left wing succumbed to both high heat and aerodynamic forces, while the right wing succumbed to aerodynamic forces only. Because the left wing was already compromised, it was the first area of the Orbiter to fail structurally. Pieces were exposed to higher heating for a longer period, resulting in more heat damage and ablation of left wing structural material. The left wing was also subjected to superheated air that penetrated directly into the mid-body of the wing for a substantial period. This pre-heating likely rendered those components unable to absorb much, if any, of the post-breakup heating. Those internal and external structures were likely vaporized during post-breakup re-entry. Finally, the left wing likely lost significant amounts of the Thermal Protection System prior to breakup due to the effect of internal wing heating on the Thermal Protection System bonding materials, and this further degraded the left wing's ability to resist the high heat of re-entry after it broke up.

Tile Slumping and External Patterns of Tile Loss

Tiles recovered from the lower left wing yielded their own interesting clues. The left wing lower carrier panel 9 tiles sustained extreme heat damage (slumping) and showed more signs of erosion than any other tiles. This severe heat erosion damage was likely caused by an *outflow* of superheated air and molten material from behind RCC panel 8 through a U-shaped design gap in the panel (see Figure 3.7-2) that allows room for the T-seal attachment. Effluents from the back side of panel 8 would directly impact this area of lower carrier panel 9 and its tiles. In addition, flow lines in these tiles (see Figure 3.7-3) exhibit evidence of superheated airflow across their surface from the area of the RCC panel

Figure 3.7-3. Superheated airflow caused erosion in tiles around the RCC panel 8 and 9 interface. The tiles shown are from behind the area where the superheated air exited from the slot in Figure 3.7-2. These tiles showed much greater thermal damage than other tiles in this area and chemical analysis showed the presence of metals only found in wing leading edge components.

8 and 9 interface. Chemical analysis shows that these carrier panel tiles were covered with molten Inconel, which is found in wing leading edge attachment fittings, and other metals coming from inside the RCC cavity. Slumping and heavy erosion of this magnitude is not noted on tiles from anywhere else on the Orbiter.

Failure modes of recovered tiles from the left and the right wing also differ. Most right wing tiles were simply broken off the wing due to aerodynamic forces, which indicates that they failed due to physical overload at breakup, not because of heat. Most of the tiles on the left wing behind RCC panels 8 and 9 show significant evidence of backside heating of the wing skin and failure of the adhesive that held the tiles on the wing. This pattern of failure suggests that heat penetrated the left wing cavity and then heated the aluminum skin *from the inside out*. As the aluminum skin was heated, the strength of the tile bond degraded, and tiles separated from the Orbiter.

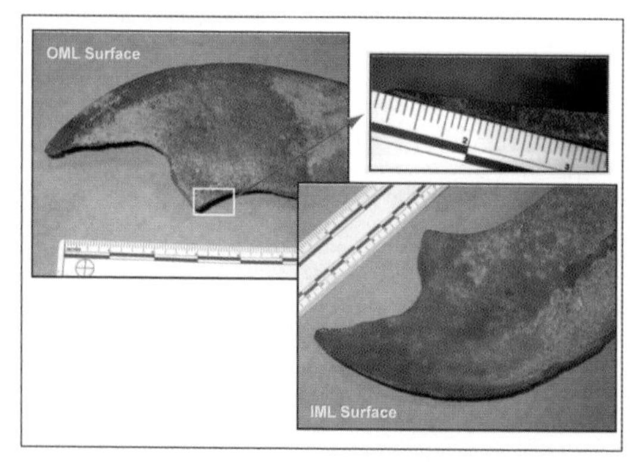

Figure 3.7-4. The outboard rib of panel 8 and the inboard rib of panel 9 showed signs of extreme heating and erosion. RCC erosion of this magnitude was not observed in any other location on the Orbiter.

Erosion of Left Wing Reinforced Carbon-Carbon

Several pieces of left wing RCC showed unique signs of heavy erosion from exposure to extreme heat. There was erosion on two rib panels on the left wing leading edge in the RCC panel 8 and 9 interface. Both the outboard rib of panel 8 and the inboard rib of panel 9 showed signs of extreme heating and erosion (see Figure 3.7-4). This erosion indicates that there was extreme heat behind RCC panels 8 and 9. This type of RCC erosion was not seen on any other part of the left or right wing.

Locations of Reinforced Carbon-Carbon Debris

The location of debris on the ground also provided evidence of where the initial breach occurred. The location of every piece of recovered RCC was plotted on a map and labeled according to the panel the piece originally came from. Two distinct patterns were immediately evident. First, it was clear that pieces from left wing RCC panels 9 through 22 had fallen the farthest west, and that RCC from left wing panels 1 through 7 had fallen considerably farther east (see Figure 3.7-5). Second, pieces from left wing panel 8 were

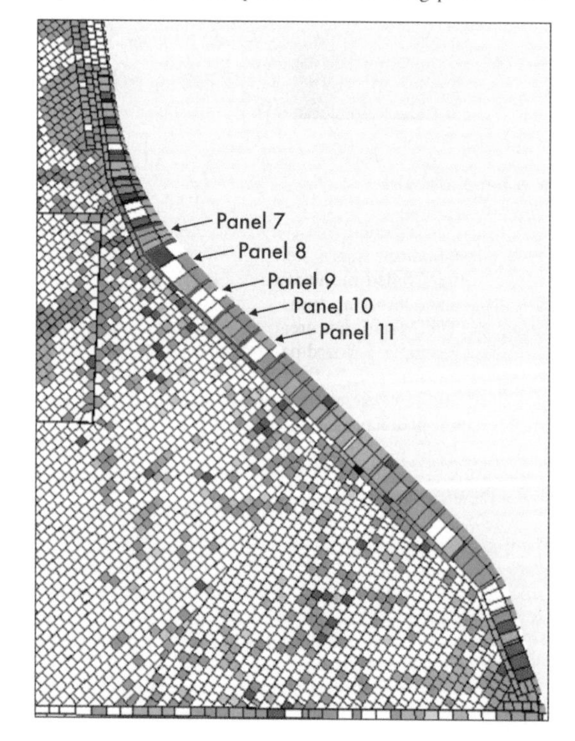

Figure 3.7-6. The tiles recovered farthest west all came from the area immediately behind left wing RCC panels 8 and 9. In the figure, each small box represents an individual tile on the lower surface of the left wing. The more red an individual tile appears, the farther west it was found.

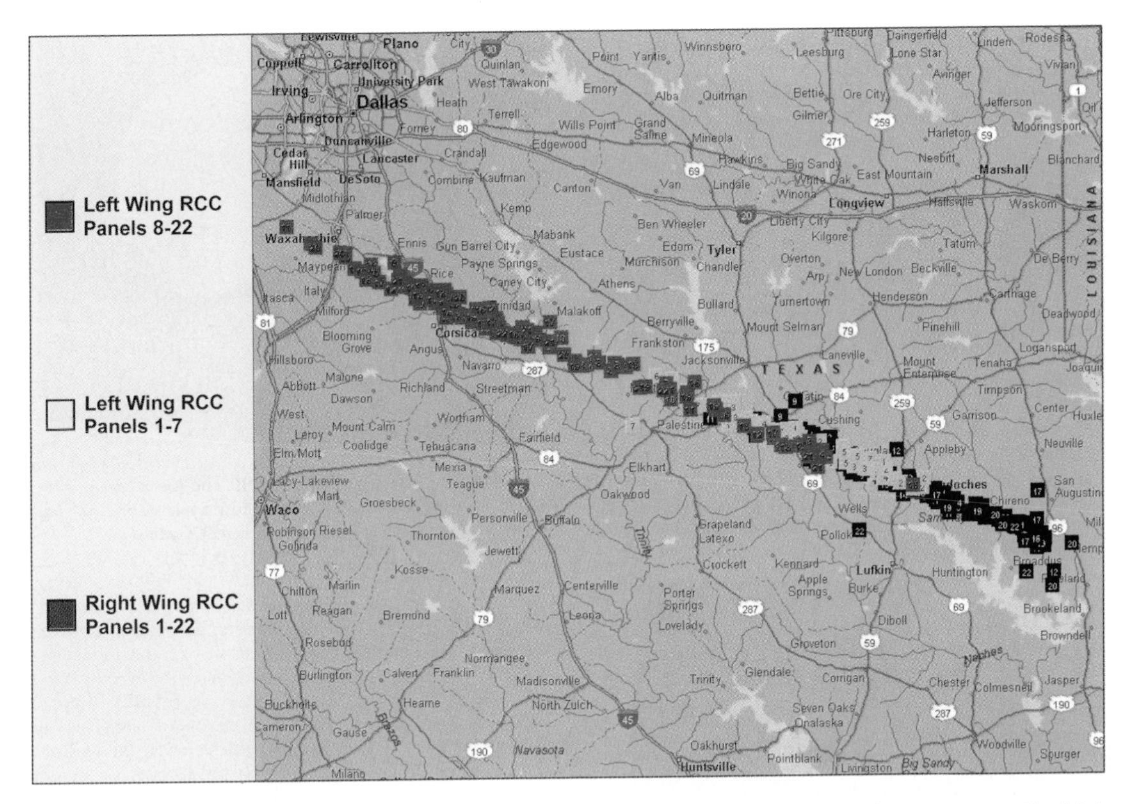

Figure 3.7-5. The location of RCC panel debris from the left and right wings, shown where it was recovered from in East Texas. The debris pattern suggested that the left wing failed before the right wing, most likely near left RCC panels 8 and 9.

found throughout the debris field, which suggested that the left wing likely failed in the vicinity of RCC panel 8. The early loss of the left wing from RCC panel 9 and outboard caused the RCC from that area to be deposited well west of the RCC from the inboard part of the wing. Since panels 1 through 7 were so much farther to the east, investigators concluded that RCC panels 1 through 7 had stayed with the Orbiter longer than had panels 8 through 22.

Tile Locations

An analysis of where tiles were found on the ground also yielded significant evidence of the breach location. Since most of the tiles are of similar size, weight, and shape, they would all have similar ballistic coefficients and would have behaved similarly after they separated from the Orbiter. By noting where each tile fell and then plotting its location on the Orbiter tile map, a distinctive pattern emerged. The tiles recovered farthest west all came from the area immediately behind the left wing RCC panel 8 and 9 (see Figure 3.7-6), which suggests that these tiles were released earlier than those from other areas of the left wing. While it is not conclusive evidence of a breach in this area, this pattern does suggest unique damage around RCC panels 8 and 9 that was not seen in other areas. Tiles from this area also showed evidence of a brown deposit that was not seen on tiles from any other part of the Orbiter. Chemical analysis revealed it was an Inconel-based deposit that had come from inside the RCC cavity on the left wing (Inconel is found in wing leading edge attachment fittings). Since the streamlines from tiles with the brown deposit originate near left RCC panels 8 and 9, this brown deposit likely originated as an outflow of superheated air and molten metal from the panel 8 and 9 area.

Molten Deposits

High heat damage to metal parts caused molten deposits to form on some Orbiter debris. Early analysis of these deposits focused on their density and location. Much of the left wing leading edge showed some signs of deposits, but the left wing RCC panels 5 to 10 had the highest levels.

Of all the debris pieces recovered, left wing panels 8 and 9 showed the largest amounts of deposits. Significant but lesser amounts of deposits were also observed on left wing RCC panels 5 and 7. Right wing RCC panel 8 was the only right-wing panel with significant deposits.

Chemical and X-Ray Analysis

Chemical analysis focused on recovered pieces of RCC panels with unusual deposits. Samples were obtained from areas in the vicinity of left wing RCC panel 8 as well as other left and right wing RCC panels. Deposits on recovered RCC debris were analyzed by cross-sectional optical and scanning electron microscopy, microprobe analysis, and x-ray diffraction to determine the content and layering of slag deposits. Slag was defined as metallic and non-metallic deposits that resulted from the melting of the internal wing structures. X-ray analysis determined the best areas to sample for chemical testing and to see if an overall flow pattern could be discerned.

The X-ray analysis of left wing RCC panel 8 (see Figure 3.7-7) showed a bottom-to-top pattern of slag deposits. In some areas, small spheroids of heavy metal were aligned vertically on the recovered pieces, which indicated a superheated airflow from the bottom of the panel toward the top in the area of RCC panel 8-left. These deposits were later determined by chemical analysis to be Inconel 718, probably from the wing leading edge attachment fittings on the spanner beams on RCC panels 8 and 9. Computational fluid dynamics modeling of the flow behind panel 8 indicated that the molten deposits would be laid down in this manner.

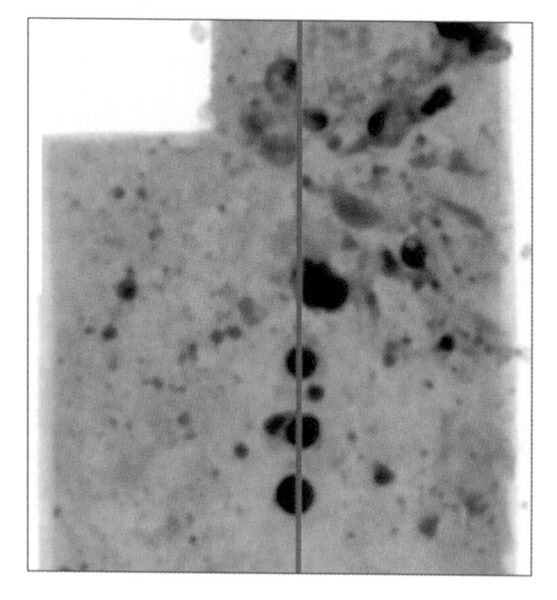

Figure 3.7-7. X-ray analysis of RCC panel 8-left showed a bottom-to-top pattern of slag deposits.

The layered deposits on panel 8 were also markedly different from those on all other left- and right-wing panels. There was much more material deposited on RCC panel 8-left. These deposits had a much rougher overall structure, including rivulets of Cerachrome slag deposited directly on the RCC. This indicated that Cerachrome, the insulation that protects the wing leading edge spar, was one of the first materials to succumb to the superheated air entering through the breach in RCC panel 8-left. Because the melting temperature of Cerachrome is greater than 3,200 degrees Fahrenheit, analysis indicated that materials in this area were exposed to extremely

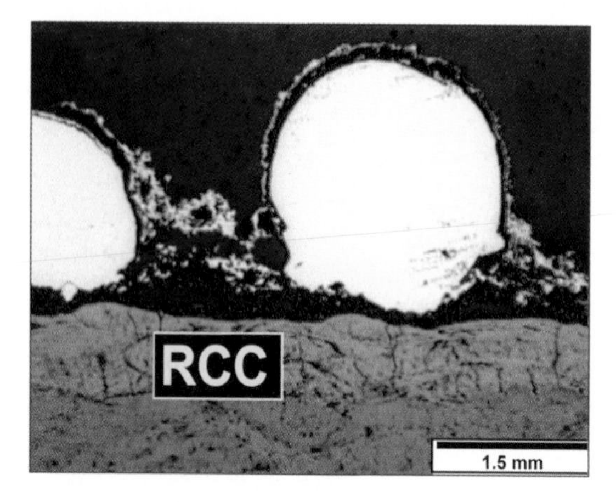

Figure 3.7-8. Spheroids of Inconel 718 and Cerachrome were deposited directly on the surface of RCC panel 8-left. This slag deposit pattern was not seen on any other RCC panels.

high temperatures for a long period. Spheroids of Inconel 718 were mixed in with the Cerachrome. Because these spheroids (see Figure 3.7-8) were directly on the surface of the RCC and also in the first layers of deposits, investigators concluded that the Inconel 718 spanner beam RCC fittings were most likely the first internal structures subjected to intense heating. No aluminum was detected in the earliest slag layers on RCC panel 8-left. Only one location on an upper corner piece, near the spar fitting attachment, contained A-286 stainless steel. This steel was not present in the bottom layer of the slag directly on the RCC surface, which indicated that the A-286 attachment fittings on the wing spar were not in the direct line of the initial plume impingement.

In wing locations other than left RCC panels 8 and 9, the deposits were generally thinner and relatively uniform. This suggests no particular breach location other than in left RCC panels 8 and 9. These other slag deposits contained primarily aluminum and aluminum oxides mixed with A-286, Inconel, and Cerachrome, with no consistent layering. This mixing of multiple metals in no apparent order suggests concurrent melting and re-depositing of all leading-edge components, which is more consistent with post-breakup damage than the organized melting and depositing of materials that occurred near the original breach at left RCC panels 8 and 9. RCC panel 9-left also differs from the rest of the locations analyzed. It was similar to panel 8-left on the inboard side, but more like the remainder of the samples analyzed on its outboard side. The deposition of molten deposits strongly suggests the original breach occurred in RCC panel 8-left.

Spanner Beams, Fittings, and Upper Carrier Panels

Spanner beams, fittings, and upper carrier panels were recovered from areas adjacent to most of the RCC panels on both wings. However, significant numbers of these items were not recovered from the vicinity of left RCC panels 6 to 10. None of the left wing upper carrier panels at positions 9, 10, or 11 were recovered. No spanner beam parts were recovered from

STS-107 Crew Survivability

At the Board's request, NASA formed a Crew Survivability Working Group within two weeks of the accident to better understand the cause of crew death and the breakup of the crew module. This group made the following observations.

Medical and Life Sciences

The Working Group found no irregularities in its extensive review of all applicable medical records and crew health data. The Armed Forces Institute of Pathology and the Federal Bureau of Investigation conducted forensic analyses on the remains of the crew of *Columbia* after they were recovered. It was determined that the acceleration levels the crew module experienced prior to its catastrophic failure were not lethal. The death of the crew members was due to blunt trauma and hypoxia. The exact time of death – sometime after 9:00:19 a.m. Eastern Standard Time – cannot be determined because of the lack of direct physical or recorded evidence.

Failure of the Crew Module

The forensic evaluation of all recovered crew module/forward fuselage components did not show any evidence of over-pressurization or explosion. This conclusion is supported by both the lack of forensic evidence and a credible source for either sort of event.[11] The failure of the crew module resulted from the thermal degradation of structural properties, which resulted in a rapid catastrophic sequential structural breakdown rather than an instantaneous "explosive" failure.

Separation of the crew module/forward fuselage assembly from the rest of the Orbiter likely occurred immediately in front of the payload bay (between Xo576 and Xo582 bulkheads). Subsequent breakup of the assembly was a result of ballistic heating and dynamic loading. Evaluations of fractures on both primary and secondary structure elements suggest that structural failures occurred at high temperatures and in some cases at high strain rates. An extensive trajectory reconstruction established the most likely breakup sequence, shown below.

The load and heat rate calculations are shown for the crew module along its reconstructed trajectory. The band superimposed on the trajectory (starting about 9:00:58 a.m. EST) represents the window where all the evaluated debris originated. It appears that the destruction of the crew module took place over a period of 24 seconds beginning at an altitude of approximately 140,000 feet and ending at 105,000 feet. These figures are consistent with the results of independent thermal re-entry and aerodynamic models. The debris footprint proved consistent with the results of these trajectory analyses and models. Approximately 40 to 50 percent, by weight, of the crew module was recovered.

The Working Group's results significantly add to the knowledge gained from the loss of *Challenger* in 1986. Such knowledge is critical to efforts to improve crew survivability when designing new vehicles and identifying feasible improvements to the existing Orbiters.

Crew Worn Equipment

Videos of the crew during re-entry that have been made public demonstrate that prescribed procedures for use of equipment such as full-pressure suits, gloves, and helmets were not strictly followed. This is confirmed by the Working Group's conclusions that three crew members were not wearing gloves, and one was not wearing a helmet. However, under these circumstances, this did not affect their chances of survival.

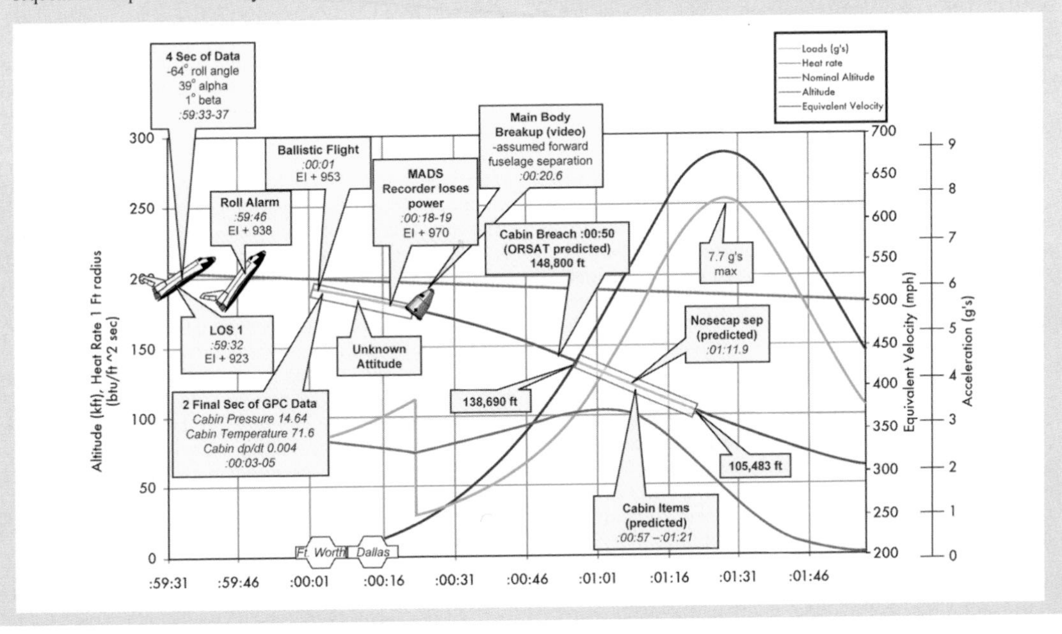

Board Testing

NASA and the Board agreed that tests would be required and a test plan developed to validate an impact/breach scenario. Initially, the Board intended to act only in an oversight role in the development and implementation of a test plan. However, ongoing and continually unresolved debate on the size and velocity of the foam projectile, largely due to the Marshall Space Flight Center's insistence that, despite overwhelming evidence to the contrary, the foam could have been no larger than 855 cubic inches, convinced the Board to take a more active role. Additionally, in its assessment of potential foam damage NASA continued to rely heavily on the Crater model, which was used during the mission to determine that the foam-shedding event was non-threatening. Crater is a semi-empirical model constructed from Apollo-era data. Another factor that contributed to the Board's decision to play an active role in the test program was the Orbiter Vehicle Engineering Working Group's requirement that the test program be used to validate the Crater model. NASA failed to focus on physics-based pre-test predictions, the schedule priorities for RCC tests that were determined by transport analysis, the addition of appropriate test instrumentation, and the consideration of additional factors such as launch loads. Ultimately, in discussions with the Orbiter Vehicle Engineering Working Group and the NASA Accident Investigation Team, the Board provided test plan requirements that outlined the template for all testing. The Board directed that a detailed written test plan, with Board-signature approval, be provided before each test.

the left RCC panel 8 to 10 area. No upper or lower RCC fittings were recovered for left panels 8, 9, or 10. Some of this debris may not have been found in the search, but it is unlikely that all of it was missed. Much of this structure probably melted, and was burned away by superheated air inside the wing. What did not melt was so hot that when it broke apart, it did not survive the heat of re-entry. This supports the theory that superheated air penetrated the wing in the general area of RCC panel 8-left and caused considerable structural damage to the left wing leading edge spar and hardware.

Debris Analysis Conclusions

A thorough analysis of left wing debris (independent of the preceding aerodynamic, aerothermal, sensor, and photo data) supports the conclusion that significant abnormalities occurred in the vicinity of left RCC panels 8 and 9. The preponderance of debris evidence alone strongly indicates that the breach occurred in the bottom of panel 8-left. The unique composition of the slag found in panels 8 and 9, and especially on RCC panel 8-left, indicates extreme and prolonged heating in these areas very early in re-entry.

The early loss of tiles in the region directly behind left RCC panels 8 and 9 also supports the conclusion that a breach through the wing leading edge spar occurred here. This allowed superheated air to flow into the wing directly behind panel 8. The heating of the aluminum wing skin degraded tile adhesion and contributed to the early loss of tiles.

Severe damage to the lower carrier panel 9-left tiles is indicative of a flow out of panel 8-left, also strongly suggesting that the breach in the RCC was through panel 8-left. It is noteworthy that it occurred only in this area and not in any other areas on either the left or the right wing lower carrier panels. There is also significant and unique evidence of severe "knife edges" erosion in left RCC panels 8 and 9. Lastly, the pattern of the debris field also suggests the left wing likely failed in the area of RCC panel 8-left.

The preponderance of unique debris evidence in and near RCC panel 8-left strongly suggests that a breach occurred here. Finally, the unique debris damage in the RCC panel 8-left area is completely consistent with other data, such as the Modular Auxiliary Data System recorder, visual imagery analysis, and the aerodynamic and aerothermal analysis.

Findings:

F3.7–1 Multiple indications from the debris analysis establish the point of heat intrusion as RCC panel 8-left.

F3.7–2 The recovery of debris from the ground and its reconstruction was critical to understanding the accident scenario.

Recommendations:

- None

3.8 Impact Analysis and Testing

The importance of understanding this potential impact damage and the need to prove or disprove the impression that foam could not break an RCC panel prompted the investigation to develop computer models for foam impacts and undertake an impact-testing program of shooting pieces of foam at a mockup of the wing leading edge to re-create, to the extent practical, the actual STS-107 debris impact event.

Based on imagery analysis conducted during the mission and early in the investigation, the test plan included impacts on the lower wing tile, the left main landing gear door, the wing leading edge, and the carrier panels.

A main landing gear door assembly was the first unit ready for testing. By the time that testing occurred, however, analysis was pointing to an impact site in RCC panels 6 through 9. After the main landing gear door tests, the analysis and testing effort shifted to the wing leading edge RCC panel assemblies. The main landing gear door testing provided valuable data on test processes, equipment, and instrumentation. Insignificant tile damage was observed at the low impact angles of less than 20 degrees (the impact angle if the foam had struck the main landing gear door would have been roughly five degrees). The apparent damage threshold was consistent with previous testing with much smaller projectiles in 1999, and with independent modeling by Southwest Research Institute. (See Appendix D.12.)

Impact Test – Wing Leading Edge Panel Assemblies

The test concept was to impact flightworthy wing leading edge RCC panel assemblies with a foam projectile fired by a compressed-gas gun. Target panel assemblies with a flight history similar to *Columbia*'s would be mounted on a support that was structurally equivalent to *Columbia*'s wing. The attaching hardware and fittings would be either flight certified or built to *Columbia* drawings. Several considerations influenced the overall RCC test design:

- RCC panel assemblies were limited, particularly those with a flight history similar to *Columbia*'s.
- The basic material properties of new RCC were known to be highly variable and were not characterized for high strain rate loadings typical of an impact.
- The influence of aging was uncertain.
- The RCC's brittleness allowed only one test impact on each panel to avoid the possibility that hidden damage would influence the results of later impacts.
- The structural system response of RCC components, their support hardware, and the wing structure was complex.
- The foam projectile had to be precisely targeted, because the predicted structural response depended on the impact point.

Because of these concerns, engineering tests with fiberglass panel assemblies from the first Orbiter, *Enterprise*,[12] were used to obtain an understanding of overall system response to various impact angles, locations, and foam orientations. The fiberglass panel impact tests were used to confirm instrumentation design and placement and the adequacy of the overall test setup.

Test projectiles were made from the same type of foam as the bipod ramp on STS-107's External Tank. The projectile's mass and velocity were determined by the previously described "best fit" image and transport analyses. Because the precise impact point was estimated, the aiming point for any individual test panel was based on structural analyses to maximize the loads in the area being assessed without producing a spray of foam over the top of the wing. The angle of impact relative to the test panel was determined from the transport analysis of the panel being tested. The foam's rotational velocity was accounted for with a three-degree increase in the impact angle.

Computer Modeling of Impact Tests

The investigation used sophisticated computer models to analyze the foam impact and to help develop an impact test program. Because an exhaustive test matrix to cover all feasible impact scenarios was not practical, these models were especially important to the investigation.

The investigation impact modeling team included members from Boeing, Glenn Research Center, Johnson Space Center, Langley Research Center, Marshall Space Flight Center, Sandia National Laboratory, and Stellingwerf Consulting. The Board also contracted with Southwest Research Institute to perform independent computer analyses because of the institute's extensive test and analysis experience with ballistic impacts, including work on the Orbiter's Thermal Protection System. (Appendix D.12 provides a complete description of Southwest's impact modeling methods and results.)

The objectives of the modeling effort included (1) evaluation of test instrumentation requirements to provide test data with which to calibrate the computer models, (2) prediction of stress, damage, and instrumentation response prior to the Test Readiness Reviews, and (3) determination of the flight conditions/loads (vibrations, aerodynamic, inertial, acoustic, and thermal) to include in the tests. In addition, the impact modeling team provided information about foam impact locations, orientation at impact, and impact angle adjustments that accounted for the foam's rotational velocity.

Flight Environment

A comprehensive consideration of the Shuttle's flight environment, including temperature, pressure, and vibration, was required to establish the experimental protocol.

Figure 3.8-1. Nitrogen-powered gun at the Southwest Research Institute used for the test series.

Based on the results of Glenn Research Center sub-scale impact tests of how various foam temperatures and pressures influence the impact force, the Board found that full-scale impact tests with foam at room temperature and pressure could adequately simulate the conditions during the foam strike on STS-107.[13]

The structure of the foam complicated the testing process. The bipod ramp foam is hand-sprayed in layers, which creates "knit lines," the boundaries between each layer, and the foam compression characteristics depend on the knit lines' orientation. The projectiles used in the full-scale impact tests had knit lines consistent with those in the bipod ramp foam.

A primary concern of investigators was that external loads present in the flight environment might add substantial extra force to the left wing. However, analysis demonstrated that the only significant external loads on the wing leading edge structural subsystem at about 82 seconds into flight are due to random vibration and the pressure differences inside and outside the leading edge. The Board concluded that the flight environment stresses in the RCC panels and the attachment fittings could be accounted for in post-impact analyses if necessary. However, the dramatic damage produced by the impact tests demonstrated that the foam strike could breach the wing leading edge structure subsystem independent of any stresses associated with the flight environment. (Appendix D.12 contains more detail.)

Test Assembly

The impact tests were conducted at a Southwest Research Institute facility. Figure 3.8-1 shows the nitrogen gas gun that had evaluated bird strikes on aircraft fuselages. The gun was modified to accept a 35-foot-long rectangular barrel, and the target site was equipped with sensors and high-speed cameras that photographed 2,000 to 7,000 frames per second, with intense light provided by theater spotlights and the sun.

Test Impact Target

The leading edge structural subsystem test target was designed to accommodate the Board's evolving determination of the most likely point of impact. Initially, analysis pointed to the main landing gear door. As the imaging and transport teams refined their assessments, the likely strike zone narrowed to RCC panels 6 through 9. Because of the long lead time to develop and produce the large complex test assemblies, investigators developed an adaptable test assembly (Figure 3.8-2) that would provide a structurally similar mounting for RCC panel assemblies 5 to 10 and would accommodate some 200 sensors, including high-speed cameras, strain and deflection gauges, accelerometers, and load cells.[14]

Test Panels

RCC panels 6 and 9, which bracketed the likely impact region, were the first identified for testing. They would also permit a comparison of the structural response of panels with and without the additional thickness at certain locations.

Panel 6 tests demonstrated the complex system response to impacts. While the initial focus of the investigation had been on single panel response, early results from the tests with fiberglass panels hinted at "boundary condition" effects. Instruments measured high stresses through panels 6, 7, and 8. With this in mind, as well as forensic and sensor evidence that panel 8 was the likeliest location of the foam strike, the Board decided that the second RCC test should target panel 8, which was placed in an assembly that included RCC panels 9 and 10 to provide high fidelity boundary conditions. The decision to impact test RCC panel 8 was complicated by the lack of spare RCC components.

The specific RCC panel assemblies selected for testing had flight histories similar to that of STS-107, which was *Columbia*'s 28th flight. Panel 6 had flown 30 missions on *Discovery*, and Panel 8 had flown 26 missions on *Atlantis*.

Test Projectile

The preparation of BX-250 foam test projectiles used the same material and preparation processes that produced the foam bipod ramp. Foam was selected as the projectile material because foam was the most likely debris, and materials other than foam would represent a greater threat.

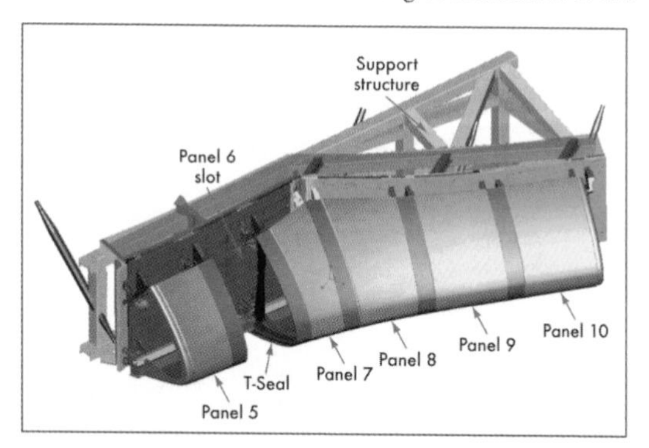

Figure 3.8-2. Test assembly that provided a structural mounting for RCC panel assemblies 5 to 10 and would accommodate some 200 sensors and other test equipment.

Figure 3.8-3. A typical foam projectile, which has marks for determining position and velocity as well as blackened outlines for indicating the impact footprint.

The testing required a projectile (see Figure 3.8-3) made from standard stock, so investigators selected a rectangular cross-section of 11.5 by 5.5 inches, which was within 15 percent of the footprint of the mean debris size initially estimated by image analysis. To account for the foam's density, the projectile length was cut to weigh 1.67 pounds, a figure determined by image and transport analysis to best represent the STS-107 projectile. For foam with a density of 2.4 pounds per cubic foot,[15] the projectile dimensions were 19 inches by 11.5 inches by 5.5 inches.

Impact Angles

The precise impact location of the foam determined the impact angle because the debris was moving almost parallel to the Orbiter's fuselage at impact. Tile areas would have been hit at very small angles (approximately five degrees), but the curvature of the leading edge created angles closer to 20 degrees (see Figure 3.4-4).

The foam that struck *Columbia* on January 16, 2003, had both a translational speed and a rotational speed relative to the Orbiter. The translational velocity was easily replicated by adjusting the gas pressure in the gun. The rotational energy could be calculated, but the impact force depends on the material composition and properties of the impacting body and how the rotating body struck the wing. Because the details of the foam contact were not available from any visual evidence, analysis estimated the increase in impact energy that would be imparted by the rotation. These analyses resulted in a three-degree increase in the angle at which the foam test projectile would hit the test panel.[16]

The "clocking angle" was an additional consideration. As shown in Figure 3.8-4, the gun barrel could be rotated to change the impact point of the foam projectile on the leading edge. Investigators conducted experiments to determine if the corner of the foam block or the full edge would impart a greater force. During the fiberglass tests, it was found that a clocking angle of 30 degrees allowed the 11.5-inch-edge to fully contact the panel at impact, resulting in a greater local force than a zero degree angle, which was achieved with the barrel aligned vertically. A zero-degree angle was used for the test on RCC panel 6, and a 30-degree angle was used for RCC panel 8.

Figure 3.8-4. The barrel on the nitrogen gun could be rotated to adjust the impact point of the foam projectile.

Test Results from Fiberglass Panel Tests 1-5

Five engineering tests on fiberglass panels (see Figure 3.8-5) established the test parameters of the impact tests on RCC panels. Details of the fiberglass tests are in Appendix D.12.

Figure 3.8-5. A typical foam strike leaves impact streaks, and the foam projectile breaks into shards and larger pieces. Here the foam is striking Panel 6 on a fiberglass test article.

Test Results from Reinforced Carbon-Carbon Panel 6 (From Discovery)

RCC panel 6 was tested first to begin to establish RCC impact response, although by the time of the test, other data had indicated that RCC panel 8-left was the most likely site of the breach. RCC panel 6 was impacted using the same parameters as the test on fiberglass panel 6 and developed a 5.5-inch crack on the outboard end of the panel that extended through the rib (see Figure 3.8-6). There was also a crack through the "web" of the T-seal between panels 6 and 7 (see Figure 3.8-7). As in the fiberglass test, the foam block deflected, or moved, the face of the RCC panel, creating a slit between the panel and the adjacent T-seal, which ripped the projectile and stuffed pieces of foam into the slit (see Figure 3.8-8). The panel rib failed at lower stresses than predicted, and the T-seal failed closer to predictions, but overall, the stress pattern was similar to what was predicted, demonstrating the need to incorporate more complete RCC failure criteria in the computational models.

Without further crack growth, the specific structural damage this test produced would probably not have allowed enough superheated air to penetrate the wing during re-entry to cause serious damage. However, the test did demonstrate that a foam impact representative of the debris strike at 81.9 seconds after launch could damage an RCC panel. Note that

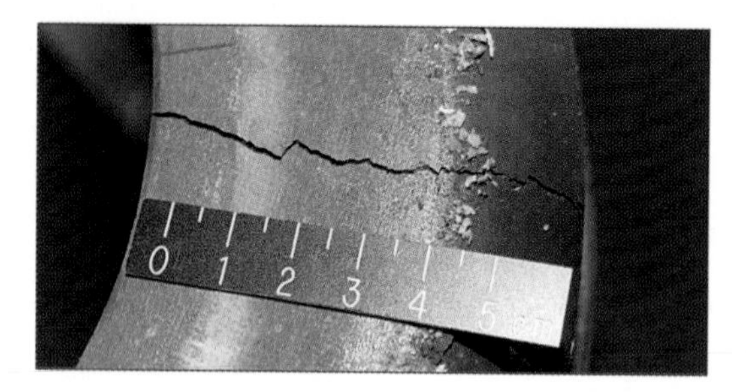

Figure 3.8-6. A 5.5-inch crack on the outboard portion of RCC Panel 6 during testing.

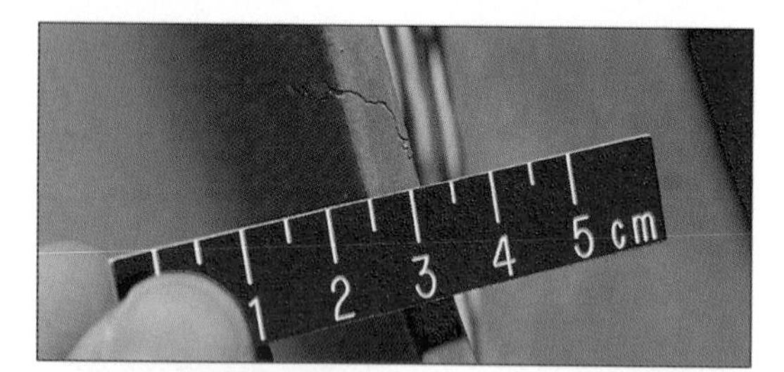

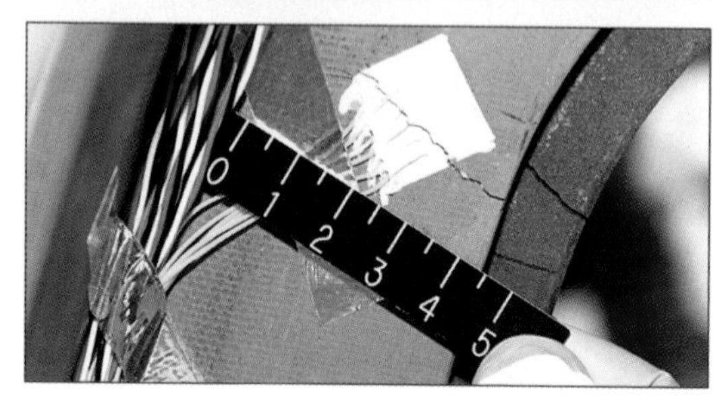

Figure 3.8-7. Two views of the crack in the T-seal between RCC Panels 6 and 7.

Figure 3.8-8. Two views of foam lodged into the slit during tests.

Figure 3.8-9. The large impact hole in Panel 8 from the final test.

Figure 3.8-10. Numerous cracks were also noted in RCC Panel 8.

the RCC panel 6-left test used fiberglass panels and T-seals in panel 7, 8, 9, and 10 locations. As seen later in the RCC panel 8-left test, this test configuration may not have adequately reproduced the flight configuration. Testing of a full RCC panel 6, 7, and 8 configuration might have resulted in more severe damage.

Test Results from Reinforced Carbon-Carbon Panel 8 (From *Atlantis*)

The second impact test of RCC material used panel 8 from *Atlantis*, which had flown 26 missions. Based on forensic evidence, sensor data, and aerothermal studies, panel 8 was considered the most likely point of the foam debris impact on *Columbia*.

Based on the system response of the leading edge in the fiberglass and RCC panel 6 impact tests, the adjacent RCC panel assemblies (9 and 10) were also flown hardware. The reference 1.67-pound foam test projectile impacted panel 8 at 777 feet per second with a clocking angle of 30 degrees and an angle of incidence of 25.1 degrees.

The impact created a hole roughly 16 inches by 17 inches, which was within the range consistent with all the findings of the investigation (see Figure 3.8-9). Additionally, cracks in the panel ranged up to 11 inches in length (Figure 3.8-10). The T-seal between panels 8 and 9 also failed at the lower outboard mounting lug.

Three large pieces of the broken panel face sheet (see Figure 3.8-11) were retained within the wing. The two largest pieces had surface areas of 86 and 75 square inches. While this test cannot exactly duplicate the damage *Columbia* incurred, pieces such as these could have remained in the wing cavity for some time, and could then have floated out of the damaged wing while the Orbiter was maneuvering in space. This scenario is consistent with the event observed on Flight Day 2 (see Section 3.5).

The test clearly demonstrated that a foam impact of the type *Columbia* sustained could seriously breach the Wing Leading Edge Structural Subsystem.

Conclusion

At the beginning of this chapter, the Board stated that the physical cause of the accident was a breach in the Thermal Protection System on the leading edge of the left wing. The breach was initiated by a piece of foam that separated from the left bipod ramp of the External Tank and struck the wing in the vicinity of the lower half of the Reinforced Carbon-Carbon (RCC) panel 8.

The conclusion that foam separated from the External Tank bipod ramp and struck the wing in the vicinity of panel 8 is documented by photographic evidence (Section 3.4). Sensor data and the aerodynamic and thermodynamic analyses (Section 3.6) based on that data led to the determination that the breach was in the vicinity of panel 8 and also accounted for the subsequent melting of the supporting structure, the spar, and the wiring behind the spar that occurred behind panel 8. The detailed examination of the debris (Section 3.7) also pointed to panel 8 as the breach site. The impact tests (Section 3.8) established that foam can breach the RCC, and also counteracted the lingering denial or discounting of the analytic evidence. Based on this evidence, the Board concluded that panel 8 was the site of the foam strike to *Columbia* during the liftoff of STS-107 on January 23, 2003.

Findings:

F3.8-1 The impact test program demonstrated that foam can cause a wide range of impact damage, from cracks to a 16- by 17-inch hole.

F3.8-2 The wing leading edge Reinforced Carbon-Carbon composite material and associated support hardware are remarkably tough and have impact capabilities that far exceed the minimal impact resistance specified in their original design requirements. Nevertheless, these tests demonstrate that this inherent toughness can be exceeded by impacts representative of those that occurred during *Columbia*'s ascent.

F3.8-3 The response of the wing leading edge to impacts is complex and can vary greatly, depending on the location of the impact, projectile mass, orientation, composition, and the material properties of the panel assembly, making analytic predictions of damage to RCC assemblies a challenge.[17]

F3.8-4 Testing indicates the RCC panels and T-seals have much higher impact resistance than the design specifications call for.

F3.8-5 NASA has an inadequate number of spare Reinforced Carbon-Carbon panel assemblies.

F3.8-6 NASA's current tools, including the Crater model, are inadequate to evaluate Orbiter Thermal Protection System damage from debris impacts during pre-launch, on-orbit, and post-launch activity.

F3.8-7 The bipod ramp foam debris critically damaged the leading edge of *Columbia*'s left wing.

Recommendations:

R3.8-1 Obtain sufficent spare Reinforced Carbon-Carbon panel assemblies and associated support components to ensure that decisions related to Reinforced Carbon-Carbon maintenance are made on the basis of component specifications, free of external pressures relating to schedules, costs, or other considerations.

R3.8-2 Develop, validate, and maintain physics-based computer models to evaluate Thermal Protection System damage from debris impacts. These tools should provide realistic and timely estimates of any impact damage from possible debris from any source that may ultimately impact the Orbiter. Establish impact damage thresholds that trigger responsive corrective action, such as on-orbit inspection and repair, when indicated.

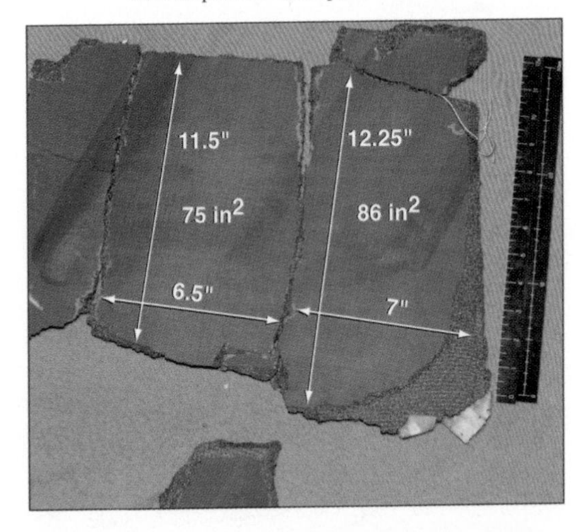

Figure 3.8-11. Three large pieces of debris from the panel face sheet were lodged within the hollow area behind the RCC panel.

ENDNOTES FOR CHAPTER 3

The citations that contain a reference to "CAIB document" with CAB or CTF followed by seven to eleven digits, such as CAB001-0010, refer to a document in the Columbia Accident Investigation Board database maintained by the Department of Justice and archived at the National Archives.

1 See Dennis R. Jenkins, *Space Shuttle: The History of the National Space Transportation System – The First 100 Missions* (Cape Canaveral, FL, Specialty Press, 2001), pp. 421-424 for a complete description of the External Tank.

2 Scotty Sparks and Lee Foster, "ET Cryoinsulation," CAIB Public Hearing, April 7, 2003. CAIB document CAB017-03140371.

3 Scotty Sparks and Steve Holmes, Presentation to the CAIB, March 27, 2003, CAIB document CTF036-02000200.

4 See the CAIB/NAIT Joint Working Scenario in Appendix D.7 of Volume II of this report.

5 Boeing Specification MJ070-0001-1E, "Orbiter End Item Specification for the Space Shuttle Systems, Part 1, Performance and Design Requirements, November 7, 2002.

6 Ibid., Paragraph 3.3.1.8.16.

7 NSTS-08171, "Operations and Maintenance Requirements and Specifications Document (OMRSD)" File II, Volume 3. CAIB document CAB033-12821997.

8 Dr. Gregory J. Byrne and Dr. Cynthia A. Evans, "STS-107 Image Analysis Team Final Report in Support of the Columbia Accident Investigation," NSTS-37384, June 2003. CAIB document CTF076-15511657. See Appendix E.2 for a copy of the report.

9 R. J. Gomex et al, "STS-107 Foam Transport Final Report," NSNS-60506, August 2003.

10 This section based on information from the following reports: MIT Lincoln Laboratory "Report on Flight Day 2 Object Analysis;" Dr. Brian M. Kent, Dr. Kueichien C. Hill, and Captain John Gulick, "An Assessment of Potential Material Candidates for the 'Flight Day 2' Radar Object Observed During the NASA Mission STS-107 (Columbia)", Air Force Research Laboratory Final Summary Report AFRL-SNS-2003-001, July 20, 2003 (see Appendix E.2); Multiple briefings to the CAIB from Dr. Brian M. Kent, AFRL/SN (CAIB document CTF076-19782017); Briefing to the CAIB from HQ AFSPC/XPY, April 18, 2003 (CAIB document CAB066-13771388).

11 The water tanks from below the mid-deck floor, along with both Forward Reaction Control System propellant tanks were recovered in good condition.

12 *Enterprise* was used for the initial Approach and Landing Tests and ground tests of the Orbiter, but was never used for orbital tests. The vehicle is now held by the National Air and Space Museum. See Jenkins, *Space Shuttle*, pp. 205-223, for more information on *Enterprise*.

13 Philip Kopfinger and Wanda Sigur, "Impact Test Results of BX-250 In Support of the Columbia Accident Investigation," ETTP-MS-03-021, July 17, 2003.

14 Details of the test instrumentation are in Appendix D.12.

15 Evaluations of the adjustments in the angle of incidence to account for rotation are in Appendix D.12.

16 The potential damage estimates had great uncertainty because the database of bending, tension, crushing, and other measures of failure were incomplete, particularly for RCC material.

CHAPTER 4

Other Factors Considered

During its investigation, the Board evaluated every known factor that could have caused or contributed to the *Columbia* accident, such as the effects of space weather on the Orbiter during re-entry and the specters of sabotage and terrorism. In addition to the analysis/scenario investigations, the Board oversaw a NASA "fault tree" investigation, which accounts for every chain of events that could possibly cause a system to fail. Most of these factors were conclusively eliminated as having nothing to do with the accident; however, several factors have yet to be ruled out. Although deemed by the Board as unlikely to have contributed to the accident, these are still open and are being investigated further by NASA. In a few other cases, there is insufficient evidence to completely eliminate a factor, though most evidence indicates that it did not play a role in the accident. In the course of investigating these factors, the Board identified several serious problems that were not part of the accident's causal chain but nonetheless have major implications for future missions.

In this chapter, a discussion of these potential causal and contributing factors is divided into two sections. The first introduces the primary tool used to assess potential causes of the breakup: the fault tree. The second addresses fault tree items and particularly notable factors that raised concerns for this investigation and, more broadly, for the future operation of the Space Shuttle.

4.1 FAULT TREE

The NASA Accident Investigation Team investigated the accident using "fault trees," a common organizational tool in systems engineering. Fault trees are graphical representations of every conceivable sequence of events that could cause a system to fail. The fault tree's uppermost level illustrates the events that could have directly caused the loss of *Columbia* by aerodynamic breakup during re-entry. Subsequent levels comprise all individual elements or factors that could cause the failure described immediately above it. In this way, all potential chains of causation that lead ultimately to the loss of *Columbia* can be diagrammed, and the behavior of every subsystem that was not a precipitating cause can be eliminated from consideration. Figure 4.1-1 depicts the fault tree structure for the *Columbia* accident investigation.

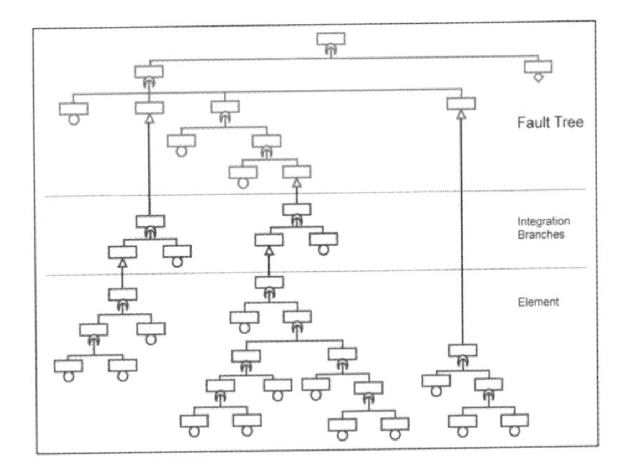

Figure 4.1-1. Accident investigation fault tree structure.

NASA chartered six teams to develop fault trees, one for each of the Shuttle's major components: the Orbiter, Space Shuttle Main Engine, Reusable Solid Rocket Motor, Solid Rocket Booster, External Tank, and Payload. A seventh "systems integration" fault tree team analyzed failure scenarios involving two or more Shuttle components. These interdisciplinary teams included NASA and contractor personnel, as well as outside experts.

Some of the fault trees are very large and intricate. For instance, the Orbiter fault tree, which only considers events on the Orbiter that could have led to the accident, includes 234 elements. In contrast, the Systems Integration fault tree, which deals with interactions among parts of the Shuttle, includes 295 unique multi-element integration faults, 128 Orbiter multi-element faults, and 221 connections to the other Shuttle components. These faults fall into three categories: induced and natural environments (such as structural interface loads and electromechanical effects); integrated vehicle mass properties; and external impacts (such as debris from the External Tank). Because the Systems Integration team considered multi-element faults – that is, scenarios involving several Shuttle components – it frequently worked in tandem with the Component teams.

In the case of the *Columbia* accident, there could be two plausible explanations for the aerodynamic breakup of the Orbiter: (1) the Orbiter sustained structural damage that undermined attitude control during re-entry; or (2) the Orbiter maneuvered to an attitude in which it was not designed to fly. The former explanation deals with structural damage initiated before launch, during ascent, on orbit, or during re-entry. The latter considers aerodynamic breakup caused by improper attitude or trajectory control by the Orbiter's Flight Control System. Telemetry and other data strongly suggest that improper maneuvering was not a factor. Therefore, most of the fault tree analysis concentrated on structural damage that could have impeded the Orbiter's attitude control, in spite of properly operating guidance, navigation, and flight control systems.

When investigators ruled out a potential cascade of events, as represented by a branch on the fault tree, it was deemed "closed." When evidence proved inconclusive, the item remained "open." Some elements could be dismissed at a high level in the tree, but most required delving into lower levels. An intact Shuttle component or system (for example, a piece of Orbiter debris) often provided the basis for closing an element. Telemetry data can be equally persuasive: it frequently demonstrated that a system operated correctly until the loss of signal, providing strong evidence that the system in question did not contribute to the accident. The same holds true for data obtained from the Modular Auxiliary Data System recorder, which was recovered intact after the accident.

The closeout of particular chains of causation was examined at various stages, culminating in reviews by the NASA Orbiter Vehicle Engineering Working Group and the NASA Accident Investigation Team. After these groups agreed to close an element, their findings were forwarded to the Board for review. At the time of this report's publication, the Board had closed more than one thousand items. A summary of fault tree elements is listed in Figure 4.1-2.

Branch	Total Number of Elements	Number of Open Elements		
		Likely	Possible	Unlikely
Orbiter	234	3	8	6
SSME	22	0	0	0
RSRM	35	0	0	0
SRB	88	0	4	4
ET	883	6	0	135
Payload	3	0	0	0
Integration	295	1	0	1

Figure 4.1-2. Summary of fault tree elements reviewed by the Board.

The open elements are grouped by their potential for contributing either directly or indirectly to the accident. The first group contains elements that may have in any way contributed to the accident. Here, "contributed" means that the element may have been an initiating event or a *likely* cause of the accident. The second group contains elements that could not be closed and may or may not have contributed to the accident. These elements are *possible* causes or factors in this accident. The third group contains elements that could not be closed, but are *unlikely* to have contributed to the accident. Appendix D.3 lists all the elements that were closed and thus eliminated from consideration as a cause or factor of this accident.

Some of the element closure efforts will continue after this report is published. Some elements will never be closed, because there is insufficient data and analysis to unconditionally conclude that they did not contribute to the accident. For instance, heavy rain fell on Kennedy Space Center prior to the launch of STS-107. Could this abnormally heavy rainfall have compromised the External Tank bipod foam? Experiments showed that the foam did not tend to absorb rain, but the rain could not be ruled out entirely as having contributed to the accident. Fault tree elements that were not closed as of publication are listed in Appendix D.4.

4.2 Remaining Factors

Several significant factors caught the attention of the Board during the investigation. Although it appears that they were not causal in the STS-107 accident, they are presented here for completeness.

Solid Rocket Booster Bolt Catchers

The fault tree review brought to light a significant problem with the Solid Rocket Booster bolt catchers. Each Solid Rocket Booster is connected to the External Tank by four separation bolts: three at the bottom plus a larger one at the top that weighs approximately 65 pounds. These larger upper (or "forward") separation bolts (one on each Solid Rocket Booster) and their associated bolt catchers on the External Tank provoked a great deal of Board scrutiny.

About two minutes after launch, the firing of pyrotechnic charges breaks each forward separation bolt into two pieces, allowing the spent Solid Rocket Boosters to separate from the External Tank (see Figure 4.2-1). Two "bolt catchers" on the External Tank each trap the upper half of a fired separation bolt, while the lower half stays attached to the Solid Rocket Booster. As a result, both halves are kept from flying free of the assembly and potentially hitting the Orbiter. Bolt catchers have a domed aluminum cover containing an aluminum honeycomb matrix that absorbs the fired bolt's energy. The two upper bolt halves and their respective catchers subsequently remain connected to the External Tank, which burns up on re-entry, while the lower halves stay with the Solid Rocket Boosters that are recovered from the ocean.

If one of the bolt catchers failed during STS-107, the resulting debris could have damaged *Columbia*'s wing leading edge. Concerns that the bolt catchers may have failed, causing metal debris to ricochet toward the Orbiter, arose because the configuration of the bolt catchers used on Shuttle missions differs in important ways from the design used in

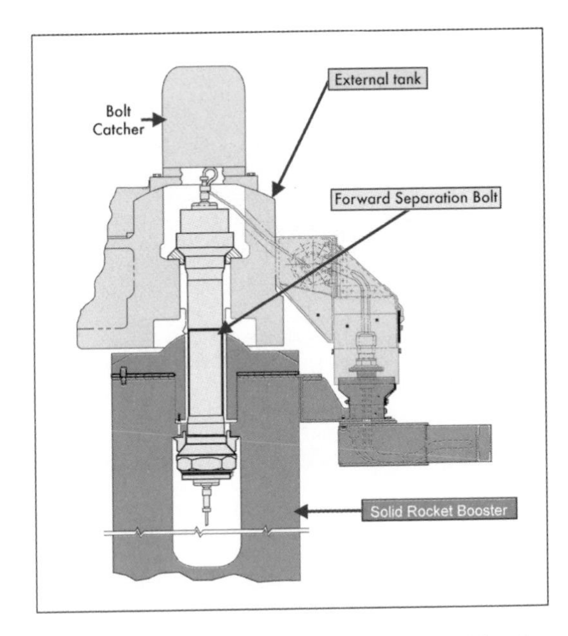

Figure 4.2-1. A cutaway drawing of the forward Solid Rocket Booster bolt catcher and separation bolt assembly.

initial qualification tests.[1] First, the attachments that currently hold bolt catchers in place use bolts threaded into inserts rather than through-bolts. Second, the test design included neither the Super Lightweight Ablative material applied to the bolt catcher apparatus for thermal protection, nor the aluminum honeycomb configuration currently used. Also, during these initial tests, temperature and pressure readings for the bolt firings were not recorded.

Instead of conducting additional tests to correct for these discrepancies, NASA engineers qualified the flight design configuration using a process called "analysis and similarity." The flight configuration was validated using extrapolated test data and redesign specifications rather than direct testing. This means that NASA's rationale for considering bolt catchers to be safe for flight is based on limited data from testing 24 years ago on a model that differs significantly from the current design.

Due to these testing deficiencies, the Board recognized that bolt catchers could have played a role in damaging *Columbia*'s left wing. The aluminum dome could have failed catastrophically, ablative coating could have come off in large quantities, or the device could have failed to hold to its mount point on the External Tank. To determine whether bolt catchers should be eliminated as a source of debris, investigators conducted tests to establish a performance baseline for bolt catchers in their current configuration and also reviewed radar data to see whether bolt catcher failure could be observed. The results had serious implications: Every bolt catcher tested failed well below the expected load range of 68,000 pounds. In one test, a bolt catcher failed at 44,000 pounds, which was two percent below the 46,000 pounds generated by a fired separation bolt. This means that the force at which a separation bolt is predicted to come apart during flight could exceed the bolt catcher's ability to safely capture the bolt. If these results are consistent with further tests, the factor of safety for the bolt catcher system would be 0.956 – far below the design requirement of 1.4 (that is, able to withstand 1.4 times the maximum load ever expected in operation).

Every bolt catcher must be inspected (via X-ray) as a final step in the manufacturing process to ensure specification compliance. There are specific requirements for film type/quality to allow sufficient visibility of weld quality (where the dome is mated to the mounting flange) and reveal any flaws. There is also a requirement to archive the film for several years after the hardware has been used. The manufacturer is required to evaluate the film, and a Defense Contract Management Agency representative certifies that requirements have been met. The substandard performance of the Summa bolt catchers tested by NASA at Marshall Space Flight Center and subsequent investigation revealed that the contractor's use of film failed to meet quality requirements and, because of this questionable quality, there were "probable" weld defects in most of the archived film. Film of STS-107's bolt catchers (serial numbers 1 and 19, both Summa-manufactured), was also determined to be substandard with "probable" weld defects (cracks, porosity, lack of penetration) on number 1 (left Solid Rocket Booster to External Tank attach point). Number 19 appeared adequate, though the substandard film quality leaves some doubt.

Further investigation revealed that a lack of qualified non-destructive inspection technicians and differing interpretations of inspection requirements contributed to this oversight. United Space Alliance, NASA's agent in procuring bolt catchers, exercises limited process oversight and delegates actual contract compliance verification to the Defense Contract Management Agency. The Defense Contract Management Agency interpreted its responsibility as limited to certifying compliance with the requirement for X-ray inspections. Since neither the Defense Contract Management Agency nor United Space Alliance had a resident non-destructive inspection specialist, they could not read the X-ray film or certify the weld. Consequently, the required inspections of weld quality and end-item certification were not properly performed. Inadequate oversight and confusion over the requirement on the parts of NASA, United Space Alliance, and the Defense Contract Management Agency all contributed to this problem.

In addition, STS-107 radar data from the U.S. Air Force Eastern Range tracking system identified an object with a radar cross-section consistent with a bolt catcher departing the Shuttle stack at the time of Solid Rocket Booster separation. The resolution of the radar return was not sufficient to definitively identify the object. However, an object that has about the same radar signature as a bolt catcher was seen on at least five other Shuttle missions. Debris shedding during Solid Rocket Booster separation is not an unusual event. However, the size of this object indicated that it could be a potential threat if it came close to the Orbiter after coming off the stack.

Although bolt catchers can be neither definitively excluded nor included as a potential cause of left wing damage to *Columbia*, the impact of such a large object would likely have registered on the Shuttle stack's sensors. The indefinite data at the time of Solid Rocket Booster separation, in tandem with overwhelming evidence related to the foam debris strike, leads the Board to conclude that bolt catchers are unlikely to have been involved in the accident.

Findings:

F4.2-1 The certification of the bolt catchers flown on STS-107 was accomplished by extrapolating analysis done on similar but not identical bolt catchers in original testing. No testing of flight hardware was performed.

F4.2-2 Board-directed testing of a small sample size demonstrated that the "as-flown" bolt catchers do not have the required 1.4 margin of safety.

F4.2-3 Quality assurance processes for bolt catchers (a Criticality 1 subsystem) were not adequate to assure contract compliance or product adequacy.

F4.2-4 An unknown metal object was seen separating from the stack during Solid Rocket Booster separation during six Space Shuttle missions. These objects were not identified, but were characterized as of little to no concern.

Recommendations:

R4.2-1 Test and qualify the flight hardware bolt catchers.

Kapton Wiring

Because of previous problems with its use in the Space Shuttle and its implication in aviation accidents, Kapton-insulated wiring was targeted as a possible cause of the *Columbia* accident. Kapton is an aromatic polyimide insulation that the DuPont Corporation developed in the 1960s. Because Kapton is lightweight, nonflammable, has a wide operating temperature range, and resists damage, it has been widely used in aircraft and spacecraft for more than 30 years. Each Orbiter contains 140 to 157 miles of Kapton-insulated wire, approximately 1,700 feet of which is inaccessible.

Despite its positive properties, decades of use have revealed one significant problem that was not apparent during its development and initial use: Kapton insulation can break down, leading to a phenomenon known as arc tracking. When arc tracking occurs, the insulation turns to carbon, or carbonizes, at temperatures of 1,100 to 1,200 degrees Fahrenheit. Carbonization is not the same as combustion. During tests unrelated to *Columbia*, Kapton wiring placed in an open flame did not continue to burn when the wiring was removed from the flame. Nevertheless, when carbonized, Kapton becomes a conductor, leading to a "soft electrical short" that causes systems to gradually fail or operate in a degraded fashion. Improper installation and mishandling during inspection and maintenance can also cause Kapton insulation to split, crack, flake, or otherwise physically degrade.[2] (Arc tracking is pictured in Figure 4.2-2.)

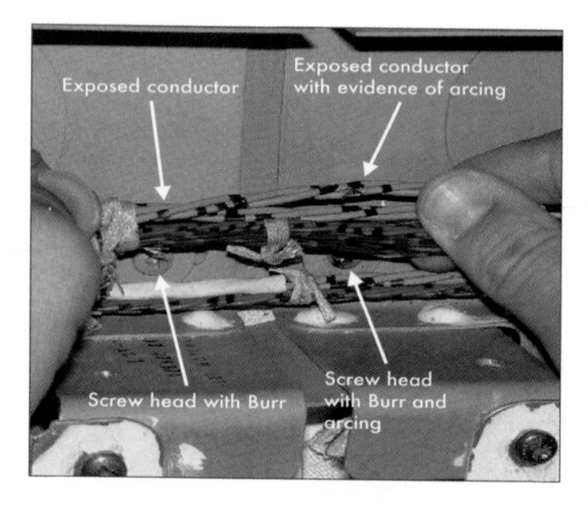

Figure 4.2-2. Arc tracking damage in Kapton wiring.

Perhaps the greatest concern is the breakdown of the wire's insulation when exposed to moisture. Over the years, the Federal Aviation Administration has undertaken extensive studies into wiring-related issues, and has issued Advisory Circulars (25-16 and 43.13-1B) on aircraft wiring that discuss using aromatic polyimide insulation. It was discovered that as long as the wiring is designed, installed, and maintained properly, it is safe and reliable. It was also discovered, however, that the aromatic polyimide insulation does not function well in high-moisture environments, or in installations that require large or frequent flexing. The military had discovered the potentially undesirable aspects of aromatic polyimide insulation much earlier, and had effectively banned its use on new aircraft beginning in 1985. These rules, however, apply only to pure polyimide insulation; various other insulations that contain polyimide are still used in appropriate areas.

The first extensive scrutiny of Kapton wiring on any of the Orbiters occurred during *Columbia*'s third Orbiter Major Modification period, after a serious system malfunction during the STS-93 launch of *Columbia* in July 1999. A short circuit five seconds after liftoff caused two of the six Main Engine Controller computers to lose power, which could have caused one or two of the three Main Engines to shut down. The ensuing investigation identified damaged Kapton wire as the cause of the malfunction. In order to identify and correct such wiring problems, all Orbiters were grounded for an initial (partial) inspection, with more extensive inspections planned during their next depot-level maintenance. During *Columbia*'s subsequent Orbiter Major Modification, wiring was inspected and redundant system wiring in the same bundles was separated to prevent arc tracking damage. Nearly 4,900 wiring nonconformances (conditions that did not meet specifications) were identified and corrected. Kapton-related problems accounted for approximately 27 percent of the nonconformances. This examination revealed a strong correlation between wire damage and the Orbiter areas that had experienced the most foot traffic during maintenance and modification.[3]

Other aspects of Shuttle operation may degrade Kapton wiring. In orbit, atomic oxygen acts as an oxidizing agent, causing chemical reactions and physical erosion that can lead to mass loss and surface property changes. Fortunately, actual exposure has been relatively limited, and inspections show that degradation is minimal. Laboratory tests on Kapton also confirm that on-orbit ultraviolet radiation can cause delamination, shrinkage, and wrinkling.

Figure 4.2-3. Typical wiring bundle inside Orbiter wing.

A typical wiring bundle is shown in Figure 4.2-3. Wiring nonconformances are corrected by rerouting, reclamping, or installing additional insulation such as convoluted tubing, insulating tape, insulating sheets, heat shrink sleeving, and abrasion pads (see Figure 4.2-4). Testing has shown that wiring bundles usually stop arc tracking when wires are physically separated from one another. Further testing under conditions simulating the Shuttle's wiring environment demonstrated that arc tracking does not progress beyond six inches. Based on these results, Boeing recommended that NASA separate all critical paths from larger wire bundles and individually protect them for a minimum of six inches beyond their separation points.[4] This recommendation is being adopted through modifications performed during scheduled Orbiter Major Modifications. For example, analysis of telemetered data from 14 of *Columbia*'s left wing sensors (hydraulic line/wing skin/wheel temperatures, tire pressures, and landing gear downlock position indication) provided failure signatures supporting the scenario of left-wing thermal intrusion, as opposed to a catastrophic failure (extensive arc tracking) of Kapton wiring. Actual NASA testing in the months following the accident, during which wiring bundles were subjected to intense heat (ovens, blowtorch, and arc jet), verified the failure signature analyses. Finally, extensive testing and analysis in years prior to STS-107 showed that, with the low currents and low voltages associated with the Orbiter's instrumentation system (such as those in the left wing), the probability of arc tracking is commensurately low.

Finding:

F4.2-5 Based on the extensive wiring inspections, maintenance, and modifications prior to STS-107, analysis of sensor/wiring failure signatures, and the alignment of the signatures with thermal intrusion into the wing, the Board found no evidence that Kapton wiring problems caused or contributed to this accident.

Recommendation:

R4.2-2 As part of the Shuttle Service Life Extension Program and potential 40-year service life, develop a state-of-the-art means to inspect all Orbiter wiring, including that which is inaccessible.

Crushed Foam

Based on the anticipated launch date of STS-107, a set of Solid Rocket Boosters had been stacked in the Vehicle Assembly Building and a Lightweight Tank had been attached to them. A reshuffling of the manifest in July 2002 resulted in a delay to the STS-107 mission.[5] It was decided to use the already-stacked Solid Rocket Boosters for the STS-113 mission to the International Space Station. All flights to the International Space Station use Super Lightweight Tanks, meaning that the External Tank already mated would need to be removed and stored pending the rescheduled STS-107 mission. Since External Tanks are not stored with the bipod struts attached, workers at the Kennedy Space Center removed the bipod strut from the Lightweight Tank before it was lifted into a storage cell.[6]

Following the de-mating of the bipod strut, an area of crushed PDL-1034 foam was found in the region beneath where the left bipod strut attached to the tank's –Y bipod fitting. The region measured about 1.5 inches by 1.25 inches by 0.187 inches and was located at roughly the five o'clock position. Foam thickness in this region was 2.187 inches.

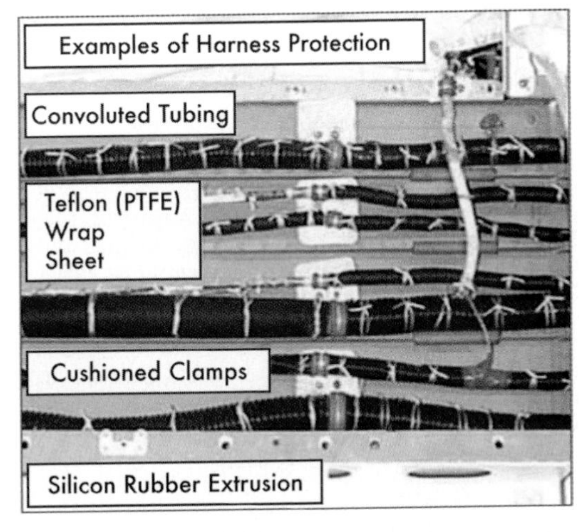

Figure 4.2-4. Typical wiring harness protection methods.

The crushed foam was exposed when the bipod strut was removed. This constituted an unacceptable condition and required a Problem Report write-up.[7]

NASA conducted testing at the Michoud Assembly Facility and at Kennedy Space Center to determine if crushed foam could have caused the loss of the left bipod ramp, and to determine if the limits specified in Problem Report procedures were sufficient for safety.[8]

Kennedy engineers decided not to take action on the crushed foam because it would be covered after the External Tank was mated to a new set of bipod struts that would connect it to *Columbia*, and the struts would sufficiently contain and shield the crushed foam.[9] An inspection after the bipod struts were attached determined that the area of crushed foam was within limits specified in the drawing for this region.[10]

STS-107 was therefore launched with crushed foam behind the clevis of the left bipod strut. Crushed foam in this region is a routine occurrence because the foam is poured and shaved so that the mating of the bipod strut to the bipod fitting results in a tight fit between the bipod strut and the foam.

Pre-launch testing showed that the extent of crushed foam did not exceed limits.[11] In these tests, red dye was wicked into the crushed (open) foam cells, and the damaged and dyed foam was then cut out and examined. Despite the effects of crushing, the foam's thickness around the bipod attach point was not substantially reduced; the foam effectively maintained insulation against ice and frost. The crushed foam was contained by the bipod struts and was subjected to little or no airflow.

Finding:

F4.2-6 Crushed foam does not appear to have contributed to the loss of the bipod foam ramp off the External Tank during the ascent of STS-107.

Recommendations:

- None

Hypergolic Fuel Spill

Concerns that hypergolic (ignites spontaneously when mixed) fuel contamination might have contributed to the accident led the Board to investigate an August 20, 1999, hydrazine spill at Kennedy Space Center that occurred while *Columbia* was being prepared for shipment to the Boeing facility in Palmdale, California. The spill occurred when a maintenance technician disconnected a hydrazine fuel line without capping it. When the fuel line was placed on a maintenance platform, 2.25 ounces of the volatile, corrosive fuel dripped onto the trailing edge of the Orbiter's left inboard elevon. After the spill was cleaned up, two tiles were removed for inspection. No damage to the control surface skin or structure was found, and the tiles were replaced.[12]

United Space Alliance briefed all employees working with these systems on procedures to prevent another spill, and on November 1, 1999, the Shuttle Operations Advisory Group was briefed on the corrective action that had been taken.

Finding:

F4.2-7 The hypergolic spill was not a factor in this accident.

Recommendations:

- None

Space Weather

Space weather refers to the action of highly energetic particles in the outer layers of Earth's atmosphere. Eruptions of particles from the sun are the primary source of space weather events, which fluctuate daily or even more frequently. The most common space weather concern is a potentially harmful radiation dose to astronauts during a mission. Particles can also cause structural damage to a vehicle, harm electronic components, and adversely affect communication links.

After the accident, several researchers contacted the Board and NASA with concerns about unusual space weather just before *Columbia* started its re-entry. A coronal mass ejection, or solar flare, of high-energy particles from the outer layers of the sun's atmosphere occurred on January 31, 2003. The shock wave from the solar flare passed Earth at about the same time that the Orbiter began its de-orbit burn. To examine the possible effects of this solar flare, the Board enlisted the expertise of the Space Environmental Center of the National Oceanic and Atmospheric Administration and the Space Vehicles Directorate of the Air Force Research Laboratory at Hanscom Air Force Base in Massachusetts.

Measurements from multiple space- and ground-based systems indicate that the solar flare occurred near the edge of the sun (as observed from Earth), reducing the impact of the subsequent shock wave to a glancing blow. Most of the effects of the solar flare were not observed on Earth until six or more hours after *Columbia* broke up. See Appendix D.5 for more on space weather effects.

Finding:

F4.2-8 Space weather was not a factor in this accident.

Recommendations:

- None

Asymmetric Boundary Layer Transition

Columbia had recently been through a complete refurbishment, including detailed inspection and certification of all lower wing surface dimensions. Any grossly protruding gap fillers would have been observed and repaired. Indeed, though investigators found that *Columbia*'s reputation for a rough left wing was well deserved prior to STS-75, quantitative measurements show that the measured wing roughness was below the fleet average by the launch of STS-107.[13]

Finding:

F4.2-9 A "rough wing" was not a factor in this accident.

Recommendations:

- None

Training and On-Orbit Performance

All mission-specific training requirements for STS-107 launch and flight control operators were completed before launch with no performance problems. However, seven flight controllers assigned to the mission did not have current recertifications at the time of the Flight Readiness Review, nor were they certified by the mission date. (Most flight controllers must recertify for their positions every 18 months.) The Board has determined that this oversight had no bearing on mission performance (see Chapter 6). The Launch Control Team and crew members held a full "dress rehearsal" of the launch day during the Terminal Countdown Demonstration Test. See Appendix D.1 for additional details on training for STS-107.

Because the majority of the mission was completed before re-entry, an assessment of the training preparation and flight readiness of the crew, launch controllers, and flight controllers was based on the documented performance of mission duties. All STS-107 personnel performed satisfactorily during the launch countdown, launch, and mission. Crew and mission controller actions were consistent with re-entry procedures.

There were a few incorrect switch movements by the crew during the mission, including the configuration of an inter-communications switch and an accidental bump of a rotational hand controller (which affected the Orbiter's attitude) after the de-orbit burn but prior to Entry Interface. The inter-communications switch error was identified and then corrected by the crew; both the crew and Mission Control noticed the bump and took the necessary steps to place the Orbiter in the correct attitude. Neither of these events was a factor in the accident, nor are they considered training or performance issues. Details on STS-107 on-orbit operations are in Appendix D.2.

Finding:

F4.2-10 The Board concludes that training and on-orbit considerations were not factors in this accident.

Recommendations:

- None

Payloads

To ensure that a payload malfunction did not cause or contribute to the *Columbia* accident, the Board conducted a thorough examination of all payloads and their integration with the Orbiter's systems. The Board reviewed all downlinked payload telemetry data during the mission, as well as all payload hardware technical documentation. Investigators assessed every payload readiness review, safety review, and payload integration process used by NASA, and interviewed individuals involved in the payload process at both Johnson and Kennedy Space Centers.

The Board's review of the STS-107 Flight Readiness Review, Payload Readiness Review, Payload Safety Review Panel, and Integrated Safety Assessments of experiment payloads on STS-107 found that all payload-associated hazards were adequately identified, accounted for, and appropriately mitigated. Payload integration engineers encountered no unique problems during SPACEHAB integration, there were no payload constraints on the launch, and there were no guideline violations during the payload preparation process.

The Board evaluated 11 payload anomalies, one of which was significant. A SPACEHAB Water Separator Assembly leak under the aft sub-floor caused an electrical short and subsequent shutdown of both Water Separator Assemblies. Ground and flight crew responses sufficiently addressed these anomalies during the mission. Circuit protection and telemetry data further indicate that during re-entry, this leak could not have produced a similar electrical short in SPACEHAB that might have affected the main Orbiter power supply.

The Board determined that the powered payloads aboard STS-107 were performing as expected when the Orbiter's signal was lost. In addition, all potential "fault-tree" payload failures that could have contributed to the Orbiter breakup were evaluated using real-time downlinked telemetry, debris analysis, or design specification analysis. These analyses indicate that no such failures occurred.

Several experiments within SPACEHAB were flammable, used flames, or involved combustible materials. All downlinked SPACEHAB telemetry was normal through re-entry, indicating no unexpected rise in temperature within the module and no increases in atmospheric or hull pressures. All fire alarms and indicators within SPACEHAB were operational, and they detected no smoke or fire. Gas percentages within SPACEHAB were also within limits.

Because a major shift in the Orbiter's center of gravity could potentially cause flight-control or heat management problems, researchers investigated a possible shifting of equipment in the payload bay. Telemetry during re-entry indicated that all payload cooling loops, electrical wiring, and communications links were functioning as expected, supporting the conclusion that no payload came loose during re-entry. In addition, there are no indications from the Orbiter's telemetry that any flight control adjustments were made to compensate for a change in the Orbiter's center of gravity, which indicates that the center of gravity in the payload bay did not shift during re-entry.

The Board explored whether the pressurized SPACEHAB module may have ruptured during re-entry. A rupture could breach the fuselage of the Orbiter or force open the payload bay doors, allowing hot gases to enter the Orbiter. All downlinked payload telemetry indicates that there was no decompression of SPACEHAB prior to loss of signal, and

(Above) The SPACEHAB Research Double Module (left) and Hitchhiker Carrier are lowered toward Columbia's payload bay on May 23, 2002. The Fast Reaction Experiments Enabling Science, Technology, Applications and Research (FREESTAR) is on the Hitchhiker Carrier.

(Below) Columbia's payload bay doors are ready to be closed over the SPACEHAB Research Double Module on June 14, 2002.

no dramatic increase in internal temperature or change in the air composition. This analysis suggests that the pressurized SPACEHAB module did not rupture during re-entry (see Appendix D.6.).

Finding:

F4.2-11 The payloads *Columbia* carried were not a factor in this accident.

Recommendations:

- None

Willful Damage and Security

During the Board's investigation, suggestions of willful damage, including the possibility of a terrorist act or sabotage by a disgruntled employee, surfaced in the media and on various Web sites. The Board assessed such theories, giving particular attention to the unprecedented security precautions taken during the launch of STS-107 because of prevailing national security concerns and the inclusion of an Israeli crew member.

Speculation that *Columbia* was shot down by a missile was easily dismissed. The Orbiter's altitude and speed prior to breakup was far beyond the reach of any air-to-air or surface-to-air missile, and telemetry and Orbiter support system data demonstrate that events leading to the breakup began at even greater altitudes.

The Board's evaluation of whether sabotage played any role included several factors: security planning and countermeasures, personnel and facility security, maintenance and processing procedures, and debris analysis.

To rule out an act of sabotage by an employee with access to these facilities, maintenance and processing procedures were thoroughly reviewed. The Board also interviewed employees who had access to the Orbiter.

The processes in place to detect anything unusual on the Orbiter, from a planted explosive to a bolt incorrectly torqued, make it likely that anything unusual would be caught during the many checks that employees perform as the Orbiter nears final closeout (closing and sealing panels that have been left open for inspection) prior to launch. In addition, the process of securing various panels before launch and taking closeout photos of hardware (see Figure 4.2-5) almost always requires the presence of more than one person, which means a saboteur would need the complicity of at least one other employee, if not more.

Debris from *Columbia* was examined for traces of explosives that would indicate a bomb onboard. Federal Bureau of Investigation laboratories provided analysis. Laboratory technicians took multiple samples of debris specimens and compared them with swabs from *Atlantis* and *Discovery*. Visual examination and gas chromatography with chemiluminescence detection found no explosive residues on any specimens that could not be traced to the Shuttle's pyrotechnic devices. Additionally, telemetry and other data indicate these pyrotechnic devices operated normally.

Figure 4.2-5. At left, a wing section open for inspection; at right, wing access closed off after inspection.

In its review of willful damage scenarios mentioned in the press or submitted to the investigation, the Board could not find any that were plausible. Most demonstrated a basic lack of knowledge of Shuttle processing and the physics of explosives, altitude, and thermodynamics, as well as the processes of maintenance documentation and employee screening.

NASA and its contractors have a comprehensive security system, outlined in documents like NASA Policy Directive 1600.2A. Rules, procedures, and guidelines address topics ranging from foreign travel to information security, from security education to investigations, and from the use of force to security for public tours.

The Board examined security at NASA and its related facilities through a combination of employee interviews, site visits, briefing reviews, and discussions with security personnel. The Board focused primarily on reviewing the capability of unauthorized access to Shuttle system components. Facilities and programs examined for security and sabotage potential included ATK Thiokol in Utah and its Reusable Solid Rocket Motor production, the Michoud Assembly Facility in Louisiana and its External Tank production, and the Kennedy Space Center in Florida for its Orbiter and overall integration responsibilities.

The Board visited the Boeing facility in Palmdale, California; Edwards Air Force Base in California; Stennis Space Center in Bay St. Louis, Mississippi; Marshall Space Flight Center near Huntsville, Alabama; and Cape Canaveral Air Force Station in Florida. These facilities exhibited a variety of security processes, according to each site's unique demands. At Kennedy, access to secure areas requires a series of identification card exchanges that electronically record each entry. The Michoud Assembly Facility employs similar measures, with additional security limiting access to a completed External Tank. The use of closed-circuit television systems complemented by security patrols is universal.

Employee screening and tracking measures appear solid across NASA and at the contractors examined by the Board. The agency relies on standard background and law enforcement checks to prevent the hiring of applicants with questionable records and the dismissal of those who may accrue such a record.

It is difficult for anyone to access critical Shuttle hardware alone or unobserved by a responsible NASA or contractor employee. With the exception of two processes when foam is applied to the External Tank at the Michoud Assembly Facility, there are no known final closeouts of any Shuttle component that can be completed with fewer than two people. Most closeouts involve at least five to eight employees before the component is sealed and certified for flight. All payloads also undergo an extensive review to ensure proper processing and to verify that they pose no danger to the crew or the Orbiter.

Security reviews also occur at locations such as the Trans-oceanic Abort Landing facilities. These sites are assessed prior to launch, and appropriate measures are taken to guarantee they are secure in case an emergency landing is required. NASA also has contingency plans in place, including dealing with bioterrorism.

Both daily and launch-day security at the Kennedy Space Center has been tightened in recent years. Each Shuttle launch has an extensive security countdown, with a variety of checks to guarantee that signs are posted, beaches are closed, and patrols are deployed. K-9 patrols and helicopters guard the launch area against intrusion.

Because the STS-107 manifest included Israel's first astronaut, security measures, developed with National Security Council approval, went beyond the normally stringent precautions, including the development of a Security Support Plan.

Military aircraft patrolled a 40-mile Federal Aviation Administration-restricted area starting nine hours before the launch of STS-107. Eight Coast Guard vessels patrolled a three-nautical-mile security zone around Kennedy Space Center and Cape Canaveral Air Force Station, and Coast Guard and NASA boats patrolled the inland waterways. Security forces were doubled on the day of the launch.

Findings:

F4.2-12 The Board found no evidence that willful damage was a factor in this accident.

F4.2-13 Two close-out processes at the Michoud Assembly Facility are currently able to be performed by a single person.

F4.2-14 Photographs of every close out activity are not routinely taken.

Recommendation:

R4.2-3 Require that at least two employees attend all final closeouts and intertank area hand-spraying procedures.

Micrometeoroids and Orbital Debris Risks

Micrometeoroids and space debris (often called "space junk") are among the most serious risk factors in Shuttle missions. While there is little evidence that micrometeoroids or space debris caused the loss of *Columbia*, and in fact a review of on-board accelerometer data rules out a major strike, micrometeoroids or space debris cannot be entirely ruled out as a potential or contributing factor.

Micrometeoroids, each usually no larger than a grain of sand, are numerous and particularly dangerous to orbiting spacecraft. Traveling at velocities that can exceed 20,000 miles per hour, they can easily penetrate the Orbiter's skin. In contrast to micrometeoroids, orbital debris generally comes from destroyed satellites, payload remnants, exhaust from solid rockets, and other man-made objects, and typically travel at far lower velocities. Pieces of debris four inches or larger are catalogued and tracked by the U.S. Air Force Space Command so they can be avoided during flight.

NASA has developed computer models to predict the risk of impacts. The Orbital Debris Model 2000 (ORDEM2000) database is used to predict the probability of a micrometeoroid or space debris collision with an Orbiter, based on its flight trajectory, altitude, date, and duration. Development of the database was based on radar tracking of debris and satellite experiments, as well as inspections of returned Orbiters. The computer code BUMPER translates expected debris hits from ORDEM2000 into an overall risk probability for each flight. The worst-case scenario during orbital debris strikes is known as the Critical Penetration Risk, which can include the depressurization of the crew module, venting or explosion of pressurized systems, breaching of the Thermal Protection System, and damage to control surfaces.

NASA guidelines require the Critical Penetration Risk to be better than 1 in 200, a number that has been the subject of several reviews. NASA has made changes to reduce the probability. For STS-107, the estimated risk was 1 in 370, though the actual as-flown value turned out to be 1 in 356. The current risk guideline of 1 in 200 makes space debris or micrometeoroid strikes by far the greatest risk factor in the Probabilistic Risk Assessment used for missions. Although 1-in-200 flights may seem to be long odds, and many flights have exceeded the guideline, the cumulative risk for such a strike over the 113-flight history of the Space Shuttle Program is calculated to be 1 in 3. The Board considers this probability of a critical penetration to be unacceptably high. The Space Station's micrometeoroid and space debris protection system reduces its critical penetration risk to five percent or less over 10 years, which translates into a per-mission risk of 1 in 1,200 with 6 flights per year, or 60 flights over 10 years.

To improve crew and vehicle safety over the next 10 to 20 years, the Board believes risk guidelines need to be changed to compel the Shuttle Program to identify and, more to the point, reduce the micrometeoroid and orbital debris threat to missions.

Findings:

F4.2-15 There is little evidence that *Columbia* encountered either micrometeoroids or orbital debris on this flight.

F4.2-16 The Board found markedly different criteria for margins of micrometeoroid and orbital debris safety between the International Space Station and the Shuttle.

Recommendation:

R4.2-4 Require the Space Shuttle to be operated with the same degree of safety for micrometeoroid and orbital debris as the degree of safety calculated for the International Space Station. Change the micrometeoroid and orbital debris safety criteria from guidelines to requirements.

Orbiter Major Modification

The Board investigated concerns that mistakes, mishaps, or human error during *Columbia*'s last Orbiter Major Modification might have contributed to the accident. Orbiters are removed from service for inspection, maintenance, and modification approximately every eight flights or three years. *Columbia* began its last Orbiter Major Modification in September 1999, completed it in February 2001, and had flown once before STS-107. Several aspects of the Orbiter Major Modification process trouble the Board, and need to be addressed for future flights. These concerns are discussed in Chapter 10.

Findings:

F4.2-17 Based on a thorough investigation of maintenance records and interviews with maintenance personnel, the Board found no errors during *Columbia*'s most recent Orbiter Major Modification that contributed to the accident.

Recommendations:

- None

Foreign Object Damage Prevention

Problems with the Kennedy Space Center and United Space Alliance Foreign Object Damage Prevention Program, which in the Department of Defense and aviation industry typically falls under the auspices of Quality Assurance, are related to changes made in 2001. In that year, Kennedy and Alliance redefined the single term "Foreign Object Damage" – an industry-standard blanket term – into two terms: "Processing Debris" and "Foreign Object Debris."

Processing Debris then became:

> *Any material, product, substance, tool or aid generally used during the processing of flight hardware that remains in the work area when not directly in use, or that is left unattended in the work area for any length of time during the processing of tasks, or that is left remaining or forgotten in the work area after the completion of a task or at the end of a work shift. Also any item, material or substance in the work area that should be found and removed as part of standard housekeeping, Hazard Recognition and Inspection Program (HRIP) walkdowns, or as part of "Clean As You Go" practices.*[14]

Foreign Object Debris then became:

> *Processing debris becomes FOD when it poses a potential risk to the Shuttle or any of its components, and only occurs when the debris is found during or subsequent to a final/flight Closeout Inspection, or subsequent to OMI S0007 ET Load SAF/FAC walkdown.*[15]

These definitions are inconsistent with those of other NASA centers, Naval Reactor programs, the Department of Defense, commercial aviation, and National Aerospace FOD Prevention Inc. guidelines.[16] They are unique to Kennedy Space Center and United Space Alliance.

Because debris of any kind has critical safety implications, these definitions are important. The United Space Alliance Foreign Object Program includes daily debris checks by management to ensure that workers comply with United Space Alliance's "clean as you go" policy, but United Space Alliance statistics reveal that the success rate of daily debris checks is between 70 and 86 percent.[17]

The perception among many interviewees is that these novel definitions mitigate the impact of Kennedy Mission Assurance-found Foreign Object Debris on the United Space Alliance award fee. This is because "Processing Debris" statistics do not directly affect the award fee. Simply put, in splitting "Foreign Object Damage" into two categories, many of the violations are tolerated. Indeed, with 18 problem reports generated on "lost items" during the processing of STS-107 alone, the need for an ongoing, thorough, and stringent Foreign Object Debris program is indisputable. However, with two definitions of foreign objects – Processing Debris and Foreign Object Debris – the former is portrayed as less significant and dangerous than the latter. The assumption that all debris will be found before flight fails to underscore the destructive potential of Foreign Object Debris, and creates an incentive to simply accept "Processing Debris."

Finding:

F4.2-18 Since 2001, Kennedy Space Center has used a non-standard approach to define foreign object debris. The industry standard term "Foreign Object Damage" has been divided into two categories, one of which is much more permissive.

Recommendation:

R4.2-5 Kennedy Space Center Quality Assurance and United Space Alliance must return to the straightforward, industry-standard definition of "Foreign Object Debris," and eliminate any alternate or statistically deceptive definitions like "processing debris."

ENDNOTES FOR CHAPTER 4

The citations that contain a reference to "CAIB document" with CAB or CTF followed by seven to eleven digits, such as CAB001-0010, refer to a document in the Columbia Accident Investigation Board database maintained by the Department of Justice and archived at the National Archives.

1. SRB Forward Separation Bolt Test Plan, Document Number 90ENG-00XX, April 2, 2003. CAIB document CTF044-62496260.
2. Cynthia Furse and Randy Haupt, "Down to the Wire," in the online version of the IEEE Spectrum magazine, accessed at http://www.spectrum.ieee.org/WEBONLY/publicfeature/feb01/wire.html on 2 August 2002.
3. Boeing Inspection Report, OV-102 J3, V30/V31 (Wire) Inspection Report, September 1999-February 2001. CAIB document CTF070-34793501.
4. Boeing briefing, "Arc Tracking Separation of Critical Wiring Redundancy Violations", present to NASA by Joe Daileda and Bill Crawford, April 18, 2001. CAIB document CAB033-43774435.
5. E-mail message from Jim Feeley, Lockheed Martin, Michoud Assembly Facility, April 24, 2003. This External Tank (ET-93) was originally mated to the Solid Rocket Boosters and bipod struts in anticipation of an earlier launch date for mission STS-107. Since Space Station missions require the use of a Super Light Weight Tank, ET-93 (which is a Light Weight Tank) had to be de-mated from the Solid Rocket Boosters so that they could be mated to such a Super Light Weight Tank. The mating of the bipod struts to ET-93 was performed in anticipation of an Orbiter mate. Once STS-107 was delayed and ET-93 had to be de-mated from the Solid Rocket Boosters, the bipod struts were also de-mated, since they are not designed to be attached to the External Tank during subsequent Solid Rocket Booster de-mate/mate operations.
6. "Production Info – Splinter Meeting," presented at Michoud Assembly Facility, March 13, 2002. TSPB ET-93-ST-003, "Bipod Strut Removal," August 1, 2002.
7. PR ET-93-TS-00073, "There Is An Area Of Crushed Foam From The Installation Of The –Y Bipod," August 8, 2002.
8. "Crushed Foam Testing." CAIB document CTF059-10561058.
9. PR ET-93-TS-00073, "There Is An Area Of Crushed Foam From The Installation Of The –Y Bipod," August 8, 2002; Meeting with John Blue, USA Engineer, Kennedy Space Center, March 10, 2003.
10. Lockheed Martin drawing 80911019109-509, "BIPOD INSTL,ET/ORB,FWD"
11. "Crushed Foam Testing." CAIB document CTF059-10561058.
12. Minutes of Orbiter Structures Telecon meeting, June 19, 2001, held with NASA, KSC, USA, JSC, BNA-Downey, Huntington Beach and Palmdale. CAIB document CAB033-38743888.
13. NASA Report NSTS-37398.
14. Standard Operating Procedure, Foreign Object Debris (FOD) Reporting, Revision A, Document Number SOP-O-0801-035, October 1, 2002, United Space Alliance, Kennedy Space Center, pg. 3.
15. Ibid, pg. 2.
16. "An effective FOD prevention program identifies potential problems, corrects negative factors, provides awareness, effective employee training, and uses industry "lessons learned" for continued improvement. There is no mention of Processing Debris, but the guidance does address potential Foreign Object Damage and Foreign Object Debris. While NASA has done a good job of complying with almost every area of this guideline, the document addresses Foreign Object investigations in a *singular* sense: "All incidents of actual or potential FOD should be reported and investigated. These reports should be directed to the FOD Focal Point who should perform tracking and trending analysis. The focal point should also assure all affected personnel are aware of all potential (near mishap) and actual FOD reports to facilitate feedback ('lessons learned')."
17. Space Flight Operations Contract, Performance Measurement System Reports for January 2003, February 2003, USA004840, issue 014, contract NAS9-2000.

Part Two

Why The Accident Occurred

Many accident investigations do not go far enough. They identify the technical cause of the accident, and then connect it to a variant of "operator error" – the line worker who forgot to insert the bolt, the engineer who miscalculated the stress, or the manager who made the wrong decision. But this is seldom the entire issue. When the determinations of the causal chain are limited to the technical flaw and individual failure, typically the actions taken to prevent a similar event in the future are also limited: fix the technical problem and replace or retrain the individual responsible. Putting these corrections in place leads to another mistake – the belief that the problem is solved. The Board did not want to make these errors.

Attempting to manage high-risk technologies while minimizing failures is an extraordinary challenge. By their nature, these complex technologies are intricate, with many interrelated parts. Standing alone, the components may be well understood and have failure modes that can be anticipated. Yet when these components are integrated into a larger system, unanticipated interactions can occur that lead to catastrophic outcomes. The risk of these complex systems is increased when they are produced and operated by complex organizations that also break down in unanticipated ways.

In our view, the NASA organizational culture had as much to do with this accident as the foam. Organizational culture refers to the basic values, norms, beliefs, and practices that characterize the functioning of an institution. At the most basic level, organizational culture defines the assumptions that employees make as they carry out their work. It is a powerful force that can persist through reorganizations and the change of key personnel. It can be a positive or a negative force.

In a report dealing with nuclear wastes, the National Research Council quoted Alvin Weinberg's classic statement about the "Faustian bargain" that nuclear scientists made with society. "The price that we demand of society for this magical energy source is both a vigilance and a longevity of our social institutions that we are quite unaccustomed to." This is also true of the space program. At NASA's urging, the nation committed to building an amazing, if compromised, vehicle called the Space Shuttle. When the agency did this, it accepted the bargain to operate and maintain the vehicle in the safest possible way. The Board is not convinced that NASA has completely lived up to the bargain, or that Congress and the Administration has provided the funding and support necessary for NASA to do so. This situation needs to be addressed – if the nation intends to keep conducting human space flight, it needs to live up to its part of the bargain.

Part Two of this report examines NASA's organizational, historical, and cultural factors, as well as how these factors contributed to the accident. As in Part One, this part begins with history. Chapter 5 examines the post-*Challenger* history of NASA and its Human Space Flight Program. This includes reviewing the budget as well as organizational and management history, such as shifting management systems and locations. Chapter 6 documents management performance related to *Columbia* to establish events analyzed in later chapters. The chapter reviews the foam strikes, intense schedule pressure driven by an artificial requirement to deliver Node 2 to the International Space Station by a certain date, and NASA management's handling of concerns regarding *Columbia* during the STS-107 mission.

In Chapter 7, the Board presents its views of how high-risk activities should be managed, and lists the characteristics of institutions that emphasize high-reliability results over economic efficiency or strict adherence to a schedule. This chapter measures the Space Shuttle Program's organizational and management practices against these principles and finds them wanting. Chapter 7 defines the organizational cause and offers recommendations. Chapter 8 draws from the previous chapters on history, budgets, culture, organization, and safety practices, and analyzes how all these factors contributed to this accident. This chapter captures the Board's views of the need to adjust management to enhance safety margins in Shuttle operations, and reaffirms the Board's position that without these changes, we have no confidence that other "corrective actions" will improve the safety of Shuttle operations. The changes we recommend will be difficult to accomplish – and will be internally resisted.

NASA

CHAPTER 5

From *Challenger* to *Columbia*

The Board is convinced that the factors that led to the *Columbia* accident go well beyond the physical mechanisms discussed in Chapter 3. The causal roots of the accident can also be traced, in part, to the turbulent post-Cold War policy environment in which NASA functioned during most of the years between the destruction of *Challenger* and the loss of *Columbia*. The end of the Cold War in the late 1980s meant that the most important political underpinning of NASA's Human Space Flight Program – U.S.-Soviet space competition – was lost, with no equally strong political objective to replace it. No longer able to justify its projects with the kind of urgency that the superpower struggle had provided, the agency could not obtain budget increases through the 1990s. Rather than adjust its ambitions to this new state of affairs, NASA continued to push an ambitious agenda of space science and exploration, including a costly Space Station Program.

If NASA wanted to carry out that agenda, its only recourse, given its budget allocation, was to become more efficient, accomplishing more at less cost. The search for cost reductions led top NASA leaders over the past decade to downsize the Shuttle workforce, outsource various Shuttle Program responsibilities – including safety oversight – and consider eventual privatization of the Space Shuttle Program. The program's budget was reduced by 40 percent in purchasing power over the past decade and repeatedly raided to make up for Space Station cost overruns, even as the Program maintained a launch schedule in which the Shuttle, a developmental vehicle, was used in an operational mode. In addition, the uncertainty of top policymakers in the White House, Congress, and NASA as to how long the Shuttle would fly before being replaced resulted in the delay of upgrades needed to make the Shuttle safer and to extend its service life.

The Space Shuttle Program has been transformed since the late 1980s implementation of post-*Challenger* management changes in ways that raise questions, addressed here and in later chapters of Part Two, about NASA's ability to safely operate the Space Shuttle. While it would be inaccurate to say that NASA managed the Space Shuttle Program at the time of the *Columbia* accident in the same manner it did prior to *Challenger*, there are unfortunate similarities between the agency's performance and safety practices in both periods.

5.1 THE *CHALLENGER* ACCIDENT AND ITS AFTERMATH

The inherently vulnerable design of the Space Shuttle, described in Chapter 1, was a product of policy and technological compromises made at the time of its approval in 1972. That approval process also produced unreasonable expectations, even myths, about the Shuttle's future performance that NASA tried futilely to fulfill as the Shuttle became "operational" in 1982. At first, NASA was able to maintain the image of the Shuttle as an operational vehicle. During its early years of operation, the Shuttle launched satellites, performed on-orbit research, and even took members of Congress into orbit. At the beginning of 1986, the goal of "routine access to space" established by President Ronald Reagan in 1982 was ostensibly being achieved. That appearance soon proved illusory. On the cold morning of January 28, 1986, the Shuttle *Challenger* broke apart 73 seconds into its climb towards orbit. On board were Francis R. Scobee, Michael J. Smith, Ellison S. Onizuka, Judith A. Resnick, Ronald E. McNair, Sharon Christa McAuliffe, and Gregory B. Jarvis. All perished.

Rogers Commission

On February 3, 1986, President Reagan created the Presidential Commission on the Space Shuttle Challenger Accident, which soon became known as the Rogers Commission after its chairman, former Secretary of State William Rogers. The Commission's report, issued on June 6, 1986, concluded that the loss of *Challenger* was caused by a failure of the joint and seal between the two lower segments of the right Solid Rocket Booster. Hot gases blew past a rubber O-ring in the joint, leading to a structural failure and the explosive burning of the Shuttle's hydrogen fuel. While the Rogers Commission identified the failure of the Solid Rocket Booster joint and seal as the physical cause of the accident, it also noted a number of NASA management failures that contributed to the catastrophe.

The Rogers Commission concluded "the decision to launch the *Challenger* was flawed." Communication failures, incomplete and misleading information, and poor management judgments all figured in a decision-making process that permitted, in the words of the Commission, "internal flight safety problems to bypass key Shuttle managers." As a result, if those making the launch decision "had known all the facts, it is highly unlikely that they would have decided to launch." Far from meticulously guarding against potential problems, the Commission found that NASA had required "a contractor to prove that it was not safe to launch, rather than proving it was safe."[1]

The Commission also found that NASA had missed warning signs of the impending accident. When the joint began behaving in unexpected ways, neither NASA nor the Solid Rocket Motor manufacturer Morton-Thiokol adequately tested the joint to determine the source of the deviations from specifications or developed a solution to them, even though the problems frequently recurred. Nor did they respond to internal warnings about the faulty seal. Instead, Morton-Thiokol and NASA management came to see the problems as an acceptable flight risk – a violation of a design requirement that could be tolerated.[2]

During this period of increasing uncertainty about the joint's performance, the Commission found that NASA's safety system had been "silent." Of the management, organizational, and communication failures that contributed to the accident, four related to faults within the safety system, including "a lack of problem reporting requirements, inadequate trend analysis, misrepresentation of criticality, and lack of involvement in critical discussions."[3] The checks and balances the safety system was meant to provide were not working.

Still another factor influenced the decisions that led to the accident. The Rogers Commission noted that the Shuttle's increasing flight rate in the mid-1980s created schedule pressure, including the compression of training schedules, a shortage of spare parts, and the focusing of resources on near-term problems. NASA managers "may have forgotten–partly because of past success, partly because of their own well-nurtured image of the program–that the Shuttle was still in a research and development phase."[4]

The *Challenger* accident had profound effects on the U.S. space program. On August 15, 1986, President Reagan announced that "NASA will no longer be in the business of launching private satellites." The accident ended Air Force and intelligence community reliance on the Shuttle to launch national security payloads, prompted the decision to abandon the yet-to-be-opened Shuttle launch site at Vandenberg Air Force Base, and forced the development of improved expendable launch vehicles.[6] A 1992 White House advisory committee concluded that the recovery from the *Challenger* disaster cost the country $12 billion, which included the cost of building the replacement Orbiter *Endeavour*.[7]

SELECTED ROGERS COMMISSION RECOMMENDATIONS

- "The faulty Solid Rocket Motor joint and seal must be changed. This could be a new design eliminating the joint or a redesign of the current joint and seal. No design options should be prematurely precluded because of schedule, cost or reliance on existing hardware. All Solid Rocket Motor joints should satisfy the following:
  - "The joints should be fully understood, tested and verified."
  - "The certification of the new design should include:
    - Tests which duplicate the actual launch configuration as closely as possible.
    - Tests over the full range of operating conditions, including temperature."
- "Full consideration should be given to conducting static firings of the exact flight configuration in a vertical attitude."
- "The Shuttle Program Structure should be reviewed. The project managers for the various elements of the Shuttle program felt more accountable to their center management than to the Shuttle program organization."
- "NASA should encourage the transition of qualified astronauts into agency management positions."
- "NASA should establish an Office of Safety, Reliability and Quality Assurance to be headed by an Associate Administrator, reporting directly to the NASA Administrator. It would have direct authority for safety, reliability, and quality assurance throughout the agency. The office should be assigned the work force to ensure adequate oversight of its functions and should be independent of other NASA functional and program responsibilities."
- "NASA should establish an STS Safety Advisory Panel reporting to the STS Program Manager. The charter of this panel should include Shuttle operational issues, launch commit criteria, flight rules, flight readiness and risk management."
- "The Commission found that Marshall Space Flight Center project managers, because of a tendency at Marshall to management isolation, failed to provide full and timely information bearing on the safety of flight 51-L [the *Challenger* mission] to other vital elements of Shuttle program management ... NASA should take energetic steps to eliminate this tendency at Marshall Space Flight Center, whether by changes of personnel, organization, indoctrination or all three."
- "The nation's reliance on the Shuttle as its principal space launch capability created a relentless pressure on NASA to increase the flight rate ... NASA must establish a flight rate that is consistent with its resources."[5]

It took NASA 32 months after the *Challenger* accident to redesign and requalify the Solid Rocket Booster and to return the Shuttle to flight. The first post-accident flight was launched on September 29, 1988. As the Shuttle returned to flight, NASA Associate Administrator for Space Flight

Richard Truly commented, "We will always have to treat it [the Shuttle] like an R&D test program, even many years into the future. I don't think calling it operational fooled anybody within the program ... It was a signal to the public that shouldn't have been sent."[8]

The Shuttle Program After Return to Flight

After the Rogers Commission report was issued, NASA made many of the organizational changes the Commission recommended. The space agency moved management of the Space Shuttle Program from the Johnson Space Center to NASA Headquarters in Washington, D.C. The intent of this change was to create a management structure "resembling that of the Apollo program, with the aim of preventing communication deficiencies that contributed to the *Challenger* accident."[9] NASA also established an Office of Safety, Reliability, and Quality Assurance at its Headquarters, though that office was not given the "direct authority" over all of NASA's safety operations as the Rogers Commission had recommended. Rather, NASA human space flight centers each retained their own safety organization reporting to the Center Director.

In the almost 15 years between the return to flight and the loss of *Columbia*, the Shuttle was again being used on a regular basis to conduct space-based research, and, in line with NASA's original 1969 vision, to build and service a space station. The Shuttle flew 87 missions during this period, compared to 24 before *Challenger*. Highlights from these missions include the 1990 launch, 1993 repair, and 1999 and 2002 servicing of the Hubble Space Telescope; the launch of several major planetary probes; a number of Shuttle-Spacelab missions devoted to scientific research; nine missions to rendezvous with the Russian space station *Mir*; the return of former Mercury astronaut Senator John Glenn to orbit in October 1998; and the launch of the first U.S. elements of the International Space Station.

After the *Challenger* accident, the Shuttle was no longer described as "operational" in the same sense as commercial aircraft. Nevertheless, NASA continued planning as if the Shuttle could be readied for launch at or near whatever date was set. Tying the Shuttle closely to International Space Station needs, such as crew rotation, added to the urgency of maintaining a predictable launch schedule. The Shuttle is currently the only means to launch the already-built European, Japanese, and remaining U.S. modules needed to complete Station assembly and to carry and return most experiments and on-orbit supplies.[10] Even after three occasions when technical problems grounded the Shuttle fleet for a month or more, NASA continued to assume that the Shuttle could regularly and predictably service the Station. In recent years, this coupling between the Station and Shuttle has become the primary driver of the Shuttle launch schedule. Whenever a Shuttle launch is delayed, it impacts Station assembly and operations.

In September 2001, testimony on the Shuttle's achievements during the preceding decade by NASA's then-Deputy Associate Administrator for Space Flight William Readdy indicated the assumptions under which NASA was operating during that period:

> *The Space Shuttle has made dramatic improvements in the capabilities, operations and safety of the system. The payload-to-orbit performance of the Space Shuttle has been significantly improved – by over 70 percent to the Space Station. The safety of the Space Shuttle has also been dramatically improved by reducing risk by more than a factor of five. In addition, the operability of the system has been significantly improved, with five minute launch windows – which would not have been attempted a decade ago – now becoming routine. This record of success is a testament to the quality and dedication of the Space Shuttle management team and workforce, both civil servants and contractors.*[11]

5.2 The NASA Human Space Flight Culture

Though NASA underwent many management reforms in the wake of the *Challenger* accident and appointed new directors at the Johnson, Marshall, and Kennedy centers, the agency's powerful human space flight culture remained intact, as did many institutional practices, even if in a modified form. As a close observer of NASA's organizational culture has observed, "Cultural norms tend to be fairly resilient ... The norms bounce back into shape after being stretched or bent. Beliefs held in common throughout the organization resist alteration."[12] This culture, as will become clear across the chapters of Part Two of this report, acted over time to resist externally imposed change. By the eve of the *Columbia* accident, institutional practices that were in effect at the time of the *Challenger* accident – such as inadequate concern over deviations from expected performance, a silent safety program, and schedule pressure – had returned to NASA.

Organizational Culture

Organizational culture refers to the basic values, norms, beliefs, and practices that characterize the functioning of a particular institution. At the most basic level, organizational culture defines the assumptions that employees make as they carry out their work; it defines "the way we do things here." An organization's culture is a powerful force that persists through reorganizations and the departure of key personnel.

The human space flight culture within NASA originated in the Cold War environment. The space agency itself was created in 1958 as a response to the Soviet launch of *Sputnik*, the first artificial Earth satellite. In 1961, President John F. Kennedy charged the new space agency with the task of reaching the moon before the end of the decade, and asked Congress and the American people to commit the immense resources for doing so, even though at the time NASA had only accumulated 15 minutes of human space flight experience. With its efforts linked to U.S.-Soviet competition for global leadership, there was a sense in the NASA workforce that the agency was engaged in a historic struggle central to the nation's agenda.

The Apollo era created at NASA an exceptional "can-do" culture marked by tenacity in the face of seemingly impossible challenges. This culture valued the interaction among research and testing, hands-on engineering experience, and a dependence on the exceptional quality of the its workforce and leadership that provided in-house technical capability to oversee the work of contractors. The culture also accepted risk and failure as inevitable aspects of operating in space, even as it held as its highest value attention to detail in order to lower the chances of failure.

The dramatic Apollo 11 lunar landing in July 1969 fixed NASA's achievements in the national consciousness, and in history. However, the numerous accolades in the wake of the moon landing also helped reinforce the NASA staff's faith in their organizational culture. Apollo successes created the powerful image of the space agency as a "perfect place," as "the best organization that human beings could create to accomplish selected goals."[13] During Apollo, NASA was in many respects a highly successful organization capable of achieving seemingly impossible feats. The continuing image of NASA as a "perfect place" in the years after Apollo left NASA employees unable to recognize that NASA never had been, and still was not, perfect, nor was it as symbolically important in the continuing Cold War struggle as it had been for its first decade of existence. NASA personnel maintained a vision of their agency that was rooted in the glories of an earlier time, even as the world, and thus the context within which the space agency operated, changed around them.

As a result, NASA's human space flight culture never fully adapted to the Space Shuttle Program, with its goal of routine access to space rather than further exploration beyond low-Earth orbit. The Apollo-era organizational culture came to be in tension with the more bureaucratic space agency of the 1970s, whose focus turned from designing new spacecraft at any expense to repetitively flying a reusable vehicle on an ever-tightening budget. This trend toward bureaucracy and the associated increased reliance on contracting necessitated more effective communications and more extensive safety oversight processes than had been in place during the Apollo era, but the Rogers Commission found that such features were lacking.

In the aftermath of the *Challenger* accident, these contradictory forces prompted a resistance to externally imposed changes and an attempt to maintain the internal belief that NASA was still a "perfect place," alone in its ability to execute a program of human space flight. Within NASA centers, as Human Space Flight Program managers strove to maintain their view of the organization, they lost their ability to accept criticism, leading them to reject the recommendations of many boards and blue-ribbon panels, the Rogers Commission among them.

External criticism and doubt, rather than spurring NASA to change for the better, instead reinforced the will to "impose the party line vision on the environment, not to reconsider it," according to one authority on organizational behavior. This in turn led to "flawed decision making, self deception, introversion and a diminished curiosity about the world outside the perfect place."[14] The NASA human space flight culture the Board found during its investigation manifested many of these characteristics, in particular a self-confidence about NASA possessing unique knowledge about how to safely launch people into space.[15] As will be discussed later in this chapter, as well as in Chapters 6, 7, and 8, the Board views this cultural resistance as a fundamental impediment to NASA's effective organizational performance.

5.3 AN AGENCY TRYING TO DO TOO MUCH WITH TOO LITTLE

A strong indicator of the priority the national political leadership assigns to a federally funded activity is its budget. By that criterion, NASA's space activities have not been high on the list of national priorities over the past three decades (see Figure 5.3-1). After a peak during the Apollo program, when NASA's budget was almost four percent of the federal budget, NASA's budget since the early 1970s has hovered at one percent of federal spending or less.

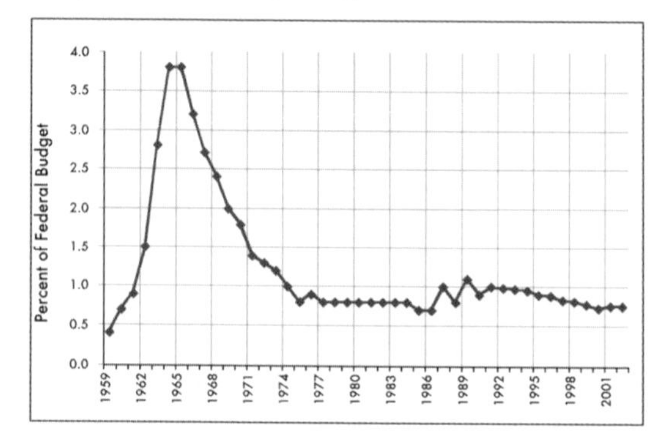

Figure 5.3-1. NASA budget as a percentage of the Federal budget. (Source: NASA History Office)

Particularly in recent years, as the national leadership has confronted the challenging task of allocating scarce public resources across many competing demands, NASA has had difficulty obtaining a budget allocation adequate to its continuing ambitions. In 1990, the White House chartered a blue-ribbon committee chaired by aerospace executive Norman Augustine to conduct a sweeping review of NASA and its programs in response to Shuttle problems and the flawed mirror on the Hubble Space Telescope.[16] The review found that NASA's budget was inadequate for all the programs the agency was executing, saying that "NASA is currently over committed in terms of program obligations relative to resources available–in short, it is trying to do too much, and allowing too little margin for the unexpected."[17] "A reinvigorated space program," the Augustine committee went on to say, "will require real growth in the NASA budget of approximately 10 percent per year (through the year 2000) reaching a peak spending level of about $30 billion per year (in constant 1990 dollars) by about the year 2000." Translated into the actual dollars of Fiscal Year 2000, that recommendation would have meant a NASA budget of over $40 billion; the actual NASA budget for that year was $13.6 billion.[18]

During the past decade, neither the White House nor Congress has been interested in "a reinvigorated space program." Instead, the goal has been a program that would continue to produce valuable scientific and symbolic payoffs for the nation without a need for increased budgets. Recent budget allocations reflect this continuing policy reality. Between 1993 and 2002, the government's discretionary spending grew in purchasing power by more than 25 percent, defense spending by 15 percent, and non-defense spending by 40 percent (see Figure 5.3-2). NASA's budget, in comparison, showed little change, going from $14.31 billion in Fiscal Year 1993 to a low of $13.6 billion in Fiscal Year 2000, and increasing to $14.87 billion in Fiscal Year 2002. This represented a loss of 13 percent in purchasing power over the decade (see Figure 5.3-3).[19]

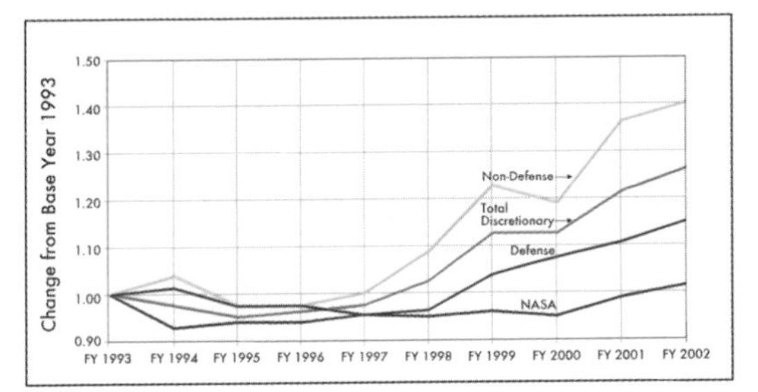

Figure 5.3-2. Changes in Federal spending from 1993 through 2002. (Source: NASA Office of Legislative Affairs)

Fiscal Year	Real Dollars (in millions)	Constant Dollars (in FY 2002 millions)
1965	5,250	24,696
1975	3,229	10,079
1985	7,573	11,643
1993	14,310	17,060
1994	14,570	16,965
1995	13,854	15,790
1996	13,884	15,489
1997	13,709	14,994
1998	13,648	14,641
1999	13,653	14,443
2000	13,601	14,202
2001	14,230	14,559
2002	14,868	14,868
2003	15,335	NA
2004	(requested) 15,255	NA

Figure 5.3-3. NASA Budget. (Source: NASA and Office of Management and Budget)

The lack of top-level interest in the space program led a 2002 review of the U.S. aerospace sector to observe that "a sense of lethargy has affected the space industry and community. Instead of the excitement and exuberance that dominated our early ventures into space, we at times seem almost apologetic about our continued investments in the space program."[20]

WHAT THE EXPERTS HAVE SAID

Warnings of a Shuttle Accident

"Shuttle reliability is uncertain, but has been estimated to range between 97 and 99 percent. If the Shuttle reliability is 98 percent, there would be a 50-50 chance of losing an Orbiter within 34 flights … The probability of maintaining at least three Orbiters in the Shuttle fleet declines to less than 50 percent after flight 113."[21]

-The Office of Technology Assessment, 1989

"And although it is a subject that meets with reluctance to open discussion, and has therefore too often been relegated to silence, the statistical evidence indicates that we are likely to lose another Space Shuttle in the next several years … probably before the planned Space Station is completely established on orbit. This would seem to be the weak link of the civil space program – unpleasant to recognize, involving all the uncertainties of statistics, and difficult to resolve."

-The Augustine Committee, 1990

Shuttle as Developmental Vehicle

"Shuttle is also a complex system that has yet to demonstrate an ability to adhere to a fixed schedule"

-The Augustine Committee, 1990

NASA Human Space Flight Culture

"NASA has not been sufficiently responsive to valid criticism and the need for change."[22]

-The Augustine Committee, 1990

Faced with this budget situation, NASA had the choice of either eliminating major programs or achieving greater efficiencies while maintaining its existing agenda. Agency leaders chose to attempt the latter. They continued to develop the space station, continued robotic planetary and scientific missions, and continued Shuttle-based missions for both scientific and symbolic purposes. In 1994 they took on the responsibility for developing an advanced technology launch vehicle in partnership with the private sector. They tried to do this by becoming more efficient. "Faster, better, cheaper" became the NASA slogan of the 1990s.[23]

The flat budget at NASA particularly affected the human space flight enterprise. During the decade before the *Columbia* accident, NASA rebalanced the share of its budget allocated to human space flight from 48 percent of agency funding in Fiscal Year 1991 to 38 percent in Fiscal Year 1999, with the remainder going mainly to other science and technology efforts. On NASA's fixed budget, that meant

EARMARKS

Pressure on NASA's budget has come not only from the White House, but also from the Congress. In recent years there has been an increasing tendency for the Congress to add "earmarks" – congressional additions to the NASA budget request that reflect targeted Members' interests. These earmarks come out of already-appropriated funds, reducing the amounts available for the original tasks. For example, as Congress considered NASA's Fiscal Year 2002 appropriation, the NASA Administrator told the House Appropriations subcommittee with jurisdiction over the NASA budget that the agency was "extremely concerned regarding the magnitude and number of congressional earmarks" in the House and Senate versions of the NASA appropriations bill.[24] He noted "the total number of House and Senate earmarks ... is approximately 140 separate items, an increase of nearly 50 percent over FY 2001." These earmarks reflected "an increasing fraction of items that circumvent the peer review process, or involve construction or other objectives that have no relation to NASA mission objectives." The potential Fiscal Year 2002 earmarks represented "a net total of $540 million in reductions to ongoing NASA programs to fund this extremely large number of earmarks."[25]

the Space Shuttle and the International Space Station were competing for decreasing resources. In addition, at least $650 million of NASA's human space flight budget was used to purchase Russian hardware and services related to U.S.-Russian space cooperation. This initiative was largely driven by the Clinton Administration's foreign policy and national security objectives of supporting the administration of Boris Yeltsin and halting the proliferation of nuclear weapons and the means to deliver them.

Space Shuttle Program Budget Patterns

For the past 30 years, the Space Shuttle Program has been NASA's single most expensive activity, and of all NASA's efforts, that program has been hardest hit by the budget constraints of the past decade. Given the high priority assigned after 1993 to completing the costly International Space Station, NASA managers have had little choice but to attempt to reduce the costs of operating the Space Shuttle. This left little funding for Shuttle improvements. The squeeze on the Shuttle budget was even more severe after the Office of Management and Budget in 1994 insisted that any cost overruns in the International Space Station budget be made up from within the budget allocation for human space flight, rather than from the agency's budget as a whole. The Shuttle was the only other large program within that budget category.

Figures 5.3-4 and 5.3-5 show the trajectory of the Shuttle budget over the past decade. In Fiscal Year 1993, the outgoing Bush administration requested $4.128 billion for the Space Shuttle Program; five years later, the Clinton Administration request was for $2.977 billion, a 27 percent reduction. By Fiscal Year 2003, the budget request had increased to $3.208 billion, still a 22 percent reduction from a decade earlier. With inflation taken into account, over the past decade, there has been a reduction of approximately 40 percent in the purchasing power of the program's budget, compared to a reduction of 13 percent in the NASA budget overall.

Fiscal Year	President's Request to Congress	Congressional Appropriation	Change	NASA Operating Plan*	Change
1993	4,128.0	4,078.0	-50.0	4,052.9	-25.1
1994	4,196.1	3,778.7	-417.4**	3,772.3	-6.4
1995	3,324.0	3,155.1	-168.9	3,155.1	0.0
1996	3,231.8	3,178.8	-53.0	3,143.8	-35.0
1997	3,150.9	3,150.9	0.0	2,960.9	-190.0
1998	2,977.8	2,927.8	-50.0	2,912.8	-15.0
1999	3,059.0	3,028.0	-31.0	2,998.3	-29.7
2000	2,986.2	3,011.2	+25.0	2,984.4	-26.8
2001	3,165.7	3,125.7	-40.0	3,118.8	-6.9
2002	3,283.8	3,278.8	-5.0	3,270.0	-8.9
2003	3,208.0	3,252.8	+44.8		

Figure 5.3-4. Space Shuttle Program Budget (in millions of dollars). (Source: NASA Office of Space Flight)
** NASA's operating plan is the means for adjusting congressional appropriations among various activities during the fiscal year as changing circumstances dictate. These changes must be approved by NASA's appropriation subcommittees before they can be put into effect.*
***This reduction primarily reflects the congressional cancellation of the Advanced Solid Rocket Motor Program*

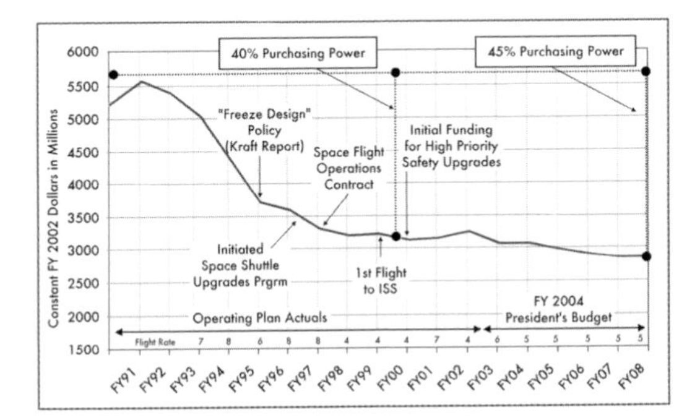

Figure 5.3-5. NASA budget as a percentage of the Federal budget from 1991 to 2008. (Source: NASA Office of Space Flight)

This budget squeeze also came at a time when the Space Shuttle Program exhibited a trait common to most aging systems: increased costs due to greater maintenance requirements, a declining second- and third-tier contractor support base, and deteriorating infrastructure. Maintaining the Shuttle was becoming more expensive at a time when Shuttle budgets were decreasing or being held constant. Only in the last few years have those budgets begun a gradual increase.

As Figure 5.3-5 indicates, most of the steep reductions in the Shuttle budget date back to the first half of the 1990s. In the second half of the decade, the White House Office of Management and Budget and NASA Headquarters held the Shuttle budget relatively level by deferring substantial funding for Shuttle upgrades and infrastructure improvements, while keeping pressure on NASA to limit increases in operating costs.

5.4 Turbulence in NASA Hits the Space Shuttle Program

In 1992 the White House replaced NASA Administrator Richard Truly with aerospace executive Daniel S. Goldin, a self-proclaimed "agent of change" who held office from April 1, 1992, to November 17, 2001 (in the process becoming the longest-serving NASA Administrator). Seeing "space exploration (manned and unmanned) as NASA's principal purpose with Mars as a destiny," as one management scholar observed, and favoring "administrative transformation" of NASA, Goldin engineered "not one or two policy changes, but a torrent of changes. This was not evolutionary change, but radical or discontinuous change."[26] His tenure at NASA was one of continuous turmoil, to which the Space Shuttle Program was not immune.

Of course, turbulence does not necessarily degrade organizational performance. In some cases, it accompanies productive change, and that is what Goldin hoped to achieve. He believed in the management approach advocated by W. Edwards Deming, who had developed a series of widely acclaimed management principles based on his work in Japan during the "economic miracle" of the 1980s. Goldin attempted to apply some of those principles to NASA, including the notion that a corporate headquarters should not attempt to exert bureaucratic control over a complex organization, but rather set strategic directions and provide operating units with the authority and resources needed to pursue those directions. Another Deming principle was that checks and balances in an organization were unnecessary

Congressional Budget Reductions

In most years, Congress appropriates slightly less for the Space Shuttle Program than the President requested; in some cases, these reductions have been requested by NASA during the final stages of budget deliberations. After its budget was passed by Congress, NASA further reduced the Shuttle budget in the agency's operating plan–the plan by which NASA actually allocates its appropriated budget during the fiscal year to react to changing program needs. These released funds were allocated to other activities, both within the human space flight program and in other parts of the agency. Changes in recent years include:

Fiscal Year 1997

- NASA transferred $190 million to International Space Station (ISS).

Fiscal Year 1998

- At NASA's request, Congress transferred $50 million to ISS.
- NASA transferred $15 million to ISS.

Fiscal Year 1999

- At NASA's request, Congress reduced Shuttle $31 million so NASA could fund other requirements.
- NASA reduced Shuttle $32 million by deferring two flights; funds transferred to ISS.
- NASA added $2.3 million from ISS to previous NASA request.

Fiscal Year 2000

- Congress added $25 million to Shuttle budget for upgrades and transferred $25 million from operations to upgrades.
- NASA reduced Shuttle $11.5 million per government-wide rescission requirement and transferred $15.3 million to ISS.

Fiscal Year 2001

- At NASA's request, Congress reduced Shuttle budget by $40 million to fund Mars initiative.
- NASA reduced Shuttle $6.9 million per rescission requirement.

Fiscal Year 2002

- Congress reduced Shuttle budget $50 million to reflect cancellation of electric Auxiliary Power Unit and added $20 million for Shuttle upgrades and $25 million for Vehicle Assembly Building repairs.
- NASA transferred $7.6 million to fund Headquarters requirements and cut $1.2 million per rescission requirement.

[Source: Marcia Smith, Congressional Research Service, Presentation at CAIB Public Hearing, June 12, 2003] and sometimes counterproductive, and those carrying out the work should bear primary responsibility for its quality. It is arguable whether these business principles can readily be applied to a government agency operating under civil service rules and in a politicized environment. Nevertheless, Goldin sought to implement them throughout his tenure.[27]

Goldin made many positive changes in his decade at NASA. By bringing Russia into the Space Station partnership in 1993, Goldin developed a new post-Cold War rationale for the agency while managing to save a program that was politically faltering. The International Space Station became NASA's premier program, with the Shuttle serving in a supporting role. Goldin was also instrumental in gaining acceptance of the "faster, better, cheaper"[28] approach to the planning of robotic missions and downsizing "an agency that was considered bloated and bureaucratic when he took it over."[29]

Goldin described himself as "sharp-edged" and could often be blunt. He rejected the criticism that he was sacrificing safety in the name of efficiency. In 1994 he told an audience at the Jet Propulsion Laboratory, "When I ask for the budget to be cut, I'm told it's going to impact safety on the Space Shuttle ... I think that's a bunch of crap."[30]

One of Goldin's high-priority objectives was to decrease involvement of the NASA engineering workforce with the Space Shuttle Program and thereby free up those skills for finishing the space station and beginning work on his preferred objective–human exploration of Mars. Such a shift would return NASA to its exploratory mission. He was often at odds with those who continued to focus on the centrality of the Shuttle to NASA's future.

Initial Shuttle Workforce Reductions

With NASA leadership choosing to maintain existing programs within a no-growth budget, Goldin's "faster, better, cheaper" motto became the agency's slogan of the 1990s.[31] NASA leaders, however, had little maneuvering room in which to achieve efficiency gains. Attempts by NASA Headquarters to shift functions or to close one of the three human space flight centers were met with strong resistance from the Centers themselves, the aerospace firms they used as contractors, and the congressional delegations of the states in which the Centers were located. This alliance resembles the classic "iron triangle" of bureaucratic politics, a conservative coalition of bureaucrats, interest groups, and congressional subcommittees working together to promote their common interests.[32]

With Center infrastructure off-limits, this left the Space Shuttle budget as an obvious target for cuts. Because the Shuttle required a large "standing army" of workers to

	1993	1994	1995	1996	1997	1998	1999	2000	2001	2002
Total Workforce	30,091	27,538	25,346	23,625	19,476	18,654	18,068	17,851	18,012	17,462
Total Civil Service Workforce	3,781	3,324	2,959	2,596	2,195	1,954	1,777	1,786	1,759	1,718
JSC	1,330	1,304	1,248	1,076	958	841	800	798	794	738
KSC	1,373	1,104	1,018	932	788	691	613	626	614	615
MSFC	874	791	576	523	401	379	328	336	327	337
Stennis/Dryden	84	64	55	32	29	27	26	16	14	16
Headquarters	120	61	62	32	20	16	10	10	10	12
Total Contractor Workforce	26,310	24,214	22,387	21,029	17,281	16,700	16,291	16,065	16,253	15,744
JSC	7,487	6,805	5,887	5,442	*10,556	10,525	10,733	10,854	11,414	11,445
KSC	9,173	8,177	7,691	7,208	539	511	430	436	439	408
MSFC	9,298	8,635	8,210	7,837	5,650	5,312	4,799	4,444	4,197	3,695
Stennis/Dryden	267	523	529	505	536	453	329	331	203	196
Headquarters	85	74	70	37	0	0	0	0	0	0

Figure 5.4-1. Space Shuttle Program workforce. [Source: NASA Office of Space Flight]
** Because Johnson Space Center manages the Space Flight Operations Contract, all United Space Alliance employees are counted as working for Johnson.*

keep it flying, reducing the size of the Shuttle workforce became the primary means by which top leaders lowered the Shuttle's operating costs. These personnel reduction efforts started early in the decade and continued through most of the 1990s. They created substantial uncertainty and tension within the Shuttle workforce, as well as the transitional difficulties inherent in any large-scale workforce reassignment.

In early 1991, even before Goldin assumed office and less than three years after the Shuttle had returned to flight after the *Challenger* accident, NASA announced a goal of saving three to five percent per year in the Shuttle budget over five years. This move was in reaction to a perception that the agency had overreacted to the Rogers Commission recommendations – for example, the notion that the many layers of safety inspections involved in preparing a Shuttle for flight had created a bloated and costly safety program.

From 1991 to 1994, NASA was able to cut Shuttle operating costs by 21 percent. Contractor personnel working on the Shuttle declined from 28,394 to 22,387 in these three years, and NASA Shuttle staff decreased from 4,031 to 2,959.[33] Figure 5.4-1 shows the changes in Space Shuttle workforce over the past decade. A 1994 National Academy of Public Administration review found that these cuts were achieved primarily through "operational and organizational efficiencies and consolidations, with resultant reductions in staffing levels and other actions which do not significantly impact basic program content or capabilities."[34]

NASA considered additional staff cuts in late 1994 and early 1995 as a way of further reducing the Space Shuttle Program budget. In early 1995, as the national leadership focused its attention on balancing the federal budget, the projected five-year Shuttle budget requirements exceeded by $2.5 billion the budget that was likely to be approved by the White House Office of Management and Budget.[35] Despite its already significant progress in reducing costs, NASA had to make further workforce cuts.

Anticipating this impending need, a 1994-1995 NASA "Functional Workforce Review" concluded that removing an additional 5,900 people from the NASA and contractor Shuttle workforce – just under 13 percent of the total – could be done without compromising safety.[36] These personnel cuts were made in Fiscal Years 1996 and 1997. By the end of 1997, the NASA Shuttle civilian workforce numbered 2,195, and the contractor workforce 17,281.

Shifting Shuttle Management Arrangements

Workforce reductions were not the only modifications to the Shuttle Program in the middle of the decade. In keeping with Goldin's philosophy that Headquarters should concern itself primarily with strategic issues, in February 1996 Johnson Space Center was designated as "lead center" for the Space Shuttle Program, a role it held prior to the *Challenger* accident. This shift was part of a general move of all program management responsibilities from NASA Headquarters to the agency's field centers. Among other things, this change meant that Johnson Space Center managers would have authority over the funding and management of Shuttle activities at the Marshall and Kennedy Centers. Johnson and Marshall had been rivals since the days of Apollo, and long-term Marshall employees and managers did not easily accept the return of Johnson to this lead role.

The shift of Space Shuttle Program management to Johnson was worrisome to some. The head of the Space Shuttle Program at NASA Headquarters, Bryan O'Connor, argued that transfer of the management function to the Johnson Space Center would return the Shuttle Program management to the flawed structure that was in place before the *Challenger* accident. "It is a safety issue," he said, "we ran it that way [with program management at Headquarters, as recommended by the Rogers Commission] for 10 years without a mishap and I didn't see any reason why we should go back to the way we operated in the pre-*Challenger* days."[37] Goldin gave O'Connor several opportunities to present his arguments against a transfer of management responsibility, but ultimately decided to proceed. O'Connor felt he had no choice but to resign.[38] (O'Connor returned to NASA in 2002 as Associate Administrator for Safety and Mission Assurance.)

In January 1996, Goldin appointed as Johnson's director his close advisor, George W.S. Abbey. Abbey, a space program veteran, was a firm believer in the values of the original human space flight culture, and as he assumed the directorship, he set about recreating as many of the positive features of that culture as possible. For example, he and Goldin initiated, as a way for young engineers to get hands-on experience, an in-house X-38 development program as a prototype for a space station crew rescue vehicle. Abbey was a powerful leader, who through the rest of the decade exerted substantial control over all aspects of Johnson Space Center operations, including the Space Shuttle Program.

Space Flight Operations Contract

By the middle of the decade, spurred on by Vice President Al Gore's "reinventing government" initiative, the goal of balancing the federal budget, and the views of a Republican-led House of Representatives, managers throughout the government sought new ways of making public sector programs more efficient and less costly. One method considered was transferring significant government operations and responsibilities to the private sector, or "privatization." NASA led the way toward privatization, serving as an example to other government agencies.

In keeping with his philosophy that NASA should focus on its research-and-development role, Goldin wanted to remove NASA employees from the repetitive operations of various systems, including the Space Shuttle. Giving primary responsibility for Space Shuttle operations to the private sector was therefore consistent with White House and congressional priorities and attractive to Goldin on its own terms. Beginning in 1994, NASA considered the feasibility of consolidating many of the numerous Shuttle operations contracts under a single prime contractor. At that time, the Space Shuttle Program was managing 86 separate contracts held by 56 different firms. Top NASA managers thought that consolidating these contracts could reduce the amount of redundant overhead, both for NASA and for the contractors themselves. They also wanted to explore whether there were functions being carried out by NASA that could be more effectively and inexpensively carried out by the private sector.

An advisory committee headed by early space flight veteran Christopher Kraft recommended such a step in its March 1995 report, which became known as the "Kraft Report."[39] (The report characterized the Space Shuttle in a way that the Board judges to be at odds with the realities of the Shuttle Program).

The report made the following findings and recommendations:

- "The Shuttle has become a mature and reliable system ... about as safe as today's technology will provide."
- "Given the maturity of the vehicle, a change to a new mode of management with considerably less NASA oversight is possible at this time."
- "Many inefficiencies and difficulties in the current Shuttle Program can be attributed to the diffuse and fragmented NASA and contractor structure. Numerous contractors exist supporting various program elements, resulting in ambiguous lines of communication and diffused responsibilities."
- NASA should "consolidate operations under a single-business entity."
- "The program remains in a quasi-development mode and yearly costs remain higher than required," and NASA should "freeze the current vehicle configuration, minimizing future modifications, with such modifications delivered in block updates. Future block updates should implement modifications required to make the vehicle more re-usable and operational."
- NASA should "restructure and reduce the overall Safety, Reliability, and Quality Assurance elements – without reducing safety."[40]

When he released his committee's report, Kraft said that "if NASA wants to make more substantive gains in terms of efficiency, cost savings and better service to its customers, we think it's imperative they act on these recommendations ... And we believe that these savings are real, achievable, and can be accomplished with no impact to the safe and successful operation of the Shuttle system."[41]

Although the Kraft Report stressed that the dramatic changes it recommended could be made without compromising safety, there was considerable dissent about this claim. NASA's Aerospace Safety Advisory Panel – independent, but often not very influential – was particularly critical. In May 1995, the Panel noted that "the assumption [in the Kraft Report] that the Space Shuttle systems are now 'mature' smacks of a complacency which may lead to serious mishaps. The fact is that the Space Shuttle may never be mature enough to totally freeze the design." The Panel also noted that "the report dismisses the concerns of many credible sources by labeling honest reservations and the people who have made them as being partners in an unneeded 'safety shield' conspiracy. Since only one more accident would kill the program and destroy far more than the spacecraft, it is extremely callous" to make such an accusation.[42]

The notion that NASA would further reduce the number of civil servants working on the Shuttle Program prompted senior Kennedy Space Center engineer José Garcia to send to President Bill Clinton on August 25, 1995, a letter that stated, "The biggest threat to the safety of the crew since the Challenger disaster is presently underway at NASA." Garcia's particular concern was NASA's "efforts to delete the 'checks and balances' system of processing Shuttles as a way of saving money ... Historically NASA has employed two engineering teams at KSC, one contractor and one government, to cross check each other and prevent catastrophic errors ... although this technique is expensive, it is effective, and it is the single most important factor that sets the Shuttle's success above that of any other launch vehicle ... Anyone who doesn't have a hidden agenda or fear of losing his job would admit that you can't delete NASA's checks and balances system of Shuttle processing without affecting the safety of the Shuttle and crew."[43]

NASA leaders accepted the advice of the Kraft Report and in August 1995 solicited industry bids for the assignment of Shuttle prime contractor. In response, Lockheed Martin and Rockwell, the two major Space Shuttle operations contractors, formed a limited liability corporation, with each firm a 50 percent owner, to compete for what was called the Space Flight Operations Contract. The new corporation would be known as United Space Alliance.

In November 1995, NASA awarded the operations contract to United Space Alliance on a sole source basis. (When Boeing bought Rockwell's aerospace group in December 1996, it also took over Rockwell's 50 percent ownership of United Space Alliance.) The company was responsible for 61 percent of the Shuttle operations contracts. Some in Congress were skeptical that safety could be maintained under the new arrangement, which transferred significant NASA responsibilities to the private sector. Despite these concerns, Congress ultimately accepted the reasoning behind the contract.[44] NASA then spent much of 1996 negotiating the contract's terms and conditions with United Space Alliance.

The Space Flight Operations Contract was designed to reward United Space Alliance for performance successes and penalize its performance failures. Before being eligible for any performance fees, United Space Alliance would have to meet a series of safety "gates," which were intended to ensure that safety remained the top priority in Shuttle operations. The contract also rewarded any cost reductions that United Space Alliance was able to achieve, with NASA taking 65 percent of any savings and United Space Alliance 35 percent.[45]

NASA and United Space Alliance formally signed the Space Flight Operations Contract on October 1, 1996. Initially, only the major Lockheed Martin and Rockwell Shuttle contracts and a smaller Allied Signal Unisys contract were transferred to United Space Alliance. The initial contractual period was six years, from October 1996 to September 2002. NASA exercised an option for a two-year extension in 2002, and another two-year option exists. The total value of the contract through the current extension is estimated at $12.8 billion. United Space Alliance currently has approximately 10,000 employees.

SPACE FLIGHT OPERATIONS CONTRACT

The Space Flight Operations Contract has two major areas of innovation:

- It replaced the previous "cost-plus" contracts (in which a firm was paid for the costs of its activity plus a negotiated profit) with a complex contract structure that included performance-based and cost reduction incentives. Performance measures include safety, launch readiness, on-time launch, Solid Rocket Booster recovery, proper orbital insertion, and successful landing.
- It gave additional responsibilities for Shuttle operation, including safety and other inspections and integration of the various elements of the Shuttle system, to United Space Alliance. Many of those responsibilities were previously within the purview of NASA employees.

Under the Space Flight Operations Contract, United Space Alliance had overall responsibility for processing selected Shuttle hardware, including:

- Inspecting and modifying the Orbiters
- Installing the Space Shuttle Main Engines on the Orbiters
- Assembling the sections that make up the Solid Rocket Boosters
- Attaching the External Tank to the Solid Rocket Boosters, and then the Orbiter to the External Tank
- Recovering expended Solid Rocket boosters

In addition to processing Shuttle hardware, United Space Alliance is responsible for mission design and planning, astronaut and flight controller training, design and integration of flight software, payload integration, flight operations, launch and recovery operations, vehicle-sustaining engineering, flight crew equipment processing, and operation and maintenance of Shuttle-specific facilities such as the Vehicle Assembly Building, the Orbiter Processing Facility, and the launch pads. United Space Alliance also provides spare parts for the Orbiters, maintains Shuttle flight simulators, and provides tools and supplies, including consumables such as food, for Shuttle missions.

Under the Space Flight Operations Contract, NASA has the following responsibilities and roles:

- Maintaining ownership of the Shuttles and all other assets of the Shuttle program
- Providing to United Space Alliance the Space Shuttle Main Engines, the External Tanks, and the Redesigned Solid Rocket Motor segments for assembly into the Solid Rocket Boosters
- Managing the overall process of ensuring Shuttle safety
- Developing requirements for major upgrades to all assets
- Participating in the planning of Shuttle missions, the directing of launches, and the execution of flights
- Performing surveillance and audits and obtaining technical insight into contractor activities
- Deciding if and when to "commit to flight" for each mission[46]

The contract provided for additional consolidation and then privatization, when all remaining Shuttle operations would be transferred from NASA. Phase 2, scheduled for 1998-2000, called for the transfer of Johnson Space Center-managed flight software and flight crew equipment contracts and the Marshall Space Center-managed contracts for the External Tank, Space Shuttle Main Engine, Reusable Solid Rocket Motor, and Solid Rocket Booster.

However, Marshall and its contractors, with the concurrence of the Space Shuttle Program Office at Johnson Space Center, successfully resisted the transfer of its contracts. Therefore, the Space Flight Operations Contract's initial efficiency and integrated management goals have not been achieved.

The major annual savings resulting from the Space Flight Operations Contract, which in 1996 were touted to be some $500 million to $1 billion per year by the early 2000s, have not materialized. These projections assumed that by 2002, NASA would have put all Shuttle contracts under the auspices of United Space Alliance, and would be moving toward Shuttle privatization. Although the Space Flight Operations Contract has not been as successful in achieving cost efficiencies as its proponents hoped, it has reduced some Shuttle operating costs and other expenses. By one estimate, in its first six years the contract has saved NASA a total of more than $1 billion.[47]

Privatizing the Space Shuttle

To its proponents, the Space Flight Operations Contract was only a beginning. In October 1997, United Space Alliance submitted to the Space Shuttle Program Office a contractually required plan for privatizing the Shuttle, which the program did not accept. But the notion of Shuttle privatization lingered at NASA Headquarters and in Congress, where some members advocated a greater private sector role in the space program. Congress passed the Commercial Space Act of 1998, which directed the NASA Administrator to "plan for the eventual privatization of the Space Shuttle Program."[48]

By August 2001, NASA Headquarters prepared for White House consideration a "Privatization White Paper" that called for transferring all Shuttle hardware, pilot and commander astronauts, and launch and operations teams to a private operator.[49] In September 2001, Space Shuttle Program Manager Ron Dittemore released his report on a "Concept of Privatization of the Space Shuttle Program,"[50] which argued that for the Space Shuttle "to remain safe and viable, it is necessary to merge the required NASA and contractor skill bases" into a single private organization that would manage human space flight. This perspective reflected Dittemore's belief that the split of responsibilities between NASA and United Space Alliance was not optimal, and that it was unlikely that NASA would ever recapture the Shuttle responsibilities that were transferred in the Space Flight Operations Contract.

Dittemore's plan recommended transferring 700 to 900 NASA employees to the private organization, including:

- Astronauts, including the flight crew members who operate the Shuttle

- Program and project management, including Space Shuttle Main Engine, External Tank, Redesigned Solid Rocket Booster, and Extravehicular Activity
- Mission operations, including flight directors and flight controllers
- Ground operations and processing, including launch director, process engineering, and flow management
- Responsibility for safety and mission assurance

After such a shift occurred, according to the Dittemore plan, "the primary role for NASA in Space Shuttle operations ... will be to provide an SMA [Safety and Mission Assurance] independent assessment ... utilizing audit and surveillance techniques."[51]

With a change in NASA Administrators at the end of 2001 and the new Bush Administration's emphasis on "competitive sourcing" of government operations, the notion of wholesale privatization of the Space Shuttle was replaced with an examination of the feasibility of both public- and private-sector Program management. This competitive sourcing was under examination at the time of the *Columbia* accident.

Workforce Transformation and the End of Downsizing

Workforce reductions instituted by Administrator Goldin as he attempted to redefine the agency's mission and its overall organization also added to the turbulence of his reign. In the 1990s, the overall NASA workforce was reduced by 25 percent through normal attrition, early retirements, and buyouts – cash bonuses for leaving NASA employment. NASA operated under a hiring freeze for most of the decade, making it difficult to bring in new or younger people. Figure 5.4-2 shows the downsizing of the overall NASA workforce during this period as well as the associated shrinkage in NASA's technical workforce.

NASA Headquarters was particularly affected by workforce reductions. More than half its employees left or were transferred in parallel with the 1996 transfer of program management responsibilities back to the NASA centers. The Space Shuttle Program bore more than its share of Headquarters personnel cuts. Headquarters civil service staff working on the Space Shuttle Program went from 120 in 1993 to 12 in 2003.

While the overall workforce at the NASA Centers involved in human space flight was not as radically reduced, the combination of the general workforce reduction and the introduction of the Space Flight Operations Contract significantly impacted the Centers' Space Shuttle Program civil service staff. Johnson Space Center went from 1,330 in 1993 to 738 in 2002; Marshall Space Flight Center, from 874 to 337; and Kennedy Space Center from 1,373 to 615. Kennedy Director Roy Bridges argued that personnel cuts were too deep, and threatened to resign unless the downsizing of his civil service workforce, particularly those involved with safety issues, was reversed.[52]

By the end of the decade, NASA realized that staff reductions had gone too far. By early 2000, internal and external

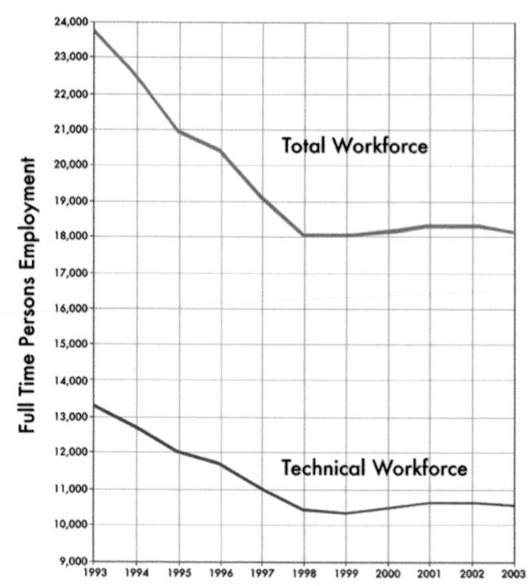

Figure 5.4-2. Downsizing of the overall NASA workforce and the NASA technical workforce.

studies convinced NASA leaders that the workforce needed to be revitalized. These studies noted that "five years of buyouts and downsizing have led to serious skill imbalances and an overtaxed core workforce. As more employees have departed, the workload and stress [on those] remaining have increased, with a corresponding increase in the potential for impacts to operational capacity and safety."[53] NASA announced that NASA workforce downsizing would stop short of the 17,500 target, and that its human space flight centers would immediately hire several hundred workers.

5.5 When to Replace the Space Shuttle?

In addition to budget pressures, workforce reductions, management changes, and the transfer of government functions to the private sector, the Space Shuttle Program was beset during the past decade by uncertainty about when the Shuttle might be replaced. National policy has vacillated between treating the Shuttle as a "going out of business" program and anticipating two or more decades of Shuttle use. As a result, limited and inconsistent investments have been made in Shuttle upgrades and in revitalizing the infrastructure to support the continued use of the Shuttle.

Even before the 1986 *Challenger* accident, when and how to replace the Space Shuttle with a second generation reusable launch vehicle was a topic of discussion among space policy leaders. In January 1986, the congressionally chartered National Commission on Space expressed the need for a Shuttle replacement, suggesting that "the Shuttle fleet will become obsolescent by the turn of the century."[54] Shortly after the *Challenger* accident (but not as a reaction to it), President Reagan announced his approval of "the new Orient Express" (see Figure 5.5-1). This reusable launch vehicle, later known as the National Aerospace Plane, "could, by the end of the decade, take off from Dulles Airport, accelerate up to 25 times the speed of sound attaining low-Earth orbit, or fly to Tokyo within two hours."[55] This goal proved too ambitious, particularly without substantial funding. In 1992, after a $1.7 billion government investment, the National Aerospace Plane project was cancelled.

This pattern – optimistic pronouncements about a revolutionary Shuttle replacement followed by insufficient government investment, and then program cancellation due to technical difficulties – was repeated again in the 1990s.

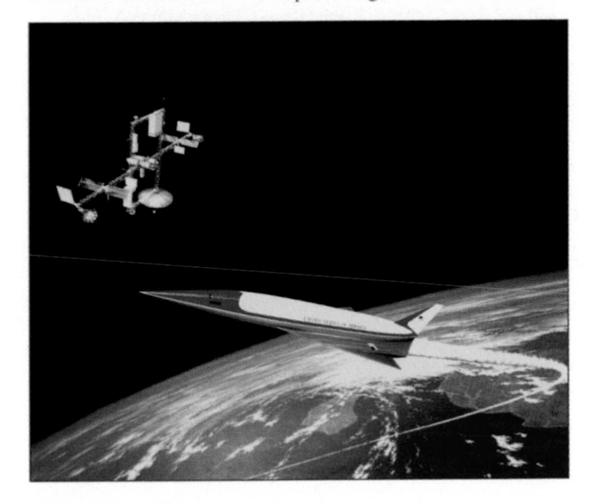

Figure 5.5-1. A 1986 artist's conception of the National Aerospace Plane on a mission to the Space Station.

In 1994, NASA listed alternatives for access to space through 2030.

- Upgrade the Space Shuttle to enable flights through 2030
- Develop a new expendable launcher
- Replace the Space Shuttle with a "leapfrog" next-generation advanced technology system that would achieve order-of-magnitude improvements in the cost effectiveness of space transportation.[56]

Figure 5.5-2. The VentureStar was intended to replace the Space Shuttle based on technology developed for the X-33.

Reflecting its leadership's preference for bold initiatives, NASA chose the third alternative. With White House support,[57] NASA began the X-33 project in 1996 as a joint effort with Lockheed Martin. NASA also initiated the less ambitious X-34 project with Orbital Sciences Corporation. At the time, the future of commercial space launches was bright, and political sentiment in the White House and Congress encouraged an increasing reliance on private-sector solutions for limiting government expenditures. In this context, these unprecedented joint projects appeared less risky than they actually were. The hope was that NASA could replace the Shuttle through private investments, without significant government spending.

Both the X-33 and X-34 incorporated new technologies. The X-33 was to demonstrate the feasibility of an aerospike engine, new Thermal Protection Systems, and composite rather than metal propellant tanks. These radically new technologies were in turn to become the basis for a new orbital vehicle called VentureStar™ that could replace the Space Shuttle by 2006 (see Figure 5.5-2). The X-33 and X-34 ran into technical problems and never flew. In 2001, after spending $1.3 billion, NASA abandoned both projects.

In all three projects – National Aerospace Plane, X-33, and X-34 – national leaders had set ambitious goals in response to NASA's ambitious proposals. These programs relied on the invention of revolutionary technology, had run into major technical problems, and had been denied the funds needed to overcome these problems – assuming they could be solved. NASA had spent nearly 15 years and several billion dollars, and yet had made no meaningful progress toward a Space Shuttle replacement.

In 2000, as the agency ran into increasing problems with the X-33, NASA initiated the Space Launch Initiative, a $4.5 billion multi-year effort to develop new space launch technologies. By 2002, after spending nearly $800 million, NASA again changed course. The Space Launch Initiative failed to find technologies that could revolutionize space launch, forcing NASA to shift its focus to an Orbital Space Plane, developed with existing technology, that would complement the Shuttle by carrying crew, but not cargo, to and from orbit. Under a new Integrated Space Transportation Plan, the Shuttle might continue to fly until 2020 or beyond. (See Section 5.6 for a discussion of this plan.)

As a result of the haphazard policy process that created these still-born developmental programs, the uncertainty over Shuttle replacement persisted. Between 1986 and 2002, the planned replacement date for the Space Shuttle was consistent only in its inconsistency: it changed from 2002 to 2006 to 2012, and before the *Columbia* accident, to 2020 or later.

Safety Concerns and Upgrading the Space Shuttle

This shifting date for Shuttle replacement has severely complicated decisions on how to invest in Shuttle Program upgrades. More often than not, investments in upgrades were delayed or deferred on the assumption they would be a waste of money if the Shuttle were to be retired in the near future (see Figure 5.5-3).

Past Reports Reviewed

During the course of the investigation, more than 50 past reports regarding NASA and the Space Shuttle Program were reviewed. The principal purpose of these reviews what factors those reports examined, what findings were made, and what response, if any, NASA may have made to the findings. Board members then used these findings and responses as a benchmark during their investigation to compare to NASA's current programs. In addition to an extensive 300-page examination of every Aerospace Safety Advisory Panel report (see Appendix D.18), the reports listed on the accompanying chart were examined for specific factors related to the investigation. A complete listing of those past reports' findings, is contained in Appendix D.18.

Report Reviewed	*Topic Examined*								
	Infrastructure	*Communica-tions*	*Contracts*	*Risk Management*	*Quality Assurance*	*Safety Programs*	*Maintenance*	*Security*	*Workforce Issues*
Rogers Commission Report – 1986	•	•	•	•	•	•	•		
STS-29R Prelaunch Assessment – 1989				•					
"Augustine Report" – 1990		•	•	•			•		•
Paté-Cornell Report – 1990				•	•				
"Aldridge Report" – 1992				•					
GAO: NASA Infrastructure – 1996	•								•
GAO: NASA Workforce Reductions – 1996	•								•
Super Light Weight Tank Independent Assessment – 1997				•	•				
Process Readiness Review – 1998				•			•		•
S&MA Ground Operations Report – 1998					•				
GAO: NASA Management Challenges – 1999		•	•	•					
Independent Assessment JS-9047 – 1999				•					
Independent Assessment JS-9059 – 1999				•					
Independent Assessment JS-9078 – 1999		•					•		
Independent Assessment JS-9083 – 1999					•				
S&MA Ground Operations Report – 1999					•	•			
Space Shuttle Independent Assessment Team – 1999			•	•		•	•		•
Space Shuttle Ground Operations Report – 1999						•			
Space Shuttle Program (SSP) Annual Report – 1999						•			

	Infrastructure	Communica-tions	Contracts	Risk Management	Quality Assurance	Safety Programs	Maintenance	Security	Workforce Issues
GAO: Human Capital & Safety – 2000						•			
Independent Assessment JS-0032 – 2000					•				
Independent Assessment JS-0034 – 2000	•								
Independent Assessment JS-0045 – 2000					•				
IG Audit Report 00-039 – 2000			•						
NASA Independent Assessment Team – 2000		•		•	•	•			•
Space Shuttle Program Annual Report – 2000	•		•	•		•			
ASAP Report – 2001	•		•	•		•	•	•	•
GAO: NASA Critical Areas – 2001				•					
GAO: Space Shuttle Safety – 2001									•
Independent Assessment JS-1014 – 2001		•		•	•	•			
Independent Assessment JS-1024 – 2001		•			•				•
Independent Assessment KS-0003 – 2001		•			•				•
Independent Assessment KS-1001 – 2001			•	•			•		
Workforce Survey-KSC – 2001				•	•				
Space Shuttle Program Annual Report – 2001		•		•					
SSP Processing Independent Assessment – 2001				•	•	•			•
ASAP Report – 2002	•		•	•	•	•			•
GAO: Lessons Learned Process – 2002		•							
Independent Assessment KS-1002 – 2002					•				
Selected NASA Lessons Learned – 1992-2002		•		•	•	•	•		•
NASA/Navy Benchmarking Exchange – 2002		•	•	•		•			•
Space Shuttle Program Annual Report – 2002	•			•		•			•
ASAP Leading Indicators – 2003		•		•		•			
NASA Quality Management System – 2003					•				
QAS Tiger Team Report – 2003					•				
Shuttle Business Environment – 2003			•						

Fiscal Year	Upgrades
1994	$454.5
1995	$247.2
1996	$224.5
1997	$215.9
1998	$206.7
1999	$175.2
2000	$239.1
2001	$289.3
2002	$379.5
2003	$347.5

Figure 5.5-3. Shuttle Upgrade Budgets (in millions of dollars). (Source: NASA)

In 1995, for instance, the Kraft Report embraced the principle that NASA should "freeze the design" of the Shuttle and defer upgrades due to the vehicle's "mature" status and the need for NASA to "concentrate scarce resources on developing potential replacements for the Shuttle."[58] NASA subsequently halted a number of planned upgrades, only to reverse course a year later to "take advantage of technologies to improve Shuttle safety and the need for a robust Space Shuttle to assemble the ISS."[59]

In a June 1999 letter to the White House, NASA Administrator Daniel Goldin declared that the nation faced a "Space Launch Crisis." He reported on a NASA review of Shuttle safety that indicated the budget for Shuttle upgrades in Fiscal year 2000 was "inadequate to accommodate upgrades necessary to yield significant safety improvements."[60] After two "close calls" during STS-93 in July 1999 Goldin also chartered a Shuttle Independent Assessment Team (SIAT) chaired by Harry McDonald, Director of NASA Ames Research Center. Among the team's findings, reported in March 2000:[61]

- "Over the course of the Shuttle Program ... processes, procedures and training have continuously been improved and implemented to make the system safer. The SIAT has a major concern ... that this critical feature of the Shuttle Program is being eroded." The major factor leading to this concern "is the reduction in allocated resources and appropriate staff ... There are important technical areas that are 'one-deep.' " Also, "the SIAT feels strongly that workforce augmentation must be realized principally with NASA personnel rather than with contractor personnel."
- The SIAT was concerned with "success-engendered safety optimism ... The SSP must rigorously guard against the tendency to accept risk solely because of prior success."
- "The SIAT was very concerned with what it perceived as Risk Management process erosion created by the desire to reduce costs ... The SIAT feels strongly that NASA Safety and Mission Assurance should be restored to its previous role of an independent oversight body, and not be simply a 'safety auditor.' "
- "The size and complexity of the Shuttle system and of NASA/contractor relationships place extreme importance on understanding, communication, and information handling ... Communication of problems and concerns upward to the SSP from the 'floor' also appeared to leave room for improvement."[62]

The Shuttle Independent Assessment Team report also stated that the Shuttle "clearly cannot be thought of as 'operational' in the usual sense. Extensive maintenance, major amounts of 'touch labor' and a high degree of skill and expertise will always be required." However, "the workforce has received a conflicting message due to the emphasis on achieving cost and staff reductions, and the pressures placed on increasing scheduled flights as a result of the Space Station."[63]

Responding to NASA's concern that the Shuttle required safety-related upgrades, the President's proposed NASA budget for Fiscal Year 2001 proposed a "safety upgrades initiative." That initiative had a short life span. In its Fiscal Year 2002 budget request, NASA proposed to spend $1.836 billion on Shuttle upgrades over five years. A year later, the Fiscal Year 2003 request contained a plan to spend $1.220 billion – a 34 percent reduction. The reductions were primarily a response to rising Shuttle operating costs and the need to stay within a fixed Shuttle budget. Cost growth in Shuttle operations forced NASA to "use funds intended for Space Shuttle safety upgrades to address operational, supportability, obsolescence, and infrastructure needs."[64]

At its March 2001 meeting, NASA's Space Flight Advisory Committee advised that "the Space Shuttle Program must make larger, more substantial safety upgrades than currently planned ... a budget on the order of three times the budget currently allotted for improving the Shuttle systems" was needed.[65] Later that year, five Senators complained that "the Shuttle program is being penalized, despite its outstanding performance, in order to conform to a budget strategy that is dangerously inadequate to ensure safety in America's human space flight program."[66] (See Chapter 7 for additional discussion of Shuttle safety upgrades.)

Deteriorating Shuttle Infrastructure

The same ambiguity about investing in Shuttle upgrades has also affected the maintenance of Shuttle Program ground infrastructure, much of which dates to Project Apollo and 1970s Shuttle Program construction. Figure 5.5-4 depicts the age of the Shuttle's infrastructure as of 2000. Most ground infrastructure was not built for such a protracted lifespan. Maintaining infrastructure has been particularly difficult at Kennedy Space Center, where it is constantly exposed to a salt water environment.

Board investigators have identified deteriorating infrastructure associated with the launch pads, Vehicle Assembly Building, and the crawler transporter. Figures 5.5-5 and 5.5-6 depict some of this deterioration. For example, NASA has installed nets, and even an entire sub-roof, inside the Vehicle Assembly Building to prevent concrete from the building's ceiling from hitting the Orbiter and Shuttle stack. In addition, the corrosion-control challenge results in zinc primer

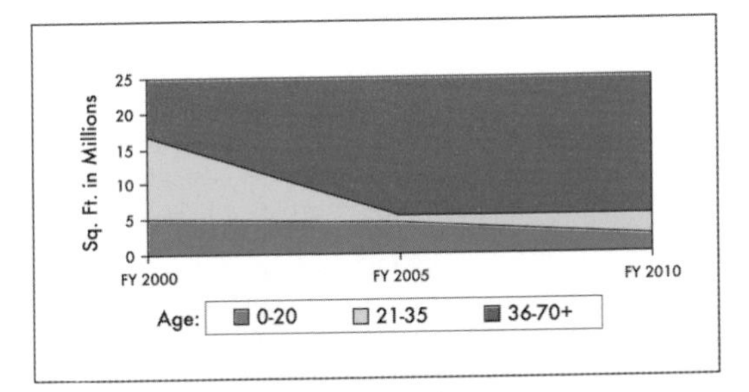

Figure 5.5-4. Age of the Space Shuttle infrastructure. (Source: Connie Milton to Space Flight Advisory Council, 2000.

on certain launch pad areas being exposed to the elements. When rain falls on these areas, it carries away zinc, runs onto the leading edge of the Orbiter's wings, and causes pinholes in the Reinforced Carbon-Carbon panels (see Chapter 3).

In 2000, NASA identified 100 infrastructure items that demanded immediate attention. NASA briefed the Space Flight Advisory Committee on this "Infrastructure Revitalization" initiative in November of that year. The Committee concluded that "deteriorating infrastructure is a serious, major problem," and, upon touring several Kennedy Space Center facilities, declared them "in deplorable condition."[67] NASA subsequently submitted a request to the White House Office of Management and Budget during Fiscal Year 2002 budget deliberations for $600 million to fund the infrastructure initiative. No funding was approved.

In Fiscal Year 2002, Congress added $25 million to NASA's budget for Vehicle Assembly Building repairs. NASA has reallocated limited funds from the Shuttle budget to pressing infrastructure repairs, and intends to take an integrated look at infrastructure as part of its new Shuttle Service Life Extension Program. Nonetheless, like Space Shuttle upgrades, infrastructure revitalization has been mired by the uncertainty surrounding the Shuttle Program's lifetime. Considering that the Shuttle will likely be flying for many years to come, NASA, the White House, and Congress alike now face the specter of having to deal with years of infrastructure neglect.

5.6 A Change in NASA Leadership

Daniel Goldin left NASA in November 2001 after more than nine years as Administrator. The White House chose Sean O'Keefe, the Deputy Director of the White House Office of Management and Budget, as his replacement. O'Keefe stated as he took office that he was not a "rocket scientist," but rather that his expertise was in the management of large government programs. His appointment was an explicit acknowledgement by the new Bush administration that NASA's primary problems were managerial and financial.

By the time O'Keefe arrived, NASA managers had come to recognize that 1990s funding reductions for the Space Shuttle Program had resulted in an excessively fragile program, and also realized that a Space Shuttle replacement was not on the horizon. In 2002, with these issues in mind, O'Keefe made a number of changes to the Space Shuttle Program. He transferred management of both the Space Shuttle Program and the International Space Station from Johnson Space Center to NASA Headquarters. O'Keefe also began considering whether to expand the Space Flight Operations Contract to cover additional Space Shuttle elements, or to pursue "competitive sourcing," a Bush administration initiative that encouraged government agencies to compete with the private sector for management responsibilities of publicly funded activities. To research whether competitive sourcing would be a viable approach for the Space Shuttle Program, NASA chartered the Space Shuttle Competitive Sourcing Task Force through the RAND Corporation, a federally funded think tank. In its report, the Task Force recognized the many obstacles to transferring the Space Shuttle to non-NASA management, primarily NASA's reticence to relinquish control, but concluded that "NASA must pursue competitive sourcing in one form or another."[68]

NASA began a "Strategic Management of Human Capital" initiative to ensure the quality of the future NASA workforce. The goal is to address the various external and internal challenges that NASA faces as it tries to ensure an appropriate mix and depth of skills for future program requirements. A number of aspects to its Strategic Human Capital Plan require legislative approval and are currently before the Congress.

Figure 5.5-5 and 5.5-6. Examples of the seriously deteriorating infrastructure used to support the Space Shuttle Program. At left is Launch Complex 39A, and at right is the Vehicle Assembly building, both at the Kennedy Space Center.

The new NASA leadership also began to compare Space Shuttle program practices with the practices of similar high-technology, high-risk enterprises. The Navy nuclear submarine program was the first enterprise selected for comparative analysis. An interim report on this "benchmarking" effort was presented to NASA in December 2002.[69]

In November 2002, NASA made a fundamental change in strategy. In what was called the Integrated Space Transportation Plan (see Figure 5.6-1), NASA shifted money from the Space Launch Initiative to the Space Shuttle and International Space Station programs. The plan also introduced the Orbital Space Plane as a complement to the Shuttle for the immediate future. Under this strategy, the Shuttle is to fly through at least 2010, when a decision will be made on how long to extend Shuttle operations – possibly through 2020 or even beyond.

As a step in implementing the plan, NASA included $281.4 million in its Fiscal Year 2004 budget submission to begin a Shuttle Service Life Extension Program,[70] which NASA describes as a "strategic and proactive program designed to keep the Space Shuttle flying safely and efficiently." The program includes "high priority projects for safety, supportability, and infrastructure" in order to "combat obsolescence of vehicle, ground systems, and facilities."[71]

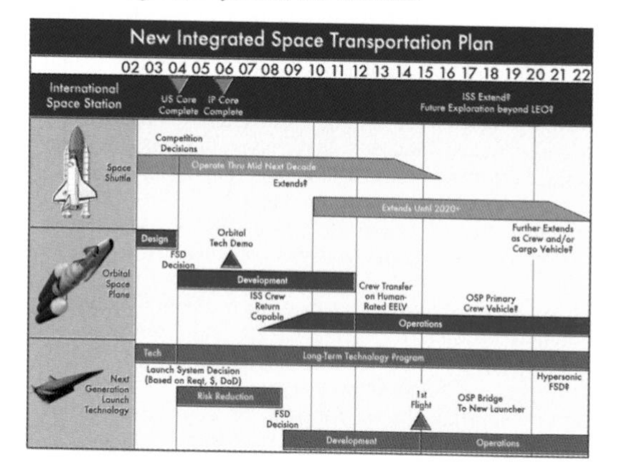

Figure 5.6-1. The Integrated Space Transportation Plan.

5.7 THE RETURN OF SCHEDULE PRESSURE

The International Space Station has been the centerpiece of NASA's human space flight program in the 1990s. In several instances, funds for the Shuttle Program have paid for various International Space Station items. The Space Station has also affected the Space Shuttle Program schedule. By the time the functional cargo block *Zarya,* the Space Station's first element, was launched from the Baikonur Cosmodrome in Kazakhstan in November 1998, the Space Station was two years behind schedule. The launch of STS-88, the first of many Shuttle missions assigned to station assembly, followed a month later. Another four assembly missions in 1999 and 2000 readied the station for its first permanent crew, Expedition 1, which arrived in late 2000.

When the Bush Administration came to the White House in January 2001, the International Space Station program was $4 billion over its projected budget. The Administration's Fiscal Year 2002 budget, released in February 2001, declared that the International Space Station would be limited to a "U.S Core Complete" configuration, a reduced design that could accommodate only three crew members. The last step in completing the U.S. portion of this configuration would be the addition of the Italian-supplied but U.S.-owned "Node 2," which would allow Europe and Japan to connect their laboratory modules to the Station. Launching Node 2 and thereby finishing "core complete" configuration became an important political and programmatic milestone (see Figure 5.7-1).

Figure 5.7-1. The "Core Complete" configuration of the International Space Station.

During congressional testimony in May of 2001, Sean O'Keefe, who was then Deputy Director of the White House Office of Management and Budget, presented the Administration's plan to bring International Space Station costs under control. The plan outlined a reduction in assembly and logistics flights to reach "core complete" configuration from 36 to 30. It also recommended redirecting about $1 billion in funding by canceling U.S. elements not yet completed, such as the habitation module and the X-38 Crew Return Vehicle. The X-38 would have allowed emergency evacuation and landing capability for a seven-member station crew. Without it, the crew was limited to three, the number that could fit into a Russian Soyuz crew rescue vehicle.

In his remarks, O'Keefe stated:

> *NASA's degree of success in gaining control of cost growth on Space Station will not only dictate the capabilities that the Station will provide, but will send a strong signal about the ability of NASA's Human Space Flight program to effectively manage large development programs. NASA's credibility with the Administration and the Congress for delivering on what is promised and the longer-term implications that such credibility may have on the future of Human Space Flight hang in the balance.*[72]

At the request of the White House Office of Management and Budget, in July 2001 NASA Administrator Dan Goldin formed an International Space Station Management and Cost Evaluation Task Force. The International Space Station Management and Cost Evaluation Task Force was to assist NASA in identifying the reforms needed to restore the Station Program's fiscal and management credibility.

While the primary focus of the Task Force was on the Space Station Program management, its November 2001 report issued a general condemnation of how NASA, and particularly Johnson Space Center, had managed the International Space Station, and by implication, NASA's overall human space flight effort. [73] The report noted "existing deficiencies in management structure, institutional culture, cost estimating, and program control," and that "the institutional needs of the [human space flight] Centers are driving the Program, rather than Program requirements being served by the Centers." The Task Force suggested that as a cost control measure, the Space Shuttle be limited to four flights per year and that NASA revise the station crew rotation period to six months. The cost savings that would result from eliminating flights could be used to offset cost overruns.

NASA accepted a reduced flight rate. The Space Shuttle Program office concluded that, based on a rate of four flights a year, Node 2 could be launched by February 19, 2004.

In testimony before the House Committee on Science on November 7, 2001, Task Force Chairman Thomas Young identified what became known as a "performance gate." He suggested that over the next two years, NASA should plan and implement a credible "core complete" program. In Fall 2003, "an assessment would be made concerning the ISS program performance and NASA's credibility. If satisfactory, resource needs would be assessed and an [ISS] 'end state' that realized the science potential would become the baseline. If unsatisfactory, the core complete program would become the 'end state.' "[74]

Testifying the same day, Office of Management and Budget Deputy Director Sean O'Keefe indicated the Administration's agreement with the planned performance gate:

> *The concept presented by the task force of a decision gate in two years that could lead to an end state other than the U.S. core complete Station is an innovative approach, and one the Administration will adopt. It calls for NASA to make the necessary management reforms to successfully build the core complete Station and operate it within the $8.3 billion available through FY 2006 plus other human space flight resources ... If NASA fails to meet the standards, then an end-state beyond core complete is not an option. The strategy places the burden of proof on NASA performance to ensure that NASA fully implements the needed reforms.*[75]

Mr. O'Keefe added in closing:

> *A most important next step – one on which the success of all these reforms hinges – is to provide new leadership for NASA and its Human Space Flight activities. NASA has been well-served by Dan Goldin. New leadership is now necessary to continue moving the ball down the field with the goal line in sight. The Administration recognizes the importance of getting the right leaders in place as soon as possible, and I am personally engaged in making sure that this happens.*

A week later, Sean O'Keefe was nominated by President Bush as the new NASA Administrator.

To meet the new flight schedule, in 2002 NASA revised its Shuttle manifest, calling for a docking adaptor to be installed in *Columbia* after the STS-107 mission so that it could make an October 2003 flight to the International Space Station. *Columbia* was not optimal for Station flights – the Orbiter could not carry enough payload – but it was assigned to this flight because *Discovery* was scheduled for 18 months of major maintenance. To ensure adequate Shuttle availability for the February 2004 Node 2 launch date, *Columbia* would fly an International Space Station resupply mission.

The White House and Congress had put the International Space Station Program, the Space Shuttle Program, and indeed NASA on probation. NASA had to prove it could meet schedules within cost, or risk halting Space Station construction at core complete – a configuration far short of what NASA anticipated. The new NASA management viewed the achievement of an on-schedule Node 2 launch as an endorsement of its successful approach to Shuttle and Station Programs. Any suggestions that it would be difficult to meet that launch date were brushed aside.

This insistence on a fixed launch schedule was worrisome. The International Space Station Management and Cost Evaluation Task Force, in particular, was concerned with the emphasis on a specific launch date. It noted in its 2002 review of progress toward meeting its recommendations that "significant progress has been made in nearly all aspects of the ISS Program," but that there was "significant risk with the Node 2 (February '04) schedule."[76]

By November 2002, NASA had flown 16 Space Shuttle missions dedicated to Station assembly and crew rotation. Five crews had lived onboard the Station, the last four of them delivered via Space Shuttles. As the Station had grown, so had the complexity of the missions required to complete it. With the International Space Station assembly more than half complete, the Station and Shuttle programs had become irreversibly linked. Any problems with or perturbations to the planned schedule of one program reverberated through both programs. For the Shuttle program, this meant that the conduct of all missions, even non-Station missions like STS-107, would have an impact on the Node 2 launch date.

In 2002, this reality, and the events of the months that would follow, began to place additional schedule pressures on the Space Shuttle Program. Those pressures are discussed in Section 6.2.

5.8 CONCLUSION

Over the last decade, the Space Shuttle Program has operated in a challenging and often turbulent environment. As discussed in this chapter, there were at least three major contributing factors to that environment:

- Throughout the decade, the Shuttle Program has had to function within an increasingly constrained budget. Both the Shuttle budget and workforce have been reduced by over 40 percent during the past decade. The White House, Congress, and NASA leadership exerted constant pressure to reduce or at least freeze operating costs. As a result, there was little margin in the budget to deal with unexpected technical problems or make Shuttle improvements.
- The Shuttle was mischaracterized by the 1995 Kraft Report as "a mature and reliable system … about as safe as today's technology will provide." Based on this mischaracterization, NASA believed that it could turn increased responsibilities for Shuttle operations over to a single prime contractor and reduce its direct involvement in ensuring safe Shuttle operations, instead monitoring contractor performance from a more detached position. NASA also believed that it could use the "mature" Shuttle to carry out operational missions without continually focusing engineering attention on understanding the mission-by-mission anomalies inherent in a developmental vehicle.
- In the 1990s, the planned date for replacing the Shuttle shifted from 2006 to 2012 and then to 2015 or later. Given the uncertainty regarding the Shuttle's service life, there has been policy and budgetary ambivalence on investing in the vehicle. Only in the past year has NASA begun to provide the resources needed to sustain extended Shuttle operations. Previously, safety and support upgrades were delayed or deferred, and Shuttle infrastructure was allowed to deteriorate.

The Board observes that this is hardly an environment in which those responsible for safe operation of the Shuttle can function without being influenced by external pressures. It is to the credit of Space Shuttle managers and the Shuttle workforce that the vehicle was able to achieve its program objectives for as long as it did.

An examination of the Shuttle Program's history from *Challenger* to *Columbia* raises the question: Did the Space Shuttle Program budgets constrained by the White House and Congress threaten safe Shuttle operations? There is no straightforward answer. In 1994, an analysis of the Shuttle budget concluded that reductions made in the early 1990s represented a "healthy tightening up" of the program.[77] Certainly those in the Office of Management and Budget and in NASA's congressional authorization and appropriations subcommittees thought they were providing enough resources to operate the Shuttle safely, while also taking into account the expected Shuttle lifetime and the many other demands on the Federal budget. NASA Headquarters agreed, at least until Administrator Goldin declared a "space launch crisis" in June 1999 and asked that additional resources for safety upgrades be added to the NASA budget. By 2001, however, one experienced observer of the space program described the Shuttle workforce as "The Few, the Tired," and suggested that "a decade of downsizing and budget tightening has left NASA exploring the universe with a less experienced staff and older equipment."[78]

It is the Board's view that this latter statement is an accurate depiction of the Space Shuttle Program at the time of STS-107. The Program was operating too close to too many margins. The Board also finds that recent modest increases in the Shuttle Program's budget are necessary and overdue steps toward providing the resources to sustain the program for its now-extended lifetime. Similarly, NASA has recently recognized that providing an adequately sized and appropriately trained workforce is critical to the agency's future success.

An examination of the Program's management changes also leads to the question: Did turmoil in the management structure contribute to the accident? The Board found no evidence that the transition from many Space Shuttle contractors to a partial consolidation of contracts under a single firm has by itself introduced additional technical risk into the Space Shuttle Program. The transfer of responsibilities that has accompanied the Space Flight Operations Contract has, however, complicated an already complex Program structure and created barriers to effective communication. Designating the Johnson Space Center as the "lead center" for the Space Shuttle Program did resurrect some of the Center rivalries and communication difficulties that existed before the *Challenger* accident. The specific ways in which this complexity and lack of an integrated approach to Shuttle management impinged on NASA's performance during and before the flight of STS-107 are discussed in Chapters 6 and 7.

As the 21st century began, NASA's deeply ingrained human space flight culture – one that has evolved over 30 years as the basis for a more conservative, less technically and organizationally capable organization than the Apollo-era NASA – remained strong enough to resist external pressures for adaptation and change. At the time of the launch of STS-107, NASA retained too many negative (and also many positive) aspects of its traditional culture: "flawed decision making, self deception, introversion and a diminished curiosity about the world outside the perfect place."[79] These characteristics were reflected in NASA's less than stellar performance before and during the STS-107 mission, which is described in the following chapters.

ENDNOTES FOR CHAPTER 5

The citations that contain a reference to "CAIB document" with CAB or CTF followed by seven to eleven digits, such as CAB001-0010, refer to a document in the Columbia Accident Investigation Board database maintained by the Department of Justice and archived at the National Archives.

1. *Report of the Presidential Commission on the Space Shuttle Challenger Accident*, June 6, 1986, (Washington: Government Printing Office, 1986), Vol. I, p. 82, 118.
2. *Report of the Presidential Commission*, Vol. I, p. 48.
3. *Report of the Presidential Commission*, Vol. I, p. 52.
4. *Report of the Presidential Commission*, Vol. I, pp. 164-165.
5. *Report of the Presidential Commission*, Vol. I, pp. 198-201.
6. *Report of The National Commission for the Review of the National Reconnaissance Office: The NRO at the Crossroads*, November 2000, p. 66. Roger Guillemette, "Vandenberg: Space Shuttle Launch and Landing Site, Part 1," *Spaceflight*, October 1994, pp. 354-357, and Roger Guillemette, "Vandenberg: Space Shuttle Launch and Landing Site, Part 2," *Spaceflight*, November 1994, pp. 378-381; Dennis R. Jenkins, *Space Shuttle: The History of the National Space Transportation System – The First 100 Missions* (Cape Canaveral, FL, Specialty Press, 2001), pp. 467-476.
7. Vice President's Space Policy Advisory Board, *A Post Cold War Assessment of U.S. Space Policy*, December 1992, p. 6.
8. Quoted in John M. Logsdon, "Return to Flight: Richard H. Truly and the Recovery from the Challenger Accident," in Pamela E. Mack, editor, *From Engineering to Big Science: The NACA and NASA Collier Trophy Research Project Winners*, NASA SP-4219 (Washington: Government Printing Office, 1998), p. 363.
9. *Aviation Week & Space Technology*, November 10, 1986, p. 30.
10. There are proposals for using other U.S. systems, in development but not yet ready for flight, to provide an alternate U.S. means of station access. These "Alternate Access to Space" proposals have not been evaluated by the Board.
11. Testimony of William F. Readdy to the Subcommittee on Science, Technology and Space, U.S. Senate, September 6, 2001.
12. Howard E. McCurdy, *Inside NASA: High Technology and Organizational Change in the U.S. Space Program* (Baltimore: The Johns Hopkins University Press, 1993), p. 24.
13. Garry D. Brewer, "Perfect Places: NASA as an Idealized Institution," in Radford Byerly, Jr., ed., *Space Policy Reconsidered* (Boulder, CO: Westview Press, 1989), p. 158. Brewer, when he wrote these words, was a professor of organizational behavior at Yale University with no prior exposure to NASA. For first-hand discussions of NASA's Apollo-era organizational culture, see Christopher Kraft, *Flight: My Life in Mission Control* (New York: E.P. Dutton, 2001); Gene Kranz, *Failure is Not an Option: Mission Control from Mercury to Apollo 13* (New York: Simon & Schuster, 2000); and Thomas J. Kelly, *Moon Lander: How We Developed the Apollo Lunar Module* (Washington: Smithsonian Institution Press, 2001).
14. Brewer, "Perfect Places," pp. 159-165.
15. As NASA human space flight personnel began to become closely involved with their counterparts in the Russian space program after 1992, there was grudging acceptance that Russian human space flight personnel were also skilled in their work, although they carried it out rather differently than did NASA.
16. Bush administration space policy is discussed in Dan Quayle, *Standing Firm: A Vice-Presidential Memoir* (New York: Harper Collins, 1994), pp. 185-190.
17. *Report of the Advisory Committee on the Future of the U.S. Space Program*, December 1990. The quotes are from p. 2 of the report's executive summary.
18. *Report of the Advisory Committee on the Future of the U.S. Space Program*. Measured in terms of total national spending, the report's recommendations would have returned NASA spending to 0.38 percent of U.S. Gross Domestic Product – a level of investment not seen since 1969.
19. For Fiscal Years 1965-2002 in Real and Constant Dollars, see NASA, "Space Activities of the U.S. Government – in Millions of Real Year Dollars," and "Space Activities of the U.S. Government – Adjusted for Inflation," in *Aeronautics and Space Report of the President – Fiscal Year 2002 Activity*, forthcoming. For Fiscal Years 2003-2004 in Real Dollars, see Office of Management and Budget, "Outlays By Agency: 1962-2008," in Historical Budget of the United States Government, Fiscal Year 2004, (Washington: Government Printing Office, 2003), pp. 70-75.
20. Commission on the Future of the U.S. Aerospace Industry, *Final Report*, November 18, 2002, p. 3-1.
21. U.S. Congress, Office of Technology Assessment, "Shuttle Fleet Attrition if Orbiter Recovery Reliability is 98 Percent," August 1989, p. 6. From: *Round Trip to Orbit: Human Space Flight Alternatives: Special Report*, OTS-ISC-419.
22. *Report of the Advisory Committee on the Future of the U.S. Space Program*.
23. Howard E. McCurdy, *Faster, Better, Cheaper: Low-Cost Innovation in the U.S. Space Program* (Baltimore: The Johns Hopkins University Press, 2001).
24. Letter from Daniel Goldin to Representative James T. Walsh, October 4, 2001. CAIB document CAB065-01630169.
25. Ibid.
26. W. Henry Lambright, *Transforming Government: Dan Goldin and the Remaking of NASA* (Washington: Price Waterhouse Coopers Endowment for the Business of Government, March 2001), pp. 12; 27-29.
27. Deming's management philosophy was not the only new notion that Goldin attempted to apply to NASA. He was also an advocate of the "Total Quality Management" approach and other modern management schemes. Trying to adapt to these various management theories was a source of some stress.
28. For a discussion of Goldin's approach, see Howard McCurdy, *Faster, Better, Cheaper: Low-Cost Innovation in the U.S. Space Program* (Baltimore: The Johns Hopkins University Press, 2001). It is worth noting that while the "faster, better, cheaper" approach led to many more NASA robotic missions being launched after 1992, not all of those missions were successful. In particular, there were two embarrassing failures of Mars missions in 1999.
29. Lambright, *Transforming Government*, provides an early but comprehensive evaluation of the Goldin record. The quote is from p. 28.
30. Goldin is quoted in Bill Harwood, "Pace of Cuts Fuels Concerns About Shuttle," *Space News*, December 19-25, 1994, p. 1.
31. McCurdy, *Faster, Better, Cheaper*.

[32] For two recent works that apply the "Iron Triangle" concept to other policy areas, see Randall B. Ripley and Grace A. Franklin, *Congress, the Bureaucracy and Public Policy*, 5th Edition, (Pacific Grove, CA: Brooks/Cole Publishing Company, 1991); and Paul C. Light, *Forging Legislation: The Politics of Veterans Reform*, (New York: W. W. Norton, 1992).

[33] Information obtained from Anna Henderson, NASA Office of Space Flight, to e-mail to John Logsdon, June 13, 2003.

[34] National Academy of Public Administration, *A Review of the Space Shuttle Costs, Reduction Goals, and Procedures*, December 1994, pp. 3-5. CAIB document CAB026-0313.

[35] Presentation to NASA Advisory Council by Stephen Oswald, Acting Director, Space Shuttle Requirements, "Space Flight Operations Contract (SFOC) Acquisition Status," April 23, 1996. CAIB document CTF064-1369.

[36] Bryan D. O'Connor, Status Briefing to NASA Administrator, "Space Shuttle Functional Workforce Review," February 14, 1995. CAIB document CAB015-0400.

[37] Ralph Vartabedian, "Ex-NASA Chief Hits Flight Safety," *Houston Chronicle*, March 7, 1996.

[38] Kathy Sawyer, "NASA Space Shuttle Director Resigns," *Washington Post*, February 3, 1996, p. A3. See also "Take this Job and Shuttle It: Why NASA's Space Shuttle Chief Quit," *Final Frontier*, July/August 1996, pp. 16-17; "NASA Alters Its Management, Philosophy," *Space News*, February 12-18, 1996, p. 3.

[39] *Report of the Space Shuttle Management Independent Review Team*, February 1995.

[40] Ibid, pp. 3-18.

[41] NASA News Release 95-27, "Shuttle Management Team Issues Final Report," March 15, 1995.

[42] Aerospace Safety Advisory Panel, "Review of the Space Shuttle Management Independent Review Program," May 1995. CAIB document CAB015-04120413.

[43] Jose Garcia to President William Jefferson Clinton, August 25, 1995.

[44] See, for instance: "Determinations and Findings for the Space Shuttle Program," United States House of Representatives, Subcommittee on Space, of the Committee on Science, 104 Cong., 1 Sess., November 30, 1995.

[45] See remarks by Daniel S. Goldin, Opening Remarks at the September 30, 1996, ceremony commemorating the signing of the Space Flight Operations Contract, Houston, Texas. (Videotape recording.)

[46] Congressional Budget Office, "NASA's Space Flight Operations Contract and Other Technologically Complex Government Activities Conducted by Contractors," July 29, 2003.

[47] Russell Turner, testimony at public hearing before the Columbia Accident Investigation Board, June 12, 2003.

[48] See Section 204 of Public Law 105-303, October 28, 1999.

[49] Joe Rothenberg to Dan Goldin, August 17, 2001, CAIB document CAB015-1134; "Space Shuttle Privatization," CAIB document CAB015-1135; "Space Shuttle Privatization: Options and Issues," Rev: 8/14/01, CAIB document CAB015-1147.

[50] Ron Dittemore, "Concept of Privatization of the Space Shuttle Program," September 2001. CAIB document CTF005-0283.

[51] Ibid.

[52] Roy Bridges, Testimony before the Columbia Accident Investigation Board, March 25, 2003.

[53] The quotes are taken from NASA-submitted material appended to the statement of NASA Administrator Daniel Goldin to the Senate Subcommittee on Science, Technology and Space, March 22, 2000, p. 7.

[54] National Commission on Space, *Pioneering the Space Frontier: An Exciting Vision of Our Next Fifty Years in Space, Report of the National Commission on Space* (Bantam Books, 1986).

[55] President Ronald Reagan, "Message to the Congress on America's Agenda for the Future," February 6, 1986, *Public Papers of the Presidents of the United States: Ronald Reagan: Book I-January 1 to June 27, 1986* (Washington, DC: U.S. Government Printing Office, 1982-1991), p. 159.

[56] Office of Space Systems Development, NASA Headquarters, "Access to Space Study–Summary Report," January 1994, reproduced in John M. Logsdon, *et al.* eds., *Exploring the Unknown, Volume IV: Accessing Space* NASA SP-4407 (Government Printing Office, 1999), pp. 584-604.

[57] The White House, Office of Science and Technology Policy, "Fact Sheet–National Space Transportation Policy," August 5, 1994, pp. 1-2, reprinted in Logsdon *et al.*, *Exploring the Unknown, Volume IV*, pp. 626-631.

[58] *Report of the Space Shuttle Management Independent Review Team*, pp. 3-18.

[59] "Statement of William F. Readdy, Deputy Associate Administrator, Office of Space Flight, National Aeronautics and Space Administration before the Subcommittee on Space and Aeronautics Committee on Science, House of Representatives," October 21, 1999. CAIB document CAB026-0146.

[60] Letter from Daniel Goldin to Jacob Lew, Director, Office of Management and Budget, July 6, 1999.

[61] NASA, Space Shuttle Independent Assessment Team, "Report to the Associate Administrator, Office of Space Flight, October-December 1999," March 7, 2000. CAIB document CTF017-0169.

[62] Ibid.

[63] Ibid.

[64] Dr. Richard Beck, Director, Resources Analysis Division, NASA, "Agency Budget Overview, FY 2003 Budget," February 6, 2002, p. 20. CAIB document CAB070-0001.

[65] Space Flight Advisory Committee, NASA Office of Space Flight, Meeting Report, May 1-2, 2001, p. 7. CAIB document CTF017-0034.

[66] Senators Bill Nelson, Bob Graham, Mary Landrieu, John Breaux, and Orrin Hatch to Senator Barbara Mikulski, September 18, 2001.

[67] Space Flight Advisory Committee, NASA Office of Space Flight, Meeting Report, May 1-2, 2001, p. 7. CAIB document CTF017-0034.

[68] Task Force on Space Shuttle Competitive Sourcing, *Alternate Trajectories: Options for Competitive Sourcing of the Space Shuttle Program, Executive Summary*, The RAND Corporation, 2002. CAIB document CAB003-1614.

[69] NNBE Benchmarking Team, NASA Office of Safety & Mission Assurance and NAVSEA 92Q Submarine Safety & Quality Assurance Division, "NASA/Navy Benchmarking Exchange (NNBE)," Interim Report, December 20, 2002. CAIB document CAB030-0392. The team's final report was issued in July 2003.

[70] NASA FY 2004 Congressional Budget, "Theme: Space Shuttle." [Excerpt from NASA FY 2004 budget briefing book also known as the "IBPD Narrative"]. CAIB document CAB065-04190440.

[71] NASA, "Theme: Space Shuttle." CAIB document CAB065-04190440.

[72] Testimony of Sean O'Keefe, Deputy Director, Office of Management and Budget, to the Subcommittee of the Committee on Appropriations, "Part 1, National Aeronautics and Space Administration," Hearings Before a Subcommittee of the Committee on Appropriations, United States House of Representatives, 107th Congress, 1st Sess., May 2001, p. 32.

[73] "Report by the International Space Station (ISS) Management and Cost Evaluation (IMCE) Task Force to the NASA Advisory Council," November 1, 2001, pp. 1-5. CAIB document CTF044-6016.

[74] Testimony of Tom Young, Chairman, ISS Management and Cost Evaluation (IMCE) Task Force, to the Committee on Science, U.S. House of Representatives, "The Space Station Task Force Report," Hearing Before the Committee on Science, United States House of Representatives, 107th Congress, 1st Sess., November, 2001, p. 23.

[75] Testimony of Sean O'Keefe, Deputy Director, Office of Management and Budget, to the Committee on Science, U.S. House of Representatives, "The Space Station Task Force Report," Hearing Before the Committee on Science, United States House of Representatives, 107th Congress, 1st Sess., November, 2001, p. 28.

[76] Thomas Young, IMCE Chair, "International Space Station (ISS) Management and Cost Evaluation (IMCE) Task Force Status Report to the NASA Advisory Council," (Viewgraphs) December 11, 2002, p. 11. CAIB document CAB065-0189.

[77] General Research Corporation, *Space Shuttle Budget Allocation Review, Volume 1*, July 1994, p. 7. CAIB document CAIB015-0161.

[78] Beth Dickey, "The Few, the Tired," *Government Executive*, April 2001, p. 71.

[79] Brewer, "Perfect Places," pp. 159.

CHAPTER 6

Decision Making at NASA

The dwindling post-Cold War Shuttle budget that launched NASA leadership on a crusade for efficiency in the decade before *Columbia*'s final flight powerfully shaped the environment in which Shuttle managers worked. The increased organizational complexity, transitioning authority structures, and ambiguous working relationships that defined the restructured Space Shuttle Program in the 1990s created turbulence that repeatedly influenced decisions made before and during STS-107.

This chapter connects Chapter 5's analysis of NASA's broader policy environment to a focused scrutiny of Space Shuttle Program decisions that led to the STS-107 accident. Section 6.1 illustrates how foam debris losses that violated design requirements came to be defined by NASA management as an acceptable aspect of Shuttle missions, one that posed merely a maintenance "turnaround" problem rather than a safety-of-flight concern. Section 6.2 shows how, at a pivotal juncture just months before the *Columbia* accident, the management goal of completing Node 2 of the International Space Station on time encouraged Shuttle managers to continue flying, even after a significant bipod-foam debris strike on STS-112. Section 6.3 notes the decisions made during STS-107 in response to the bipod foam strike, and reveals how engineers' concerns about risk and safety were competing with – and were defeated by – management's belief that foam could not hurt the Orbiter, as well as the need to keep on schedule. In relating a rescue and repair scenario that might have enabled the crew's safe return, Section 6.4 grapples with yet another latent assumption held by Shuttle managers during and after STS-107: that even if the foam strike had been discovered, nothing could have been done.

6.1 A HISTORY OF FOAM ANOMALIES

The shedding of External Tank foam – the physical cause of the *Columbia* accident – had a long history. Damage caused by debris has occurred on every Space Shuttle flight, and most missions have had insulating foam shed during ascent. This raises an obvious question: Why did NASA continue flying the Shuttle with a known problem that violated design requirements? It would seem that the longer the Shuttle Program allowed debris to continue striking the Orbiters, the more opportunity existed to detect the serious threat it posed. But this is not what happened. Although engineers have made numerous changes in foam design and application in the 25 years that the External Tank has been in production, the problem of foam-shedding has not been solved, nor has the Orbiter's ability to tolerate impacts from foam or other debris been significantly improved.

The Need for Foam Insulation

The External Tank contains liquid oxygen and hydrogen propellants stored at minus 297 and minus 423 degrees Fahrenheit. Were the super-cold External Tank not sufficiently insulated from the warm air, its liquid propellants would boil, and atmospheric nitrogen and water vapor would condense and form thick layers of ice on its surface. Upon launch, the ice could break off and damage the Orbiter. (See Chapter 3.)

To prevent this from happening, large areas of the External Tank are machine-sprayed with one or two inches of foam, while specific fixtures, such as the bipod ramps, are hand-sculpted with thicker coats. Most of these insulating materials fall into a general category of "foam," and are outwardly similar to hardware store-sprayable foam insulation. The problem is that foam does not always stay where the External Tank manufacturer Lockheed Martin installs it. During flight, popcorn- to briefcase-size chunks detach from the External Tank.

Original Design Requirements

Early in the Space Shuttle Program, foam loss was considered a dangerous problem. Design engineers were extremely concerned about potential damage to the Orbiter and its fragile Thermal Protection System, parts of which are so vulnerable to impacts that lightly pressing a thumbnail into them leaves a mark. Because of these concerns, the baseline design requirements in the Shuttle's "Flight and Ground System Specification-Book 1, Requirements," precluded foam-shedding by the External Tank. Specifically:

> ***3.2.1.2.14 Debris Prevention:*** *The Space Shuttle System, including the ground systems, shall be designed to preclude the shedding of ice and/or other debris from the Shuttle elements during prelaunch and flight operations that would jeopardize the flight crew, vehicle, mission success, or would adversely impact turnaround operations.*[1]

> ***3.2.1.1.17 External Tank Debris Limits:*** *No debris shall emanate from the critical zone of the External Tank on the launch pad or during ascent except for such material which may result from normal thermal protection system recession due to ascent heating.*[2]

The assumption that only tiny pieces of debris would strike the Orbiter was also built into original design requirements, which specified that the Thermal Protection System (the tiles and Reinforced Carbon-Carbon, or RCC, panels) would be built to withstand impacts with a kinetic energy less than 0.006 foot-pounds. Such a small tolerance leaves the Orbiter vulnerable to strikes from birds, ice, launch pad debris, and pieces of foam.

Despite the design requirement that the External Tank shed no debris, and that the Orbiter not be subjected to any significant debris hits, *Columbia* sustained damage from debris strikes on its inaugural 1981 flight. More than 300 tiles had to be replaced.[3] Engineers stated that had they known in advance that the External Tank "was going to produce the debris shower that occurred" during launch, "they would have had a difficult time clearing *Columbia* for flight."[4]

Discussion of Foam Strikes Prior to the Rogers Commission

Foam strikes were a topic of management concern at the time of the *Challenger* accident. In fact, during the Rogers Commission accident investigation, Shuttle Program Manager Arnold Aldrich cited a contractor's concerns about foam shedding to illustrate how well the Shuttle Program manages risk:

> *On a series of four or five external tanks, the thermal insulation around the inner tank ... had large divots of insulation coming off and impacting the Orbiter. We found significant amount of damage to one Orbiter after a flight and ... on the subsequent flight we had a camera in the equivalent of the wheel well, which took a picture of the tank after separation, and we determined that this was in fact the cause of the damage. At that time, we wanted to be able to proceed with the launch program if it was acceptable ... so we undertook discussions of what would be acceptable in terms of potential field repairs, and during those discussions, Rockwell was very conservative because, rightly, damage to the Orbiter TPS [Thermal Protection System] is damage to the Orbiter system, and it has a very stringent environment to experience during the re-entry phase.*

Aldrich described the pieces of foam as "... half a foot square or a foot by half a foot, and some of them much smaller and localized to a specific area, but fairly high up on the tank. So they had a good shot at the Orbiter underbelly, and this is where we had the damage."[5]

Continuing Foam Loss

Despite the high level of concern after STS-1 and through the *Challenger* accident, foam continued to separate from the External Tank. Photographic evidence of foam shedding exists for 65 of the 79 missions for which imagery is available. Of the 34 missions for which there are no imagery, 8 missions where foam loss is not seen in the imagery, and 6 missions where imagery is inconclusive, foam loss can be inferred from the number of divots on the Orbiter's lower surfaces. Over the life of the Space Shuttle Program, Orbiters have returned with an average of 143 divots in the upper and lower surfaces of the Thermal Protection System tiles, with 31 divots averaging over an inch in one dimension.[6] (The Orbiters' lower surfaces have an average of 101 hits, 23 of which are larger than an inch in diameter.) Though the Orbiter is also struck by ice and pieces of launch-pad hardware during launch, by micrometeoroids and orbital debris in space, and by runway debris during landing, the Board concludes that foam is likely responsible for most debris hits.

With each successful landing, it appears that NASA engineers and managers increasingly regarded the foam-shedding as inevitable, and as either unlikely to jeopardize safety or simply an acceptable risk. The distinction between foam loss and debris events also appears to have become blurred. NASA and contractor personnel came to view foam strikes not as a safety of flight issue, but rather a simple maintenance, or "turnaround" issue. In Flight Readiness Review documentation, Mission Management Team minutes, In-Flight Anomaly disposition reports, and elsewhere, what was originally considered a serious threat to the Orbiter

DEFINITIONS

In Family: A reportable problem that was previously experienced, analyzed, and understood. Out of limits performance or discrepancies that have been previously experienced may be considered as in-family when specifically approved by the Space Shuttle Program or design project.[8]

Out of Family: Operation or performance outside the expected performance range for a given parameter or which has not previously been experienced.[9]

Accepted Risk: The threat associated with a specific circumstance is known and understood, cannot be completely eliminated, and the circumstance(s) producing that threat is considered unlikely to reoccur. Hence, the circumstance is fully known and is considered a tolerable threat to the conduct of a Shuttle mission.

No Safety-of-Flight-Issue: The threat associated with a specific circumstance is known and understood and does not pose a threat to the crew and/or vehicle.

Flight	STS-7	STS-32R	STS-50	STS-52	STS-62	STS-112	STS-107
ET #	06	25	45	55	62	115	93
ET Type	SWT	LWT	LWT	LWT	LWT	SLWT	LWT
Orbiter	Challenger	Columbia	Columbia	Columbia	Columbia	Atlantis	Columbia
Inclination	28.45 deg	28.45 deg	28.45 deg	28.45 deg	39.0 deg	51.6 deg	39.0 deg
Launch Date	06/18/83	01/09/90	06/25/92	10/22/92	03/04/94	10/07/02	01/16/03
Launch Time (Local)	07:33:00 AM EDT	07:35:00 AM EST	12:12:23 PM EDT	1:09:39 PM EDT	08:53:00 AM EST	3:46:00 PM EDT	10:39:00 AM EDT

Figure 6.1-1. There have been seven known cases where the left External Tank bipod ramp foam has come off in flight.

came to be treated as "in-family,"[7] a reportable problem that was within the known experience base, was believed to be understood, and was not regarded as a safety-of-flight issue.

Bipod Ramp Foam Loss Events

Chunks of foam from the External Tank's forward bipod attachment, which connects the Orbiter to the External Tank, are some of the largest pieces of debris that have struck the Orbiter. To place the foam loss from STS-107 in a broader context, the Board examined every known instance of foam-shedding from this area. Foam loss from the left bipod ramp (called the –Y ramp in NASA parlance) has been confirmed by imagery on 7 of the 113 missions flown. However, only on 72 of these missions was available imagery of sufficient quality to determine left bipod ramp foam loss. Therefore, foam loss from the left bipod area occurred on approximately 10 percent of flights (seven events out of 72 imaged flights). On the 66 flights that imagery was available for the right bipod area, foam loss was never observed. NASA could not explain why only the left bipod experienced foam loss. (See Figure 6.1-1.)

Figure 6.1-2. The first known instance of bipod ramp shedding occurred on STS-7 which was launched on June 18, 1983.

The first known bipod ramp foam loss occurred during STS-7, *Challenger*'s second mission (see Figure 6.1-2). Images taken after External Tank separation revealed that a 19- by 12-inch piece of the left bipod ramp was missing, and that the External Tank had some 25 shallow divots in the foam just forward of the bipod struts and another 40 divots in the foam covering the lower External Tank. After the mission was completed, the Program Requirements Control Board cited the foam loss as an In-Flight Anomaly. Citing an event as an In-Flight Anomaly means that before the next launch, a specific NASA organization must resolve the problem or prove that it does not threaten the safety of the vehicle or crew.[11]

At the Flight Readiness Review for the next mission, Orbiter Project management reported that, based on the completion of repairs to the Orbiter Thermal Protection System, the bipod ramp foam loss In-Flight Anomaly was resolved, or "closed." However, although the closure documents detailed the repairs made to the Orbiter, neither the Certificate of Flight Readiness documentation nor the Flight Readiness Review documentation referenced correcting the *cause* of the damage – the shedding of foam.

Figure 6.1-3. Only three months before the final launch of Columbia, the bipod ramp foam had come off during STS-112.

Umbilical Cameras and the Statistics of Bipod Ramp Loss

Over the course of the 113 Space Shuttle missions, the left bipod ramp has shed significant pieces of foam at least seven times. (Foam-shedding from the right bipod ramp has never been confirmed. The right bipod ramp may be less subject to foam shedding because it is partially shielded from aerodynamic forces by the External Tank's liquid oxygen line.) The fact that five of these left bipod shedding events occurred on missions flown by *Columbia* sparked considerable Board debate. Although initially this appeared to be a improbable coincidence that would have caused the Board to fault NASA for improper trend analysis and lack of engineering curiosity, on closer inspection, the Board concluded that this "coincidence" is probably the result of a bias in the sample of known bipod foam-shedding. Before the *Challenger* accident, only *Challenger* and *Columbia* carried umbilical well cameras that imaged the External Tank after separation, so there are more images of *Columbia* than of the other Orbiters.[10]

The bipod was imaged 26 of 28 of *Columbia*'s missions; in contrast, *Challenger* had 7 of 10, *Discovery* had only 14 of 30, *Atlantis* only 14 of 26, and *Endeavour* 12 of 19.

The second bipod ramp foam loss occurred during STS-32R, *Columbia*'s ninth flight, on January 9, 1990. A post-mission review of STS-32R photography revealed five divots in the intertank foam ranging from 6 to 28 inches in diameter, the largest of which extended into the left bipod ramp foam. A post-mission inspection of the lower surface of the Orbiter revealed 111 hits, 13 of which were one inch or greater in one dimension. An In-Flight Anomaly assigned to the External Tank Project was closed out at the Flight Readiness Review for the next mission, STS-36, on the basis that there may have been local voids in the foam bipod ramp where it attached to the metal skin of the External Tank. To address the foam loss, NASA engineers poked small "vent holes" through the intertank foam to allow trapped gases to escape voids in the foam where they otherwise might build up pressure and cause the foam to pop off. However, NASA is still studying this hypothesized mechanism of foam loss. Experiments conducted under the Board's purview indicate that other mechanisms may be at work. (See "Foam Fracture Under Hydrostatic Pressure" in Chapter 3.) As discussed in Chapter 3, the Board notes that the persistent uncertainty about the causes of foam loss and potential Orbiter damage results from a lack of thorough hazard analysis and engineering attention.

The third bipod foam loss occurred on June 25, 1992, during the launch of *Columbia* on STS-50, when an approximately 26- by 10-inch piece separated from the left bipod ramp area. Post-mission inspection revealed a 9-inch by 4.5-inch by 0.5-inch divot in the tile, the largest area of tile damage in Shuttle history. The External Tank Project at Marshall Space Flight Center and the Integration Office at Johnson Space Center cited separate In-Flight Anomalies. The Integration Office closed out its In-Flight Anomaly two days before the next flight, STS-46, by deeming damage to the Thermal Protection System an "accepted flight risk."[12] In Integration Hazard Report 37, the Integration Office noted that the impact damage was shallow, the tile loss was not a result of excessive aerodynamic loads, and the External Tank Thermal Protection System failure was the result of "inadequate venting."[13] The External Tank Project closed out its In-Flight Anomaly with the rationale that foam loss during ascent was "not considered a flight or safety issue."[14] Note the difference in how the each program addressed the foam-shedding problem: While the Integration Office deemed it an "accepted risk," the External Tank Project considered it "not a safety-of-flight issue." Hazard Report 37 would figure in the STS-113 Flight Readiness Review, where the crucial decision was made to continue flying with the foam-loss problem. This inconsistency would reappear 10 years later, after bipod foam-shedding during STS-112.

The fourth and fifth bipod ramp foam loss events went undetected until the Board directed NASA to review all available imagery for other instances of bipod foam-shedding. This review of imagery from tracking cameras, the umbilical well camera, and video and still images from flight crew hand held cameras revealed bipod foam loss on STS-52 and STS-62, both of which were flown by *Columbia*. STS-52, launched on October 22, 1992, lost an 8- by 4-inch corner of the left bipod ramp as well as portions of foam covering the left jackpad, a piece of External Tank hardware that facilitates the Orbiter attachment process. The STS-52 post-mission inspection noted a higher-than-average 290 hits on upper and lower Thermal Protection System tiles, 16 of which were greater than one inch in one dimension. External Tank separation videos of STS-62, launched on March 4, 1994, revealed that a 1- by 3-inch piece of foam in the rear face of the left bipod ramp was missing, as were small pieces of foam around the bipod ramp. Because these incidents of missing bipod foam were not detected until after the STS-107 accident, no In-Flight Anomalies had been written. The Board concludes that NASA's failure to identify these bipod foam losses at the time they occurred means the agency must examine the adequacy of its film review, post-flight inspection, and Program Requirements Control Board processes.

The sixth and final bipod ramp event before STS-107 occurred during STS-112 on October 7, 2002 (see Figure 6.1-3). At 33 seconds after launch, when *Atlantis* was at 12,500 feet and traveling at Mach 0.75, ground cameras observed an object traveling from the External Tank that subsequently impacted the Solid Rocket Booster/External Tank Attachment ring (see Figure 6.1-4). After impact, the debris broke into multiple pieces that fell along the Solid Rocket Booster exhaust plume.[15] Post-mission inspection of the Solid Rocket Booster confirmed damage to foam on the forward face of the External Tank Attachment ring. The impact was approximately 4 inches wide and 3 inches deep. Post-External Tank separation photography by the crew showed that a 4- by 5- by 12-inch (240 cubic-inch) corner section of the left bipod ramp was missing, which exposed the super lightweight ablator coating on the bipod housing. This missing chunk of foam was believed to be the debris that impacted the External Tank Attachment ring during ascent. The post-launch review of photos and video identified these debris events, but the Mission Evaluation Room logs and Mission Management Team minutes do not reflect any discussions of them.

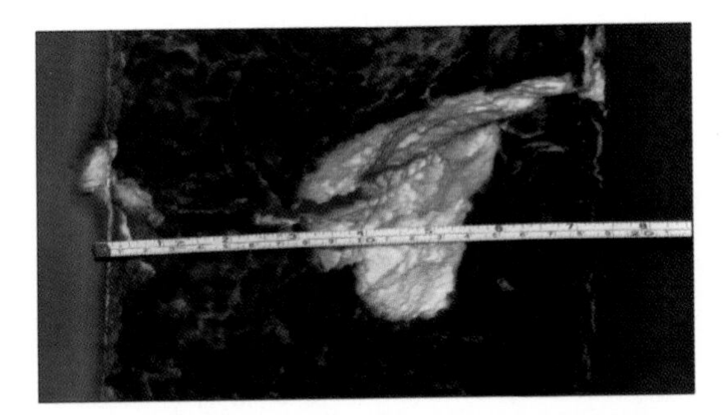

Figure 6.1-4. On STS-112, the foam impacted the External Tank Attach ring on the Solid Rocket Booster, causing this tear in the insulation on the ring.

STS-113 Flight Readiness Review: A Pivotal Decision

Because the bipod ramp shedding on STS-112 was significant, both in size and in the damage it caused, and because it occurred only two flights before STS-107, the Board investigated NASA's rationale to continue flying. This decision made by the Program Requirements Control Board at the STS-113 Flight Readiness Review is among those most directly linked to the STS-107 accident. Had the foam loss during STS-112 been classified as a more serious threat, managers might have responded differently when they heard about the foam strike on STS-107. Alternately, in the face of the increased risk, STS-107 might not have flown at all. However, at STS-113's Flight Readiness Review, managers formally accepted a flight rationale that stated it was safe to fly with foam losses. This decision enabled, and perhaps even encouraged, Mission Management Team members to use similar reasoning when evaluating whether the foam strike on STS-107 posed a safety-of-flight issue.

At the Program Requirements Control Board meeting following the return of STS-112, the Intercenter Photo Working Group recommended that the loss of bipod foam be classified as an In-Flight Anomaly. In a meeting chaired by Shuttle Program Manager Ron Dittemore and attended by many of the managers who would be actively involved with STS-107, including Linda Ham, the Program Requirements Control Board ultimately decided against such classification. Instead, after discussions with the Integration Office and the External Tank Project, the Program Requirements Control Board Chairman assigned an "action" to the External Tank Project to determine the root cause of the foam loss and to propose corrective action. This was inconsistent with previous practice, in which all other known bipod foam-shedding was designated as In-Flight Anomalies. The Program Requirements Control Board initially set December 5, 2002, as the date to report back on this action, even though STS-113 was scheduled to launch on November 10. The due date subsequently slipped until after the planned launch and return of STS-107. The Space Shuttle Program decided to fly not one but two missions before resolving the STS-112 foam loss.

The Board wondered why NASA would treat the STS-112 foam loss differently than all others. What drove managers to reject the recommendation that the foam loss be deemed an In-Flight Anomaly? Why did they take the unprecedented step of scheduling not one but eventually two missions to fly before the External Tank Project was to report back on foam losses? It seems that Shuttle managers had become conditioned over time to not regard foam loss or debris as a safety-of-flight concern. As will be discussed in Section 6.2, the need to adhere to the Node 2 launch schedule also appears to have influenced their decision. Had the STS-113 mission been delayed beyond early December 2002, the Expedition 5 crew on board the Space Station would have exceeded its 180-day on-orbit limit, and the Node 2 launch date, a major management goal, would not be met.

Even though the results of the External Tank Project engineering analysis were not due until after STS-113, the foam-shedding was reported, or "briefed," at STS-113's Flight Readiness Review on October 31, 2002, a meeting that Dittemore and Ham attended. Two slides from this brief (Figure 6.1-5) explain the disposition of bipod ramp foam loss on STS-112.

SPACE SHUTTLE PROGRAM
Space Shuttle Projects Office (MSFC)
NASA Marshall Space Flight Center, Huntsville, Alabama
Space Shuttle Program
STS-112/ET-115 Bipod Ramp Foam Loss
Presenter Jerry Smelser, NASA/MP31
Date October 31, 2002
Page 3

- **Issue**
  - Foam was lost on the STS-112/ET-115 –Y bipod ramp (≈4" X 5" X 12") exposing the bipod housing SLA closeout
- **Background**
  - ET TPS Foam loss over the life of the Shuttle Program has never been a "Safety of Flight" issue
  - More than 100 External Tanks have flown with only 3 documented instances of significant foam loss on a bipod ramp

Missing Foam on –Y Bipod Ramp

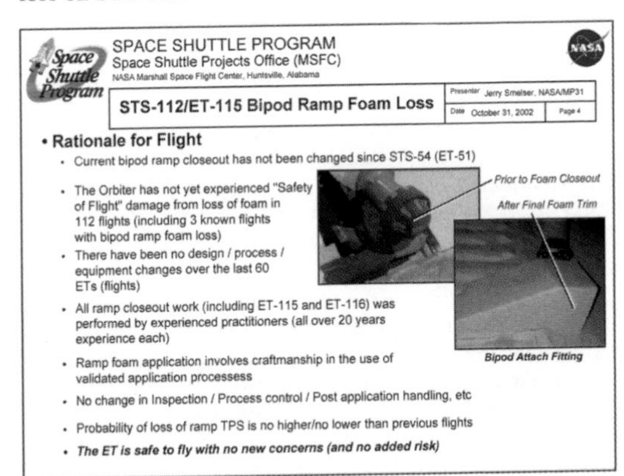

Figure 6.1-5. These two briefing slides are from the STS-113 Flight Readiness Review. The first and third bullets on the right-hand slide are incorrect since the design of the bipod ramp had changed several times since the flights listed on the slide.

This rationale is seriously flawed. The first and third statements listed under "Rationale for Flight" are incorrect. Contrary to the chart, which was presented by Jerry Smelser, the Program Manager for the External Tank Project, the bipod ramp design *had* changed, as of External Tank-76. This casts doubt on the implied argument that because the design had *not* changed, future bipod foam events were unlikely to occur. Although the other points may be factually correct, they provide an exceptionally weak rationale for safe flight. The fact that ramp closeout work was "performed by experienced practitioners" or that "application involves craftsmanship in the use of validated application processes" in no way decreases the chances of recurrent foam loss. The statement that the "probability of loss of ramp Thermal Protection System is no higher/no lower than previous flights" could be just as accurately stated "the probability of bipod foam loss on the next flight is just as high as it was on previous flights." With no engineering analysis, Shuttle managers used past success as a justification for future flights, and made no change to the External Tank configurations planned for STS-113, and, subsequently, for STS-107.

Along with this chart, the NASA Headquarters Safety Office presented a report that estimated a 99 percent probability of foam not being shed from the same area, even though no corrective action had been taken following the STS-112 foam-shedding.[16] The ostensible justification for the 99 percent figure was a calculation of the actual rate of bipod loss over 61 flights. This calculation was a sleight-of-hand effort to make the probability of bipod foam loss appear low rather than a serious grappling with the probability of bipod ramp foam separating. For one thing, the calculation equates the probability of left and right bipod loss, when right bipod loss has never been observed, and the amount of imagery available for left and right bipod events differs. The calculation also miscounts the actual number of bipod ramp losses in two ways. First, by restricting the sample size to flights between STS-112 and the last known bipod ramp loss, it excludes known bipod ramp losses from STS-7, STS-32R, and STS-50. Second, by failing to project the statistical rate of bipod loss across the many missions for which no bipod imagery is available, the calculation assumes a "what you don't see won't hurt you" mentality when in fact the reverse is true. When the statistical rate of bipod foam loss is projected across missions for which imagery is not available, and the sample size is extended to include every mission from STS-1 on, the probability of bipod loss increases dramatically. The Board's review after STS-107, which included the discovery of two additional bipod ramp losses that NASA had not previously noted, concluded that bipod foam loss occurred on approximately 10 percent of all missions.

During the brief at STS-113's Flight Readiness Review, the Associate Administrator for Safety and Mission Assurance scrutinized the Integration Hazard Report 37 conclusion that debris-shedding was an accepted risk, as well as the External Tank Project's rationale for flight. After conferring, STS-113 Flight Readiness Review participants ultimately agreed that foam shedding should be characterized as an "accepted risk" rather than a "not a safety-of-flight" issue. Space Shuttle Program management accepted this rationale, and STS-113's Certificate of Flight Readiness was signed.

The decision made at the STS-113 Flight Readiness Review seemingly acknowledged that the foam posed a threat to the Orbiter, although the continuing disagreement over whether foam was "not a safety of flight issue" versus an "accepted risk" demonstrates how the two terms became blurred over time, clouding the precise conditions under which an increase in risk would be permitted by Shuttle Program management. In retrospect, the bipod foam that caused a 4- by 3-inch gouge in the foam on one of *Atlantis*' Solid Rocket Boosters – just months before STS-107 – was a "strong signal" of potential future damage that Shuttle engineers ignored. Despite the significant bipod foam loss on STS-112, Shuttle Program engineers made no External Tank configuration changes, no moves to reduce the risk of bipod ramp shedding or potential damage to the Orbiter on either of the next two flights, STS-113 and STS-107, and did not update Integrated Hazard Report 37. The Board notes that although there is a process for conducting hazard analyses when the system is designed and a process for re-evaluating them when a design is changed or the component is replaced, no process addresses the need to update a hazard analysis when anomalies occur. A stronger Integration Office would likely have insisted that Integrated Hazard Analysis 37 be updated. In the course of that update, engineers would be forced to consider the cause of foam-shedding and the effects of shedding on other Shuttle elements, including the Orbiter Thermal Protection System.

STS-113 launched at night, and although it is occasionally possible to image the Orbiter from light given off by the Solid Rocket Motor plume, in this instance no imagery was obtained and it is possible that foam could have been shed.

The acceptance of the rationale to fly cleared the way for *Columbia*'s launch and provided a method for Mission managers to classify the STS-107 foam strike as a maintenance and turnaround concern rather than a safety-of-flight issue. It is significant that in retrospect, several NASA managers identified their acceptance of this flight rationale as a serious error.

The foam-loss issue was considered so insignificant by some Shuttle Program engineers and managers that the STS-107 Flight Readiness Review documents include no discussion of the still-unresolved STS-112 foam loss. According to Program rules, this discussion was not a requirement because the STS-112 incident was only identified as an "action," not an In-Flight Anomaly. However, because the action was still open, and the date of its resolution had slipped, the Board believes that Shuttle Program managers should have addressed it. Had the foam issue been discussed in STS-107 pre-launch meetings, Mission managers may have been more sensitive to the foam-shedding, and may have taken more aggressive steps to determine the extent of the damage.

The seventh and final known bipod ramp foam loss occurred on January 16, 2003, during the launch of *Columbia* on STS-107. After the *Columbia* bipod loss, the Program Requirements Control Board deemed the foam loss an In-Flight Anomaly to be dealt with by the External Tank Project.

Other Foam/Debris Events

To better understand how NASA's treatment of debris strikes evolved over time, the Board investigated missions where debris was shed from locations other than the External Tank bipod ramp. The number of debris strikes to the Orbiters' lower surface Thermal Protection System that resulted in tile damage greater than one inch in diameter is shown in Figure 6.1-6.[17] The number of debris strikes may be small, but a single strike could damage several tiles (see Figure 6.1-7).

One debris strike in particular foreshadows the STS-107 event. When *Atlantis* was launched on STS-27R on December 2, 1988, the largest debris event up to that time significantly damaged the Orbiter. Post-launch analysis of tracking camera imagery by the Intercenter Photo Working Group identified a large piece of debris that struck the Thermal Protection System tile at approximately 85 seconds into the flight. On Flight Day Two, Mission Control asked the flight crew to inspect *Atlantis* with a camera mounted on the remote manipulator arm, a robotic device that was not installed on *Columbia* for STS-107. Mission Commander R.L. "Hoot" Gibson later stated that *Atlantis* "looked like it had been blasted by a shotgun."[18] Concerned that the Orbiter's Thermal Protection System had been breached, Gibson ordered that the video be transferred to Mission Control so that NASA engineers could evaluate the damage.

When *Atlantis* landed, engineers were surprised by the extent of the damage. Post-mission inspections deemed it "the most severe of any mission yet flown."[19] The Orbiter had 707 dings, 298 of which were greater than an inch in one dimension. Damage was concentrated outboard of a line right of the bipod attachment to the liquid oxygen umbilical line. Even more worrisome, the debris had knocked off a tile, exposing the Orbiter's skin to the heat of re-entry. Post-flight analysis concluded that structural damage was confined to the exposed cavity left by the missing tile, which happened to be at the location of a thick aluminum plate covering an L-band navigation antenna. Were it not for the thick aluminum plate, Gibson stated during a presentation to the Board that a burn-through may have occurred.[20]

The Board notes the distinctly different ways in which the STS-27R and STS-107 debris strike events were treated. After the discovery of the debris strike on Flight Day Two of STS-27R, the crew was immediately directed to inspect the vehicle. More severe thermal damage – perhaps even a burn-through – may have occurred were it not for the aluminum plate at the site of the tile loss. Fourteen years later, when a debris strike was discovered on Flight Day Two of STS-107, Shuttle Program management declined to have the crew inspect the Orbiter for damage, declined to request on-orbit imaging, and ultimately discounted the possibility of a burn-through. In retrospect, the debris strike on STS-27R is a "strong signal" of the threat debris posed that should have been considered by Shuttle management when STS-107 suffered a similar debris strike. The Board views the failure to do so as an illustration of the lack of institutional memory in the Space Shuttle Program that supports the Board's claim, discussed in Chapter 7, that NASA is not functioning as a learning organization.

After the STS-27R damage was evaluated during a post-flight inspection, the Program Requirements Control Board assigned In-Flight Anomalies to the Orbiter and Solid Rocket Booster Projects. Marshall Sprayable Ablator (MSA-1) material found embedded in an insulation blanket on the right Orbital Maneuvering System pod confirmed that the ablator on the right Solid Rocket Booster nose cap was the most likely source of debris.[21] Because an improved ablator material (MSA-2) would now be used on the Solid Rocket Booster nose cap, the issue was considered "closed" by the time of the next mission's Flight Readiness Review. The Orbiter Thermal Protection System review team concurred with the use of the improved ablator without reservation.

An STS-27R investigation team notation mirrors a Columbia Accident Investigation Board finding. The STS-27R investigation noted: "it is observed that program emphasis

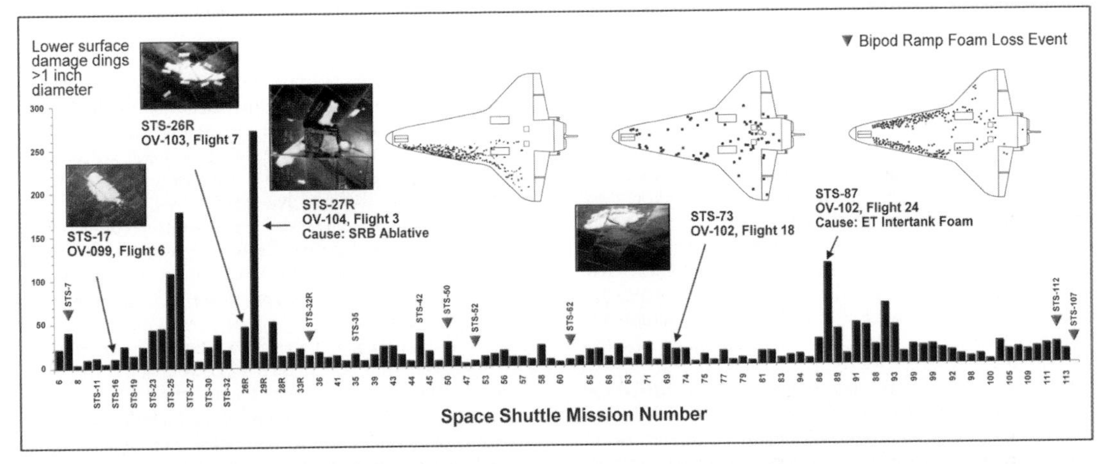

Figure 6.1-6. This chart shows the number of dings greater than one inch in diameter on the lower surface of the Orbiter after each mission from STS-6 through STS-113. Flights where the bipod ramp foam is known to have come off are marked with a red triangle.

MISSION	DATE	COMMENTS
STS-1	April 12, 1981	Lots of debris damage. 300 tiles replaced.
STS-7	June 18, 1983	First known left bipod ramp foam shedding event.
STS-27R	December 2, 1988	Debris knocks off tile; structural damage and near burn through results.
STS-32R	January 9, 1990	Second known left bipod ramp foam event.
STS-35	December 2, 1990	First time NASA calls foam debris "safety of flight issue," and "re-use or turn-around issue."
STS-42	January 22, 1992	First mission after which the next mission (STS-45) launched without debris In-Flight Anomaly closure/resolution.
STS-45	March 24, 1992	Damage to wing RCC Panel 10-right. Unexplained Anomaly, "most likely orbital debris."
STS-50	June 25, 1992	Third known bipod ramp foam event. Hazard Report 37: an "accepted risk."
STS-52	October 22, 1992	Undetected bipod ramp foam loss (Fourth bipod event).
STS-56	April 8, 1993	Acreage tile damage (large area). Called "within experience base" and considered "in family."
STS-62	October 4, 1994	Undetected bipod ramp foam loss (Fifth bipod event).
STS-87	November 19, 1997	Damage to Orbiter Thermal Protection System spurs NASA to begin 9 flight tests to resolve foam-shedding. Foam fix ineffective. In-Flight Anomaly eventually closed after STS-101 as "accepted risk."
STS-112	October 7, 2002	Sixth known left bipod ramp foam loss. First time major debris event not assigned an In-Flight Anomaly. External Tank Project was assigned an Action. Not closed out until after STS-113 and STS-107.
STS-107	January 16, 2003	*Columbia* launch. Seventh known left bipod ramp foam loss event.

Figure 6.1-7. The Board identified 14 flights that had significant Thermal Protection System damage or major foam loss. Two of the bipod foam loss events had not been detected by NASA prior to the Columbia Accident Investigation Board requesting a review of all launch images.

and *attention to tile damage assessments varies with severity* and that detailed records could be augmented to ease trend maintenance" (emphasis added).[22] In other words, Space Shuttle Program personnel knew that the monitoring of tile damage was inadequate and that clear trends could be more readily identified if monitoring was improved, but no such improvements were made. The Board also noted that an STS-27R investigation team recommendation correlated to the *Columbia* accident 14 years later: "It is recommended that the program actively solicit design improvements directed toward eliminating debris sources or minimizing damage potential."[23]

Another instance of non-bipod foam damage occurred on STS-35. Post-flight inspections of *Columbia* after STS-35 in December 1990, showed a higher-than-average amount of damage on the Orbiter's lower surface. A review of External Tank separation film revealed approximately 10 areas of missing foam on the flange connecting the liquid hydrogen tank to the intertank. An In-Flight Anomaly was assigned to the External Tank Project, which closed it by stating that there was no increase in Orbiter Thermal Protection System damage and that it was "not a safety-of-flight concern."[24] The Board notes that it was in a discussion at the STS-36 Flight Readiness Review that NASA first identified this problem as a turnaround issue.[25] Per established procedures, NASA was still designating foam-loss events as In-Flight Anomalies and continued to make various corrective actions, such as drilling more vent holes and improving the foam application process.

Discovery was launched on STS-42 on January 22, 1992. A total of 159 hits on the Orbiter Thermal Protection System were noted after landing. Two 8- to 12-inch-diameter divots in the External Tank intertank area were noted during post-External Tank separation photo evaluation, and these pieces of foam were identified as the most probable sources of the damage. The External Tank Project was assigned an

In-Flight Anomaly, and the incident was later described as an unexplained or isolated event. However, at later Flight Readiness Reviews, the Marshall Space Flight Center briefed this as being "not a safety-of-flight" concern.[26] The next flight, STS-45, would be the first mission launched before the foam-loss In-Flight Anomaly was closed.

On March 24, 1992, *Atlantis* was launched on STS-45. Post-mission inspection revealed exposed substrate on the upper surface of right wing leading edge Reinforced Carbon-Carbon (RCC) panel 10 caused by two gouges, one 1.9 inches by 1.6 inches and the other 0.4 inches by 1 inch.[27] Before the next flight, an In-Flight Anomaly assigned to the Orbiter Project was closed as "unexplained," but "most likely orbital debris."[28] Despite this closure, the Safety and Mission Assurance Office expressed concern as late as the pre-launch Mission Management Team meeting two days before the launch of STS-49. Nevertheless, the mission was cleared for launch. Later laboratory tests identified pieces of man-made debris lodged in the RCC, including stainless steel, aluminum, and titanium, but no conclusion was made about the source of the debris. (The Board notes that this indicates there were transport mechanisms available to determine the path the debris took to impact the wing leading edge. See Section 3.4.)

The Program Requirements Control Board also assigned the External Tank Project an In-Flight Anomaly after foam loss on STS-56 (*Discovery*) and STS-58 (*Columbia*), both of which were launched in 1993. These missions demonstrate the increasingly casual ways in which debris impacts were dispositioned by Shuttle Program managers. After post-flight analysis determined that on both missions the foam had come from the intertank and bipod jackpad areas, the rationale for closing the In-Flight Anomalies included notations that the External Tank foam debris was "in-family," or within the experience base.[29]

During the launch of STS-87 (*Columbia*) on November 19, 1997, a debris event focused NASA's attention on debris-shedding and damage to the Orbiter. Post-External Tank separation photography revealed a significant loss of material from both thrust panels, which are fastened to the Solid Rocket Booster forward attachment points on the intertank structure. Post-landing inspection of the Orbiter noted 308 hits, with 244 on the lower surface and 109 larger than an inch. The foam loss from the External Tank thrust panels was suspected as the most probable cause of the Orbiter Thermal Protection System damage. Based on data from post-flight inspection reports, as well as comparisons with statistics from 71 similarly configured flights, the total number of damage sites, and the number of damage sites one inch or larger, were considered "out-of-family."[30] An investigation was conducted to determine the cause of the material loss and the actions required to prevent a recurrence.

The foam loss problem on STS-87 was described as "popcorning" because of the numerous popcorn-size foam particles that came off the thrust panels. Popcorning has always occurred, but it began earlier than usual in the launch of STS-87. The cause of the earlier-than-normal popcorning (but not the fundamental cause of popcorning) was traced back to a change in foam-blowing agents that caused pressure buildups and stress concentrations within the foam. In an effort to reduce its use of chlorofluorocarbons (CFCs), NASA had switched from a CFC-11 (chlorofluorocarbon) blowing agent to an HCFC-141b blowing agent beginning with External Tank-85, which was assigned to STS-84. (The change in blowing agent affected only mechanically applied foam. Foam that is hand sprayed, such as on the bipod ramp, is still applied using CFC-11.)

The Program Requirements Control Board issued a Directive and the External Tank Project was assigned an In-Flight Anomaly to address the intertank thrust panel foam loss. Over the course of nine missions, the External Tank Project first reduced the thickness of the foam on the thrust panels to minimize the amount of foam that could be shed; and, due to a misunderstanding of what caused foam loss at that time, put vent holes in the thrust panel foam to relieve trapped gas pressure.

The In-Flight Anomaly remained open during these changes, and foam shedding occurred on the nine missions that tested the corrective actions. Following STS-101, the 10th mission after STS-87, the Program Requirements Control Board concluded that foam-shedding from the thrust panel had been reduced to an "acceptable level" by sanding and venting, and the In-Flight Anomaly was closed.[31] The Orbiter Project, External Tank Project, and Space Shuttle Program management all accepted this rationale without question. The Board notes that these interventions merely reduced foam-shedding to previously experienced levels, which have remained relatively constant over the Shuttle's lifetime.

Making the Orbiter More Resistant To Debris Strikes

If foam shedding could not be prevented entirely, what did NASA do to make the Thermal Protection System more resistant to debris strikes? A 1990 study by Dr. Elisabeth Paté-Cornell and Paul Fishback attempted to quantify the risk of a Thermal Protection System failure using probabilistic analysis.[32] The data they used included (1) the probability that a tile would become debonded by either debris strikes or a poor bond, (2) the probability of then losing adjacent tiles, (3) depending on the final size of the failed area, the probability of burn-through, and (4) the probability of failure of a critical sub-system if burn-through occurs. The study concluded that the probability of losing an Orbiter on any given mission due to a failure of Thermal Protection System tiles was approximately one in 1,000. Debris-related problems accounted for approximately 40 percent of the probability, while 60 percent was attributable to tile debonding caused by other factors. An estimated 85 percent of the risk could be attributed to 15 percent of the "acreage," or larger areas of tile, meaning that the loss of any one of a relatively small number of tiles pose a relatively large amount of risk to the Orbiter. In other words, not all tiles are equal – losing certain tiles is more dangerous. While the actual risk may be different than that computed in the 1990 study due to the limited amount of data and the underlying simplified assumptions, this type of analysis offers insight that enables management to concentrate their resources on protecting the Orbiters' critical areas.

Two years after the conclusion of that study, NASA wrote to Paté-Cornell and Fishback describing the importance of their work, and stated that it was developing a long-term effort to use probabilistic risk assessment and related disciplines to improve programmatic decisions.[33] Though NASA has taken some measures to invest in probabilistic risk assessment as a tool, it is the Board's view that NASA has not fully exploited the insights that Paté-Cornell's and Fishback's work offered.[34]

Impact Resistant Tile

NASA also evaluated the possibility of increasing Thermal Protection System tile resistance to debris hits, lowering the possibility of tile debonding, and reducing tile production and maintenance costs.[35] Indeed, tiles with a "tough" coating are currently used on the Orbiters. This coating, known as Toughened Uni-piece Fibrous Insulation (TUFI), was patented in 1992 and developed for use on high-temperature rigid insulation.[36] TUFI is used on a tile material known as Alumina Enhanced Thermal Barrier (AETB), and has a debris impact resistance that is greater than the current acreage tile's resistance by a factor of approximately 6-20.[37] At least 772 of these advanced tiles have been installed on the Orbiters' base heat shields and upper body flaps.[38] However, due to its higher thermal conductivity, TUFI-coated AETB cannot be used as a replacement for the larger areas of tile coverage. (Boeing, Lockheed Martin and NASA are developing a lightweight, impact-resistant, low-conductivity tile.[39]) Because the impact requirements for these next-generation tiles do not appear to be based on resistance to specific (and probable) damage sources, it is the Board's view that certification of the new tile will not adequately address the threat posed by debris.

Conclusion

Despite original design requirements that the External Tank not shed debris, and the corresponding design requirement that the Orbiter not receive debris hits exceeding a trivial amount of force, debris has impacted the Shuttle on each flight. Over the course of 113 missions, foam-shedding and other debris impacts came to be regarded more as a turnaround or maintenance issue, and less as a hazard to the vehicle and crew.

Assessments of foam-shedding and strikes were not thoroughly substantiated by engineering analysis, and the process for closing In-Flight Anomalies is not well-documented and appears to vary. Shuttle Program managers appear to have confused the notion of foam posing an "accepted risk" with foam not being a "safety-of-flight issue." At times, the pressure to meet the flight schedule appeared to cut short engineering efforts to resolve the foam-shedding problem.

NASA's lack of understanding of foam properties and behavior must also be questioned. Although tests were conducted to develop and qualify foam for use on the External Tank, it appears there were large gaps in NASA's knowledge about this complex and variable material. Recent testing conducted at Marshall Space Flight Center and under the auspices of the Board indicate that mechanisms previously considered a prime source of foam loss, cryopumping and cryoingestion, are not feasible in the conditions experienced during tanking, launch, and ascent. Also, dissections of foam bipod ramps on External Tanks yet to be launched reveal subsurface flaws and defects that only now are being discovered and identified as contributing to the loss of foam from the bipod ramps.

While NASA properly designated key debris events as In-Flight Anomalies in the past, more recent events indicate that NASA engineers and management did not appreciate the scope, or lack of scope, of the Hazard Reports involving foam shedding.[40] Ultimately, NASA's hazard analyses, which were based on reducing or eliminating foam-shedding, were not succeeding. Shuttle Program management made no adjustments to the analyses to recognize this fact. The acceptance of events that are not supposed to happen has been described by sociologist Diane Vaughan as the "normalization of deviance."[41] The history of foam-problem decisions shows how NASA first began and then continued flying with foam losses, so that flying with these deviations from design specifications was viewed as normal and acceptable. Dr. Richard Feynman, a member of the Presidential Commission on the Space Shuttle Challenger Accident, discusses this phenomena in the context of the *Challenger* accident. The parallels are striking:

> *The phenomenon of accepting ... flight seals that had shown erosion and blow-by in previous flights is very clear. The Challenger flight is an excellent example. There are several references to flights that had gone before. The acceptance and success of these flights is taken as evidence of safety. But erosions and blow-by are not what the design expected. They are warnings that something is wrong ... The O-rings of the Solid Rocket Boosters were not designed to erode. Erosion was a clue that something was wrong. Erosion was not something from which safety can be inferred ... If a reasonable launch schedule is to be maintained, engineering often cannot be done fast enough to keep up with the expectations of originally conservative certification criteria designed to guarantee a very safe vehicle. In these situations, subtly, and often with apparently logical arguments, the criteria are altered so that flights may still be certified in time. They therefore fly in a relatively unsafe condition, with a chance of failure of the order of a percent (it is difficult to be more accurate).*[42]

Findings

F6.1–1 NASA has not followed its own rules and requirements on foam-shedding. Although the agency continuously worked on the foam-shedding problem, the debris impact requirements have not been met on any mission.

F6.1–2 Foam-shedding, which had initially raised serious safety concerns, evolved into "in-family" or "no safety-of-flight" events or were deemed an "accepted risk."

F6.1–3 Five of the seven bipod ramp events occurred on missions flown by *Columbia*, a seemingly high number. This observation is likely due to

Columbia having been equipped with umbilical cameras earlier than other Orbiters.

F6.1–4 There is lack of effective processes for feedback or integration among project elements in the resolution of In-Flight Anomalies.

F6.1–5 Foam bipod debris-shedding incidents on STS-52 and STS-62 were undetected at the time they occurred, and were not discovered until the Board directed NASA to examine External Tank separation images more closely.

F6.1–6 Foam bipod debris-shedding events were classified as In-Flight Anomalies up until STS-112, which was the first known bipod foam-shedding event not classified as an In-Flight Anomaly.

F6.1–7 The STS-112 assignment for the External Tank Project to "identify the cause and corrective action of the bipod ramp foam loss event" was not due until after the planned launch of STS-113, and then slipped to after the launch of STS-107.

F6.1–8 No External Tank configuration changes were made after the bipod foam loss on STS-112.

F6.1–9 Although it is sometimes possible to obtain imagery of night launches because of light provided by the Solid Rocket Motor plume, no imagery was obtained for STS-113.

F6.1–10 NASA failed to adequately perform trend analysis on foam losses. This greatly hampered the agency's ability to make informed decisions about foam losses.

F6.1–11 Despite the constant shedding of foam, the Shuttle Program did little to harden the Orbiter against foam impacts through upgrades to the Thermal Protection System. Without impact resistance and strength requirements that are calibrated to the energy of debris likely to impact the Orbiter, certification of new Thermal Protection System tile will not adequately address the threat posed by debris.

Recommendations:

- None

6.2 SCHEDULE PRESSURE

Countdown to Space Station "Core Complete:" A Workforce Under Pressure

During the course of this investigation, the Board received several unsolicited comments from NASA personnel regarding pressure to meet a schedule. These comments all concerned a date, more than a year after the launch of *Columbia*, that seemed etched in stone: February 19, 2004, the scheduled launch date of STS-120. This flight was a milestone in the minds of NASA management since it would carry a section of the International Space Station called "Node 2." This would configure the International Space Station to its "U.S. Core Complete" status.

At first glance, the Core Complete configuration date seemed noteworthy but unrelated to the *Columbia* accident. However, as the investigation continued, it became apparent that the complexity and political mandates surrounding the International Space Station Program, as well as Shuttle Program management's responses to them, resulted in pressure to meet an increasingly ambitious launch schedule.

In mid-2001, NASA adopted plans to make the over-budget and behind-schedule International Space Station credible to the White House and Congress. The Space Station Program and NASA were on probation, and had to prove they could meet schedules and budgets. The plan to regain credibility focused on the February 19, 2004, date for the launch of Node 2 and the resultant Core Complete status. If this goal was not met, NASA would risk losing support from the White House and Congress for subsequent Space Station growth.

By the late summer of 2002, a variety of problems caused Space Station assembly work and Shuttle flights to slip beyond their target dates. With the Node 2 launch endpoint fixed, these delays caused the schedule to become ever more compressed.

Meeting U.S. Core Complete by February 19, 2004, would require preparing and launching 10 flights in less than 16 months. With the focus on retaining support for the Space Station program, little attention was paid to the effects the aggressive Node 2 launch date would have on the Shuttle Program. After years of downsizing and budget cuts (Chapter 5), this mandate and events in the months leading up to STS-107 introduced elements of risk to the Program. *Columbia* and the STS-107 crew, who had seen numerous launch slips due to missions that were deemed higher priorities, were further affected by the mandatory Core Complete date. The high-pressure environments created by NASA Headquarters unquestionably affected *Columbia*, even though it was not flying to the International Space Station.

February 19, 2004 – "A Line in the Sand"

Schedules are essential tools that help large organizations effectively manage their resources. Aggressive schedules by themselves are often a sign of a healthy institution. However, other institutional goals, such as safety, sometimes compete with schedules, so the effects of schedule pressure in an organization must be carefully monitored. The Board posed the question: Was there undue pressure to nail the Node 2 launch date to the February 19, 2004, signpost? The management and workforce of the Shuttle and Space Station programs each answered the question differently. Various members of NASA upper management gave a definite "no." In contrast, the workforce within both programs thought there was considerable management focus on Node 2 and resulting pressure to hold firm to that launch date, and individuals were becoming concerned that safety might be compromised. The weight of evidence supports the workforce view.

Employees attributed the Node 2 launch date to the new Administrator, Sean O'Keefe, who was appointed to execute a Space Station management plan he had proposed as Deputy Director of the White House Office of Management and Budget. They understood the scrutiny that NASA, the new Administrator, and the Space Station Program were under, but now it seemed to some that budget and schedule were of paramount concern. As one employee reflected:

> *I guess my frustration was ... I know the importance of showing that you ... manage your budget and that's an important impression to make to Congress so you can continue the future of the agency, but to a lot of people, February 19th just seemed like an arbitrary date ... It doesn't make sense to me why at all costs we were marching to this date.*

The importance of this date was stressed from the very top. The Space Shuttle and Space Station Program Managers briefed the new NASA Administrator monthly on the status of their programs, and a significant part of those briefings was the days of margin remaining in the schedule to the launch of Node 2 – still well over a year away. The Node 2 schedule margin typically accounted for more than half of the briefing slides.

Figure 6.2-1 is one of the charts presented by the Shuttle Program Manager to the NASA Administrator in December 2002. The chart shows how the days of margin in the existing schedule were being managed to meet the requirement of a Node 2 launch on the prescribed date. The triangles are events that affected the schedule (such as the slip of a Russian Soyuz flight). The squares indicate action taken by management to regain the lost time (such as authorizing work over the 2002 winter holidays).

Figure 6.2-2 shows a slide from the International Space Station Program Manager's portion of the briefing. It indicates that International Space Station Program management was also taking actions to regain margin. Over the months, the extent of some testing at Kennedy was reduced, the number of tasks done in parallel was increased, and a third shift of workers would be added in 2003 to accomplish the processing. These charts illustrate that both the Space Shuttle and Space Station Programs were being managed to a particular launch date – February 19, 2004. Days of margin in that schedule were one of the principle metrics by which both programs came to be judged.

NASA Headquarters stressed the importance of this date in other ways. A screen saver (see Figure 6.2-3) was mailed to managers in NASA's human spaceflight program that depicted a clock counting down to February 19, 2004 – U.S. Core Complete.

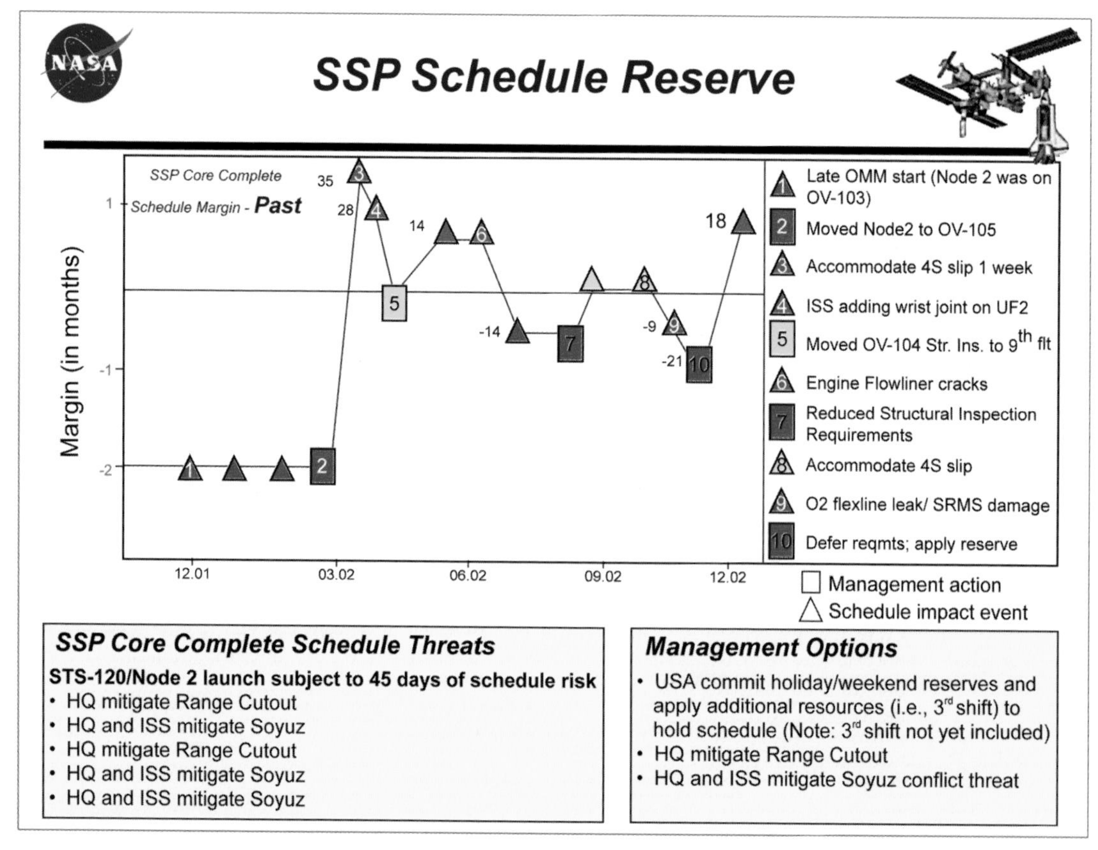

Figure 6.2-1. This chart was presented by the Space Shuttle Program Manager to the NASA Administrator in December 2002. It illustrates how the schedule was being managed to meet the Node 2 launch date of February 19, 2004.

While employees found this amusing because they saw it as a date that could not be met, it also reinforced the message that NASA Headquarters was focused on and promoting the achievement of that date. This schedule was on the minds of the Shuttle managers in the months leading up to STS-107.

The Background: Schedule Complexity and Compression

In 2001, the International Space Station Cost and Management Evaluation Task Force report recommended, as a cost-saving measure, a limit of four Shuttle flights to the International Space Station per year. To meet this requirement, managers began adjusting the Shuttle and Station manifests to "get back in the budget box." They rearranged Station assembly sequences, moving some elements forward and taking others out. When all was said and done, the launch of STS-120, which would carry Node 2 to the International Space Station, fell on February 19, 2004.

The Core Complete date simply emerged from this planning effort in 2001. By all accounts, it was a realistic and achievable date when first approved. At the time there was more concern that four Shuttle flights a year would limit the capability to carry supplies to and from the Space Station, to rotate its crew, and to transport remaining Space Station segments and equipment. Still, managers felt it was a rea-

Figure 6.2-3. NASA Headquarters distributed to NASA employees this computer screensaver counting down to February 19, 2004.

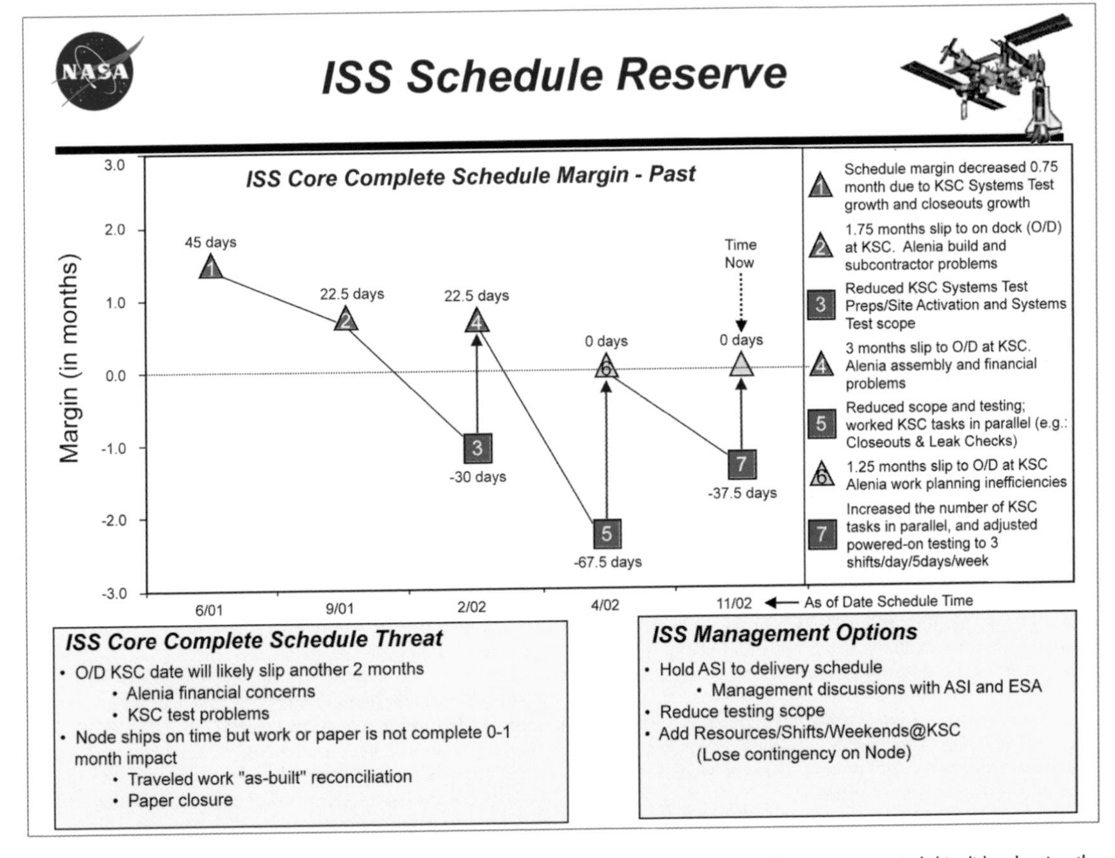

Figure 6.2-2. At the same December 2002 meeting, the International Space Station Program Manager presented this slide, showing the actions being taken to regain margin in the schedule. Note that the yellow triangles reflect zero days remaining margin.

sonable goal and assumed that if circumstances warranted a slip of that date, it would be granted.

Shuttle and Station managers worked diligently to meet the schedule. Events gradually ate away at the schedule margin. Unlike the "old days" before the Station, the Station/Shuttle partnership created problems that had a ripple effect on both programs' manifests. As one employee described it, "the serial nature" of having to fly Space Station assembly missions in a specific order made staying on schedule more challenging. Before the Space Station, if a Shuttle flight had to slip, it would; other missions that had originally followed it would be launched in the meantime. Missions could be flown in any sequence. Now the manifests were a delicate balancing act. Missions had to be flown in a certain order and were constrained by the availability of the launch site, the Russian Soyuz and Progress schedules, and a myriad of other processes. As a result, employees stated they were now experiencing a new kind of pressure. Any necessary change they made on one mission was now impacting future launch dates. They had a sense of being "under the gun."

Shuttle and Station program personnel ended up with manifests that one employee described as "changing, changing, changing" all the time. One of the biggest issues they faced entering 2002 was "up mass," the amount of cargo the Shuttle can carry to the Station. Up mass was not a new problem, but when the Shuttle flight rate was reduced to four per year, up mass became critical. Working groups were actively evaluating options in the summer of 2002 and bartering to get each flight to function as expected.

Sometimes the up mass being traded was actual Space Station crew members. A crew rotation planned for STS-118 was moved to a later flight because STS-118 was needed for other cargo. This resulted in an increase of crew duration on the Space Station, which was creeping past the 180-day limit agreed to by the astronaut office, flight surgeons, and Space Station international partners. A space station worker described how this one change created many other problems, and added: *"... we had a train wreck coming ..."* Future on-orbit crew time was being projected at 205 days or longer to maintain the assembly sequence and meet the schedule.

By July 2002, the Shuttle and Space Station Programs were facing a schedule with very little margin. Two setbacks occurred when technical problems were found during routine maintenance on *Discovery*. STS-107 was four weeks away from launch at the time, but the problems grounded the entire Shuttle fleet. The longer the fleet was grounded, the more schedule margin was lost, which further compounded the complexity of the intertwined Shuttle and Station schedules. As one worker described the situation:

> *... a one-week hit on a particular launch can start a steam roll effect including all [the] constraints and by the time you get out of here, that one-week slip has turned into a couple of months.*

In August 2002, the Shuttle Program realized it would be unable to meet the Space Station schedule with the available Shuttles. *Columbia* had never been outfitted to make a Space Station flight, so the other three Orbiters flew the Station missions. But *Discovery* was in its Orbiter Maintenance Down Period, and would not be available for another 17 months. All Space Station flights until then would have to be made by *Atlantis* and *Endeavour*. As managers looked ahead to 2003, they saw that after STS-107, these two Orbiters would have to alternate flying five consecutive missions, STS-114 through STS-118. To alleviate this pressure, and regain schedule margin, Shuttle Program managers elected to modify *Columbia* to enable it to fly Space Station missions. Those modifications were to take place immediately after STS-107 so that *Columbia* would be ready to fly its first Space Station mission eight months later. This decision put *Columbia* directly in the path of Core Complete.

As the autumn of 2002 began, both the Space Shuttle and Space Station Programs began to use what some employees termed "tricks" to regain schedule margin. Employees expressed concern that their ability to gain schedule margin using existing measures was waning.

In September 2002, it was clear to Space Shuttle and Space Station Program managers that they were not going to meet the schedule as it was laid out. The two Programs proposed a new set of launch dates, documented in an e-mail (right) that included moving STS-120, the Node 2 flight, to mid-March 2004. (Note that the first paragraph ends with *"... the 10A* [U.S. Core Complete, Node 2] *launch remains 2/19/04."*)

These launch date changes made it possible to meet the early part of the schedule, but compressed the late 2003/early 2004 schedule even further. This did not make sense to many in the program. One described the system as at *"an uncomfortable point,"* noted having to go to great lengths to reduce vehicle-processing time at Kennedy, and added:

> *... I don't know what Congress communicated to O'Keefe. I don't really understand the criticality of February 19th, that if we didn't make that date, did that mean the end of NASA? I don't know ... I would like to think that the technical issues and safely resolving the technical issues can take priority over any budget issue or scheduling issue.*

When the Shuttle fleet was cleared to return to flight, attention turned to STS-112, STS-113, and STS-107, set for October, November, and January. Workers were uncomfortable with the rapid sequence of flights.

> *The thing that was beginning to concern me ... is I wasn't convinced that people were being given enough time to work the problems correctly.*

The problems that had grounded the fleet had been handled well, but the program nevertheless lost the rest of its margin. As the pressure to keep to the Node 2 schedule continued, some were concerned that this might influence the future handling of problems. One worker expressed the concern:

> *... and I have to think that subconsciously that even though you don't want it to affect decision-making, it probably does.*

-----Original Message-----
From: THOMAS, DAWN A. (JSC-OC) (NASA)
Sent: Friday, September 20, 2002 7:10 PM
To: 'Flowers, David'; 'Horvath, Greg'; 'O'Fallon, Lee'; 'Van Scyoc, Neal'; 'Gouti, Tom'; 'Hagen, Ray'; 'Kennedy, John'; 'Thornburg, Richard'; 'Gari, Judy'; 'Dodds, Joel'; 'Janes, Lou Ann'; 'Breen, Brian'; 'Deheck-Stokes, Kristina'; 'Narita, Kaneaki (NASDA)'; 'Patrick, Penny O'; 'Michael Rasmussen (E-mail)'; DL FPWG; 'Hughes, Michael G'; 'Bennett, Patty'; 'Masazumi, Miyake'; 'Mayumi Matsuura'; NORIEGA, CARLOS I. (JSC-CB) (NASA); BARCLAY, DINA E. (JSC-DX) (NASA); MEARS, AARON (JSC-XA) (HS); BROWN, WILLIAM C. (JSC-DT) (NASA); DUMESNIL, DEANNA T. (JSC-OC) (USA); MOORE, NATHAN (JSC-REMOTE); MONTALBANO, JOEL R. (JSC-DA8) (NASA); MOORE, PATRICIA (PATTI) (JSC-DA8) (NASA); SANCHEZ, HUMBERTO (JSC-DA8) (NASA)
Subject: FPWG status - 9/20/02 OA/MA mgrs mtg results

The ISS and SSP Program Managers have agreed to proceed with the crew rotation change and the following date changes: 12A launch to 5/23/03, 12A.1 launch to 7/24/03, 13A launch to 10/2/03, and 13A.1 launch to NET 11/13/03. Please note that 10A launch remains 2/19/04.

The ISS SSCN that requests evaluation of these changes will be released Monday morning after the NASA/Russian bilateral Requirements and Increment Planning videocon. It will contain the following:

- Increments 8 and 9 redefinition - this includes baseline of ULF2 into the tactical timeframe as the new return flight for Expedition 9
- Crew size changes for 7S, 13A.1, 15A, and 10A
- Shuttle date changes as listed above
- Russian date changes for CY2003 that were removed from SSCN 6872 (11P launch/10P undock and subsequent)
- CY2004 Russian data if available Monday morning
- Duration changes for 12A and 15A
- Docking altitude update for 10A, along with "NET" TBR closure.

The evaluation due date is 10/2/02. Board/meeting dates are as follows: MIOCB status - 10/3/02; comment dispositioning - 10/3/02 FPWG (meeting date/time under review); OA/MA Program Managers status - 10/4/02; SSPCB and JPRCB - 10/8/02; MMIOCB status (under review) and SSCB - 10/10/02.

The 13A.1 date is indicated as "NET" (No Earlier Than) since SSP ability to meet that launch date is under review due to the processing flow requirements.

There is no longer a backup option to move ULF2 to OV-105: due to vehicle processing requirements, there is no launch opportunity on OV-105 past May 2004 until after OMM.

The Program Managers have asked for preparation of a backup plan in case of a schedule slip of ULF2. In order to accomplish this, the projected ISS upmass capability shortfall will be calculated as if ULF2 launch were 10/7/04, and a recommendation made for addressing the resulting shortfall and increment durations. Some methods to be assessed: manifest restructuring, fallback moves of rotation flight launch dates, LON (Launch on Need) flight on 4/29/04.

[ISS=International Space Station, SSP=Space Shuttle Program, NET=no earlier than, SSCN=Space Station Change Notice, CY=Calendar Year, TBR=To Be Revised (or Reviewed), MIOCB=Mission Integration and Operations Control Board, FPWG=Flight Planning Working Group, OA/MA=Space Station Office Symbol/Shuttle Program Office Symbol, SSPCB=Space Station Program Control Board, JPRCB=Space Shuttle/Space Station Joint Program Requirements Control Board, MMIOCB=Multi-Lateral Mission Integration and Operations Control Board, SSCB=Space Station Control Board, ULF2=U.S. Logistics Flight 2, OMM=Orbiter Major Modification, OV-105=Endeavour]

This was the environment for October and November of 2002. During this time, a bipod foam event occurred on STS-112. For the first time in the history of the Shuttle Program, the Program Requirements Control Board chose to classify that bipod foam loss as an "action" rather than a more serious In-Flight Anomaly. At the STS-113 Flight Readiness Review, managers accepted with little question the rationale that it was safe to fly with the known foam problem.

The Operations Tempo Following STS-107

After STS-107, the tempo was only going to increase. The vehicle processing schedules, training schedules, and mission control flight staffing assignments were all overburdened.

The vehicle-processing schedule for flights from February 2003, through February 2004, was optimistic. The schedule could not be met with only two shifts of workers per day. In late 2002, NASA Headquarters approved plans to hire a third shift. There were four Shuttle launches to the Space Station scheduled in the five months from October 2003, through the launch of Node 2 in February 2004. To put this in perspective, the launch rate in 1985, for which NASA was criticized by the Rogers Commission, was nine flights in 12 months – and that was accomplished with four Orbiters and a manifest that was not complicated by Space Station assembly.

Endeavour was the Orbiter on the critical path. Figure 6.2-4 shows the schedule margin for STS-115, STS-117, and STS-120 (Node 2). To preserve the margin going into 2003, the vehicle processing team would be required to work the late 2002-early 2003 winter holidays. The third shift of workers at Kennedy would be available in March 2003, and would buy eight more days of margin for STS-117 and STS-120. The workforce would likely have to work on the 2003 winter holidays to meet the Node 2 date.

Figure 6.2-5 shows the margin for each vehicle (*Discovery*, OV-103, was in extended maintenance). The large boxes indicate the "margin to critical path" (to Node 2 launch date). The three smaller boxes underneath indicate (from left to right) vehicle processing margin, holiday margin, and Dryden margin. The vehicle processing margin indicates how many days there are in addition to the days required for that mission's vehicle processing. *Endeavour* (OV-105) had zero days of margin for the processing flows for STS-115, STS-117, and STS-120. The holiday margin is the number of days that could be gained by working holidays. The Dryden margin is the six days that are always reserved to accommodate an Orbiter landing at Edwards Air Force Base in California and having to be ferried to Kennedy. If the Orbiter landed at Kennedy, those six days would automatically be regained. Note that the Dryden margin had already been surrendered in the STS-114 and STS-115 schedules. If bad weather at Kennedy forced those two flights to land at Edwards, the schedule would be directly affected.

The clear message in these charts is that any technical problem that resulted in a slip to one launch would now directly affect the Node 2 launch.

The lack of housing for the Orbiters was becoming a factor as well. Prior to launch, an Orbiter can be placed in an Orbiter Processing Facility, the Vehicle Assembly Building, or on one of the two Shuttle launch pads. Maintenance and

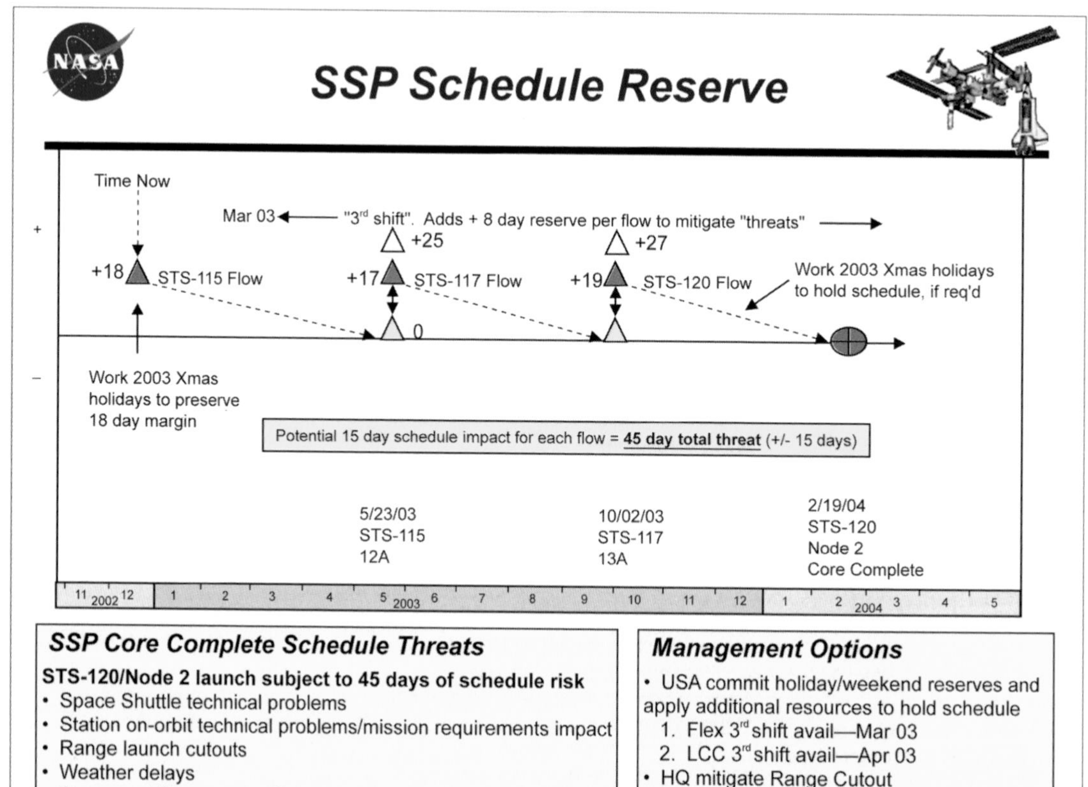

Figure 6.2-4. By late 2002, the vehicle processing team at the Kennedy Space Center would be required to work through the winter holidays, and a third shift was being hired in order to meet the February 19, 2004, schedule for U.S. Core Complete.

refurbishment is performed in the three Orbiter Processing Facilities at Kennedy. One was occupied by *Discovery* during its scheduled extended maintenance. This left two to serve the other three Orbiters over the next several months. The 2003 schedule indicated plans to move *Columbia* (after its return from STS-107) from an Orbiter Processing Facility to the Vehicle Assembly Building and back several times in order to make room for *Atlantis* (OV-104) and *Endeavour* (OV-105) and prepare them for missions. Moving an Orbiter is tedious, time-consuming, carefully orchestrated work. Each move introduces an opportunity for problems. Those 2003 moves were often slated to occur without a day of margin between them – another indication of the additional risks that managers were willing to incur to meet the schedule.

The effect of the compressed schedule was also evident in the Mission Operations Directorate. The plans for flight controller staffing of Mission Control showed that of the seven flight controllers who lacked current certifications during STS-107 (see Chapter 4), five were scheduled to work the next mission, and three were scheduled to work the next three missions (STS-114, -115, and -116). These controllers would have been constantly either supporting missions or supporting mission training, and were unlikely to have the time to complete the recertification requirements. With the pressure of the schedule, the things perceived to be less important, like recertification (which was not done before STS-107), would likely continue to be deferred. As a result of the schedule pressure, managers either were willing to delay recertification or were too busy to notice that deadlines for recertification had passed.

Columbia: Caught in the Middle

STS-112 flew in October 2002. At 33 seconds into the flight, a piece of the bipod foam from the External Tank struck one of the Solid Rocket Boosters. As described in Section 6.1, the STS-112 foam strike was discussed at the Program Requirements Control Board following the flight. Although the initial recommendation was to treat the foam loss as an In-Flight Anomaly, the Shuttle Program instead assigned it as an action, with a due date after the next launch. (This was the first instance of bipod foam loss that was not designated an In-Flight Anomaly.) The action was noted at the STS-113 Flight Readiness Review. Those Flight Readiness Review charts (see Section 6.1) provided a flawed flight rationale by concluding that the foam loss was "not a safety-of-flight" issue.

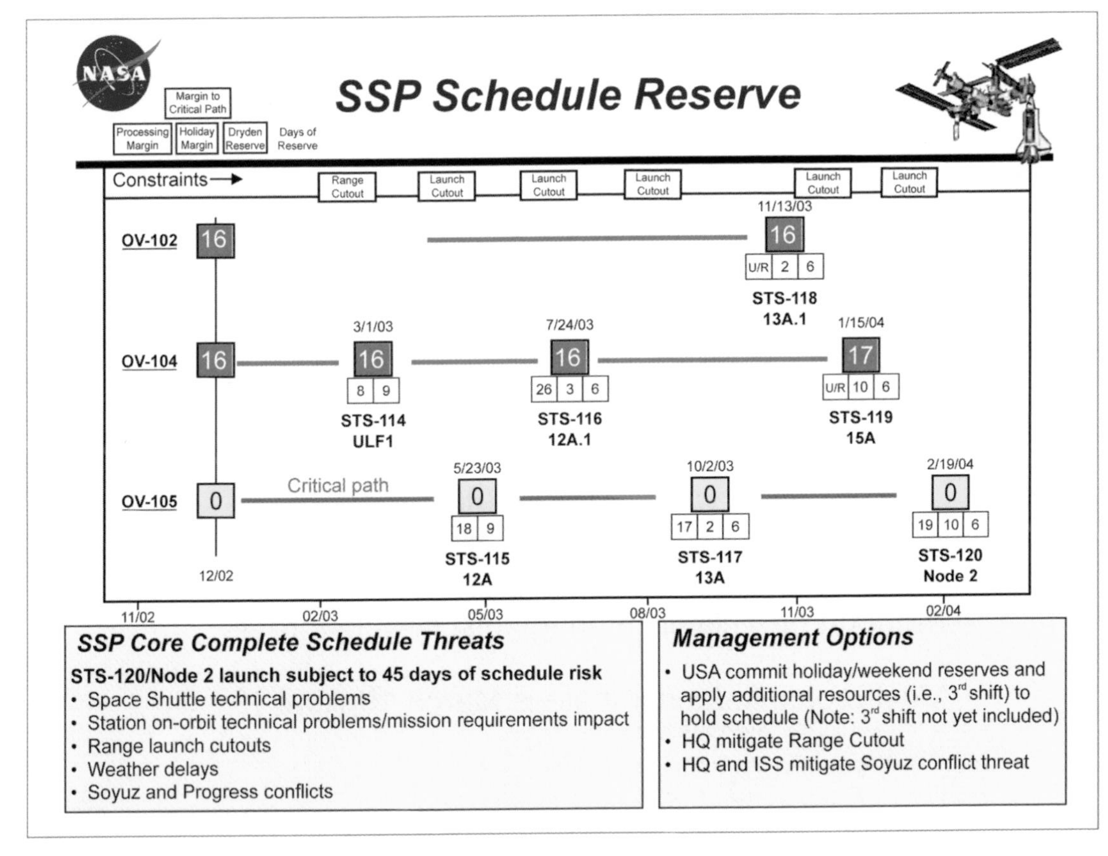

Figure 6.2-5. This slide shows the margin for each Orbiter. The large boxes show the number of days margin to the Node 2 launch date, while the three smaller boxes indicate vehicle processing margin, holiday margin, and the margin if a Dryden landing was not required.

Interestingly, during *Columbia's* mission, the Chair of the Mission Management Team, Linda Ham, would characterize that reasoning as "lousy" – though neither she nor Shuttle Program Manager Ron Dittemore, who were both present at the meeting, questioned it at the time. The pressing need to launch STS-113 to retrieve the International Space Station Expedition 5 crew before they surpassed the 180-day limit and to continue the countdown to Node 2 were surely in the back of managers' minds during these reviews.

By December 2002, every bit of padding in the schedule had disappeared. Another chart from the Shuttle and Station Program Managers' briefing to the NASA Administrator summarizes the schedule dilemma (see Figure 6.2-6).

Even with work scheduled on holidays, a third shift of workers being hired and trained, future crew rotations drifting beyond 180 days, and some tests previously deemed "requirements" being skipped or deferred, Program managers estimated that Node 2 launch would be one to two months late. They were slowly accepting additional risk in trying to meet a schedule that probably could not be met.

Interviews with workers provided insight into how this situation occurred. They noted that people who work at NASA have the legendary can-do attitude, which contributes to the agency's successes. But it can also cause problems. When workers are asked to find days of margin, they work furiously to do so and are praised for each extra day they find. But those same people (and this same culture) have difficulty admitting that something "can't" or "shouldn't" be done, that the margin has been cut too much, or that resources are being stretched too thin. No one at NASA wants to be the one to stand up and say, "We can't make that date."

STS-107 was launched on January 16, 2003. Bipod foam separated from the External Tank and struck *Columbia*'s left wing 81.9 seconds after liftoff. As the mission proceeded over the next 16 days, critical decisions about that event would be made.

The STS-107 Mission Management Team Chair, Linda Ham, had been present at the Program Requirements Control Board discussing the STS-112 foam loss and the STS-113 Flight Readiness Review. So had many of the other Shuttle Program managers who had roles in STS-107. Ham was also the Launch Integration Manager for the next mission, STS-114. In that capacity, she would chair many of the meetings leading up to the launch of that flight, and many of those individuals would have to confront *Columbia's* foam strike and its possible impact on the launch of STS-114. Would the *Columbia* foam strike be classified as an In-Flight Anomaly? Would the fact that foam had detached from the bipod ramp on two out of the last three flights have made this problem a constraint to flight that would need to be solved before the next launch? Could the Program develop a solid rationale to fly STS-114, or would additional analysis be required to clear the flight for launch?

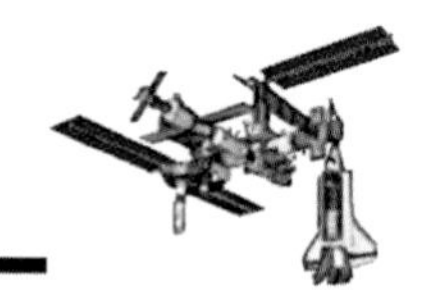

Figure 6.2-6. By December 2002, every bit of padding in the schedule had disappeared. Another chart from the Shuttle and Station Program Managers' briefing to the NASA Administrator summarizes the schedule dilemma.

-----Original Message-----
From: HAM, LINDA J. (JSC-MA2) (NASA)
Sent: Wednesday, January 22, 2003 10:16 AM
To: DITTEMORE, RONALD D. (JSC-MA) (NASA)
Subject: RE: ET Briefing - STS-112 Foam Loss

Yes, I remember....It was not good. I told Jerry to address it at the ORR next Tuesday (even though he won't have any more data and it really doesn't impact Orbiter roll to the VAB). I just want him to be thinking hard about this now, not wait until IFA review to get a formal action.

[ORR=Orbiter Rollout Review, VAB=Vehicle Assembly Building, IFA=In-Flight Anomaly]

In fact, most of Linda Ham's inquiries about the foam strike were not to determine what action to take during *Columbia's* mission, but to understand the implications for STS-114. During a Mission Management Team meeting on January 21, she asked about the rationale put forward at the STS-113 Flight Readiness Review, which she had attended. Later that morning she reviewed the charts presented at that Flight Readiness Review. Her assessment, which she e-mailed to Shuttle Program Manager Ron Dittemore on January 21, was *"Rationale was lousy then and still is …"* (See Section 6.3 for the e-mail.)

One of Ham's STS-114 duties was to chair a review to determine if the mission's Orbiter, *Atlantis*, should be rolled from the Orbiter Processing Facility to the Vehicle Assembly Building, per its pre-launch schedule. In the above e-mail to Ron Dittemore, Ham indicates a desire to have the same individual responsible for the "lousy" STS-113 flight rationale start working the foam shedding issue – and presumably present a new flight rationale – very soon.

As STS-107 prepared for re-entry, Shuttle Program managers prepared for STS-114 flight rationale by arranging to have post-flight photographs taken of *Columbia's* left wing rushed to Johnson Space Center for analysis.

As will become clear in the next section, most of the Shuttle Program's concern about *Columbia*'s foam strike were not about the threat it might pose to the vehicle in orbit, but about the threat it might pose to the schedule.

Conclusion

The agency's commitment to hold firm to a February 19, 2004, launch date for Node 2 influenced many of decisions in the months leading up to the launch of STS-107, and may well have subtly influenced the way managers handled the STS-112 foam strike and *Columbia*'s as well.

When a program agrees to spend less money or accelerate a schedule beyond what the engineers and program managers think is reasonable, a small amount of overall risk is added. These little pieces of risk add up until managers are no longer aware of the total program risk, and are, in fact, gambling. Little by little, NASA was accepting more and more risk in order to stay on schedule.

Findings

F6.2-1 NASA Headquarters' focus was on the Node 2 launch date, February 19, 2004.

F6.2-2 The intertwined nature of the Space Shuttle and Space Station programs significantly increased the complexity of the schedule and made meeting the schedule far more challenging.

F6.2-3 The capabilities of the system were being stretched to the limit to support the schedule. Projections into 2003 showed stress on vehicle processing at the Kennedy Space Center, on flight controller training at Johnson Space Center, and on Space Station crew rotation schedules. Effects of this stress included neglecting flight controller recertification requirements, extending crew rotation schedules, and adding incremental risk by scheduling additional Orbiter movements at Kennedy.

F6.2-4 The four flights scheduled in the five months from October 2003, to February 2004, would have required a processing effort comparable to the effort immediately before the *Challenger* accident.

F6.2-5 There was no schedule margin to accommodate unforeseen problems. When flights come in rapid succession, there is no assurance that anomalies on one flight will be identified and appropriately addressed before the next flight.

F6.2-6 The environment of the countdown to Node 2 and the importance of maintaining the schedule may have begun to influence managers' decisions, including those made about the STS-112 foam strike.

F6.2-7 During STS-107, Shuttle Program managers were concerned with the foam strike's possible effect on the launch schedule.

Recommendation:

R6.2-1 Adopt and maintain a Shuttle flight schedule that is consistent with available resources. Although schedule deadlines are an important management tool, those deadlines must be regularly evaluated to ensure that any additional risk incurred to meet the schedule is recognized, understood, and acceptable.

6.3 DECISION-MAKING DURING THE FLIGHT OF STS-107

Initial Foam Strike Identification

As soon as *Columbia* reached orbit on the morning of January 16, 2003, NASA's Intercenter Photo Working Group began reviewing liftoff imagery by video and film cameras on the launch pad and at other sites at and nearby the Kennedy Space Center. The debris strike was not seen during the first review of video imagery by tracking cameras, but it was noticed at 9:30 a.m. EST the next day, Flight Day Two, by Intercenter Photo Working Group engineers at Marshall Space Flight Center. Within an hour, Intercenter Photo Working Group personnel at Kennedy also identified the strike on higher-resolution film images that had just been developed.

The images revealed that a large piece of debris from the left bipod area of the External Tank had struck the Orbiter's left wing. Because the resulting shower of post-impact fragments could not be seen passing over the top of the wing, analysts concluded that the debris had apparently impacted the left wing below the leading edge. Intercenter Photo Working Group members were concerned about the size of the object and the apparent momentum of the strike. In searching for better views, Intercenter Photo Working Group members realized that none of the other cameras provided a higher-quality view of the impact and the potential damage to the Orbiter.

Of the dozen ground-based camera sites used to obtain images of the ascent for engineering analyses, each of which has film and video cameras, five are designed to track the Shuttle from liftoff until it is out of view. Due to expected angle of view and atmospheric limitations, two sites did not capture the debris event. Of the remaining three sites positioned to "see" at least a portion of the event, none provided a clear view of the actual debris impact to the wing. The first site lost track of *Columbia* on ascent, the second site was out of focus – because of an improperly maintained lens – and the third site captured only a view of the upper side of *Columbia's* left wing. The Board notes that camera problems also hindered the *Challenger* investigation. Over the years, it appears that due to budget and camera-team staff cuts, NASA's ability to track ascending Shuttles has atrophied – a development that reflects NASA's disregard of the developmental nature of the Shuttle's technology. (See recommendation R3.4-1.)

Because they had no sufficiently resolved pictures with which to determine potential damage, and having never seen such a large piece of debris strike the Orbiter so late in ascent, Intercenter Photo Working Group members decided to ask for ground-based imagery of *Columbia*.

IMAGERY REQUEST 1

To accomplish this, the Intercenter Photo Working Group's Chair, Bob Page, contacted Wayne Hale, the Shuttle Program Manager for Launch Integration at Kennedy Space Center, to request imagery of *Columbia*'s left wing on-orbit. Hale, who agreed to explore the possibility, holds a Top Secret clearance and was familiar with the process for requesting military imaging from his experience as a Mission Control Flight Director.

This would be the first of three discrete requests for imagery by a NASA engineer or manager. In addition to these three requests, there were, by the Board's count, at least eight "missed opportunities" where actions may have resulted in the discovery of debris damage.

Shortly after confirming the debris hit, Intercenter Photo Working Group members distributed a "L+1" (Launch plus one day) report and digitized clips of the strike via e-mail throughout the NASA and contractor communities. This report provided an initial view of the foam strike and served as the basis for subsequent decisions and actions.

Mission Management's Response to the Foam Strike

As soon as the Intercenter Working Group report was distributed, engineers and technical managers from NASA, United Space Alliance, and Boeing began responding. Engineers and managers from Kennedy Space Center called engineers and Program managers at Johnson Space Center. United Space Alliance and Boeing employees exchanged e-mails with details of the initial film analysis and the work in progress to determine the result of the impact. Details of the strike, actions taken in response to the impact, and records of telephone conversations were documented in the Mission Control operational log. The following section recounts in chronological order many of these exchanges and provides insight into why, in spite of the debris strike's severity, NASA managers ultimately declined to request images of *Columbia's* left wing on-orbit.

Flight Day Two, Friday, January 17, 2003

In the Mission Evaluation Room, a support function of the Shuttle Program office that supplies engineering expertise for missions in progress, a set of consoles are staffed by engineers and technical managers from NASA and contractor organizations. For record keeping, each Mission Evaluation Room member types mission-related comments into a running log. A log entry by a Mission Evaluation Room manager at 10:58 a.m. Central Standard Time noted that the vehicle may have sustained damage from a debris strike.

> *"John Disler [a photo lab engineer at Johnson Space Center] called to report a debris hit on the vehicle. The debris appears to originate from the ET Forward Bipod area...travels down the left side and hits the left wing leading edge near the fuselage...The launch video review team at KSC think that the vehicle may have been damaged by the impact. Bill Reeves and Mike Stoner (USA SAM) were notified."* [ET=External Tank, KSC=Kennedy Space Center, USA SAM=United Space Alliance Sub-system Area Manager]

At 3:15 p.m., Bob Page, Chair of the Intercenter Photo Working Group, contacted Wayne Hale, the Shuttle Program Manager for Launch Integration at Kennedy Space Center, and Lambert Austin, the head of the Space Shuttle Systems Integration at Johnson Space Center, to inform them that Boeing was performing an analysis to determine trajectories, velocities, angles, and energies for the debris impact. Page also stated that photo-analysis would continue over the Martin Luther King Jr. holiday weekend as additional film from tracking cameras was developed. Shortly thereafter, Wayne Hale telephoned Linda Ham, Chair of the Mission Management Team, and Ron Dittemore, Space Shuttle Program Manager, to pass along information about the debris strike and let them know that a formal report would be issued by the end of the day. John Disler, a member of the Intercenter Photo Working Group, notified the Mission Evaluation Room manager that a newly formed group of analysts, to be known as the Debris Assessment Team, needed the entire weekend to conduct a more thorough analysis. Meanwhile, early opinions about Reinforced Carbon-Carbon (RCC) resiliency were circulated via e-mail between United Space Alliance technical managers and NASA engineers, which may have contributed to a mindset that foam hitting the RCC was not a concern.

-----Original Message-----
From: Stoner-1, Michael D
Sent: Friday, January 17, 2003 4:03 PM
To: Woodworth, Warren H; Reeves, William D
Cc: Wilder, James; White, Doug; Bitner, Barbara K; Blank, Donald E; Cooper, Curt W; Gordon, Michael P.
Subject: RE: STS 107 Debris

Just spoke with Calvin and Mike Gordon (RCC SSM) about the impact.

Basically the RCC is extremely resilient to impact type damage. The piece of debris (most likely foam/ice) looked like it most likely impacted the WLE RCC and broke apart. It didn't look like a big enough piece to pose any serious threat to the system and Mike Gordon the RCC SSM concurs. At T +81seconds the piece wouldn't have had enough energy to create a large damage to the RCC WLE system. Plus they have analysis that says they have a single mission safe re-entry in case of impact that penetrates the system.

As far as the tile go in the wing leading edge area they are thicker than required (taper in the outer mold line) and can handle a large area of shallow damage which is what this event most likely would have caused. They have impact data that says the structure would get slightly hotter but still be OK.

Mike Stoner
USA TPS SAM

[RCC=Reinforced Carbon-Carbon, SSM=Sub-system Manager, WLE=Wing Leading Edge, TPS=Thermal Protection System, SAM= Sub-system Area Manager]

ENGINEERING COORDINATION AT NASA AND UNITED SPACE ALLIANCE

After United Space Alliance became contractually responsible for most aspects of Shuttle operations, NASA developed procedures to ensure that its own engineering expertise was coordinated with that of contractors for any "out-of-family" issue. In the case of the foam strike on STS-107, which was classified as out-of-family, clearly defined written guidance led United Space Alliance technical managers to liaise with their NASA counterparts. Once NASA managers were officially notified of the foam strike classification, and NASA engineers joined their contractor peers in an early analysis, the resultant group should, according to standing procedures, become a Mission Evaluation Room Tiger Team. Tiger Teams have clearly defined roles and responsibilities.[43] Instead, the group of analysts came to be called a Debris Assessment Team. While they were the right group of engineers working the problem at the right time, by not being classified as a Tiger Team, they did not fall under the Shuttle Program procedures described in Tiger Team checklists, and as a result were not "owned" or led by Shuttle Program managers. This left the Debris Assessment Team in a kind of organizational limbo, with no guidance except the date by which Program managers expected to hear their results: January 24th.

Already, by Friday afternoon, Shuttle Program managers and working engineers had different levels of concern about what the foam strike might have meant. After reviewing available film, Intercenter Photo Working Group engineers believed the Orbiter may have been damaged by the strike. They wanted on-orbit images of *Columbia's* left wing to confirm their suspicions and initiated action to obtain them. Boeing and United Space Alliance engineers decided to work through the holiday weekend to analyze the strike. At the same time, high-level managers Ralph Roe, head of the Shuttle Program Office of Vehicle Engineering, and Bill Reeves, from United Space Alliance, voiced a lower level of concern. It was at this point, before any analysis had started, that Shuttle Program managers officially shared their belief that the strike posed no safety issues, and that there was no need for a review to be conducted over the weekend. The following is a 4:28 p.m. Mission Evaluation Room manager log entry:

> *"Bill Reeves called, after a meeting with Ralph Roe, it is confirmed that USA/Boeing will not work the debris issue over the weekend, but will wait till Monday when the films are released. The LCC constraints on ice, the energy/speed of impact at +81 seconds, and the toughness of the RCC are two main factors for the low concern. Also, analysis supports single mission safe re-entry for an impact that penetrates the system..."* [USA=United Space Alliance, LCC=Launch Commit Criteria]

The following is a 4:37 p.m. Mission Evaluation Room manager log entry.

> *"Bob Page told MER that KSC/TPS engineers were sent by the USA SAM/Woody Woodworth to review the video and films. Indicated that Page had said that Woody had said this was an action from the MER to work this issue and a possible early landing on Tuesday. MER Manager told Bob that no official action was given by USA or Boeing and they had no concern about landing early. Woody indicated that the TPS engineers at KSC have been 'turned away' from reviewing the films. It was stated that the film reviews wouldn't be finished till Monday."* [MER=Mission Evaluation Room, KSC=Kennedy Space Center, TPS=Thermal Protection System, USA SAM=United Space Alliance Sub-system Area Manager]

The Mission Evaluation Room manager also wrote:

> *"I also confirmed that there was no rush on this issue and that it was okay to wait till the film reviews are finished on Monday to do a TPS review."*

In addition to individual log entries by Mission Evaluation Room members, managers prepared "handover" notes for delivery from one working shift to the next. Handovers from Shift 1 to 2 on January 17 included the following entry under a "problem" category.

> *"Disler Report – Debris impact on port wing edge-appears to have originated at the ET fwd bipod – foam?- if so, it shouldn't be a problem – video clip will be available on the web soon – will look at high-speed film today."* [ET=External Tank, fwd=forward]

Shortly after these entries were made, the deputy manager of Johnson Space Center Shuttle Engineering notified Rodney Rocha, NASA's designated chief engineer for the Thermal Protection System, of the strike and the approximate debris size. It was Rocha's responsibility to coordinate NASA engineering resources and work with contract engineers at United Space Alliance, who together would form a Debris Assessment Team that would be Co-Chaired by United Space Alliance engineering manager Pam Madera. The United Space Alliance deputy manager of Shuttle Engineering signaled that the debris strike was initially classified as "out-of-family" and therefore of greater concern than previous debris strikes. At about the same time, the Intercenter Photo Working Group's L+1 report, containing both video clips and still images of the debris strike, was e-mailed to engineers and technical managers both inside and outside of NASA.

Flight Days Three and Four, Saturday and Sunday, January 18 and 19, 2003

Though senior United Space Alliance Manager Bill Reeves had told Mission Evaluation Room personnel that the debris problem would not be worked over the holiday weekend, engineers from Boeing did in fact work through the weekend. Boeing analysts conducted a preliminary damage assessment on Saturday. Using video and photo images, they generated two estimates of possible debris size – 20 inches by 20 inches by 2 inches, and 20 inches by 16 inches by 6 inches – and determined that the debris was traveling at a approximately 750 feet per second, or 511 miles per hour, when it struck the Orbiter at an estimated impact angle of less than 20 degrees. These estimates later proved remarkably accurate.

To calculate the damage that might result from such a strike, the analysts turned to a Boeing mathematical modeling tool called Crater that uses a specially developed algorithm to predict the depth of a Thermal Protection System tile to which debris will penetrate. This algorithm, suitable for estimating small (on the order of three cubic inches) debris impacts, had been calibrated by the results of foam, ice, and metal debris impact testing. A similar Crater-like algorithm was also developed and validated with test results to assess the damage caused by ice projectiles impacting the RCC leading edge panels. These tests showed that within certain limits, the Crater algorithm predicted more severe damage than was observed. This led engineers to classify Crater as a "conservative" tool – one that predicts more damage than will actually occur.

Until STS-107, Crater was normally used only to predict whether small debris, usually ice on the External Tank, would pose a threat to the Orbiter during launch. The use of Crater to assess the damage caused by foam during the launch of STS-107 was the first use of the model while a mission was on orbit. Also of note is that engineers used Crater during STS-107 to analyze a piece of debris that was at maximum 640 times larger in volume than the pieces of debris used to calibrate and validate the Crater model (the Board's best estimate is that it actually was 400 times larger). Therefore, the use of Crater in this new and very different situation compromised NASA's ability to accurately predict debris damage in ways that Debris Assessment Team engineers did not fully comprehend (see Figure 6.3-1).

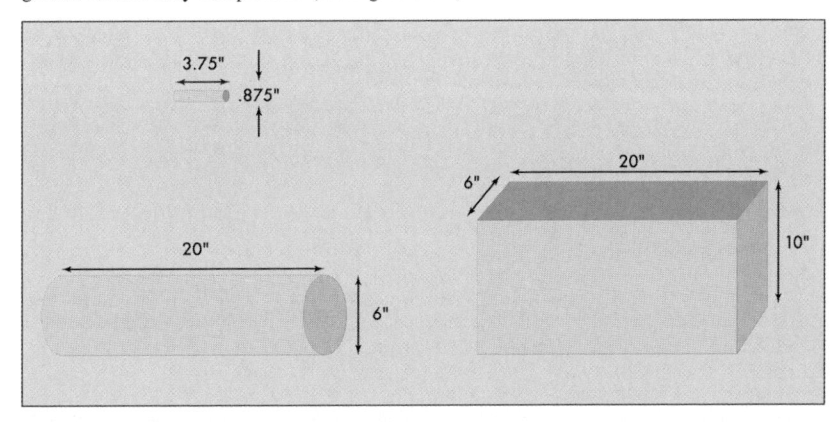

Figure 6.3-1. The small cylinder at top illustrates the size of debris Crater was intended to analyze. The larger cylinder was used for the STS-107 analysis; the block at right is the estimated size of the foam.

The Crater Model

$$p = \frac{0.0195(L/d)^{0.45}(d)(\rho_P)^{0.27}(V-V^*)^{2/3}}{(S_T)^{1/4}(\rho_T)^{1/6}}$$

p = *penetration depth*
L = *length of foam projectile*
d = *diameter of foam projectile*
ρ_P = *density of foam*
V = *component of foam velocity at right angle to foam*
V^* = *velocity required to break through the tile coating*
S_T = *compressive strength of tile*
ρ_T = *density of tile*
0.0195 = *empirical constant*

In 1966, during the Apollo program, engineers developed an equation to assess impact damage, or "cratering," by micrometeoroids.[44] The equation was modified between 1979 and 1985 to enable the analysis of impacts to "acreage" tiles that cover the lower surface of the Orbiter.[45] The modified equation, now known as Crater, predicts possible damage from sources such as foam, ice, and launch site debris, and is most often used in the day-of-launch analysis of ice debris falling off the External Tank.[46]

When used within its validated limits, Crater provides conservative predictions (that is, Crater predictions are larger than actual damage). When used outside its validated limits, Crater's precision is unknown.

Crater has been correlated to actual impact data using results from several tests. Preliminary ice drop tests were performed in 1978,[47] and additional tests using sprayed-on foam insulation projectiles were conducted in 1979 and 1999.[48] However, the test projectiles were relatively small (maximum volume of 3 cubic inches), and targeted only single tiles, not groups of tiles as actually installed on the Orbiter. No tests were performed with larger debris objects because it was not believed such debris could ever impact the Orbiter. This resulted in a very limited set of conditions under which Crater's results were empirically validated.

During 1984, tests were conducted using ice projectiles against the Reinforced Carbon-Carbon used on the Orbiters' wing leading edges.[49] These tests used an 0.875-inch diameter, 3.75-inch long ice projectile to validate an algorithm that was similar to Crater. Unlike Crater, which was designed to predict damage during a flight, the RCC predictions were intended to determine the thickness of RCC required to withstand ice impacts as an aid to design engineers. Like Crater, however, the limited set of test data significantly restricts the potential application of the model.

Other damage assessment methods available today, such as hydrodynamic structural codes, like Dyna, are able to analyze a larger set of projectile sizes and materials than Crater. Boeing and NASA did not currently sanction these finite element codes because of the time required to correlate their results in order to use the models effectively.

Although Crater was designed, and certified, for a very limited set of impact events, the results from Crater simulations can be generated quickly. During STS-107, this led to Crater being used to model an event that was well outside the parameters against which it had been empirically validated. As the accompanying table shows, many of the STS-107 debris characteristics were orders of magnitude outside the validated envelope. For instance, while Crater had been designed and validated for projectiles up to 3 cubic inches in volume, the initial STS-107 analysis estimated the piece of debris at 1,200 cubic inches – 400 times larger.

Crater parameters used during development of experimental test data versus STS-107 analysis:

Test Parameter	*Test Value*	*STS-107 Analysis*
Volume	Up to 3 cu.in	10" x 6" x 20" = 1200 cu.in. *
Length	Up to 1 inch	~ 20 inches *
Cylinder Dimensions	<= 3/8" dia x 3"	6" dia x 20"
Projectile Block Dimensions	<= 3"x 1"x 1"	6" x 10" x 20" *
Tile Material	LI-900 "acreage" tile	LI-2200 * and LI-900
Projectile Shape	Cylinder	Block

* *Outside experimental test limits*

Crater equation parameter limits:

Crater Equation Parameter	Applicable Range	STS-107 Analysis
L/d	1 - 20	3.3
L	n/a	~ 20 inches
ρ_d	1 - 3 pounds per cu.ft.	2.4 pounds per cu.ft.
d	0.4 - 2.0 inches	6 inches *
V	up to 810 fps	~ 700 fps

* *Outside validated limits*

Over the weekend, an engineer certified by Boeing to use Crater entered the two estimated debris dimensions, the estimated debris velocity, and the estimated angle of impact. The engineer had received formal training on Crater from senior Houston-based Boeing engineering staff, but he had only used the program twice before, and had reservations about using it to model the piece of foam debris that struck *Columbia*. The engineer did not consult with more experienced engineers from Boeing's Huntington Beach, California, facility, who up until the time of STS-107 had performed or overseen Crater analysis. (Boeing completed the transfer of responsibilities for Crater analysis from its Huntington Beach engineers to its Houston office in January 2003. STS-107 was the first mission that the Huntington Beach engineers were not directly involved with.)

For the Thermal Protection System tile, Crater predicted damage deeper than the actual tile thickness. This seemingly alarming result suggested that the debris that struck *Columbia* would have exposed the Orbiter's underlying aluminum airframe to extreme temperatures, resulting in a possible burn-through during re-entry. Debris Assessment Team engineers discounted the possibility of burn through for two reasons. First, the results of calibration tests with small projectiles showed that Crater predicted a deeper penetration than would actually occur. Second, the Crater equation does not take into account the increased density of a tile's lower "densified" layer, which is much stronger than tile's fragile outer layer. Therefore, engineers judged that the actual damage from the large piece of foam lost on STS-107 would not be as severe as Crater predicted, and assumed that the debris did not penetrate the Orbiter's skin. This uncertainty, however, meant that determining the precise location of the impact was paramount for an accurate damage estimate. Some areas on the Orbiter's lower surface, such as the seals around the landing gear doors, are more vulnerable than others. Only by knowing precisely where the debris struck could the analysts more accurately determine if the Orbiter had been damaged.

To determine potential RCC damage, analysts used a Crater-like algorithm that was calibrated in 1984 by impact data from ice projectiles. At the time the algorithm was empirically tested, ice was considered the only realistic threat to RCC integrity. (See Appendix E.4, RCC Impact Analysis.) The Debris Assessment Team analysis indicated that impact angles greater than 15 degrees would result in RCC penetration. A separate "transport" analysis, which attempts to determine the path the debris took, identified 15 strike regions and angles of impact. Twelve transport scenarios predicted an impact in regions of Shuttle tile. Only one scenario predicted an impact on the RCC leading edge, at a 21-degree angle. Because the foam that struck *Columbia* was less dense than ice, Debris Assessment Team analysts used a qualitative extrapolation of the test data and engineering judgment to conclude that a foam impact angle up to 21 degrees would not penetrate the RCC. Although some engineers were uncomfortable with this extrapolation, no other analyses were performed to assess RCC damage. The Debris Assessment Team focused on analyzing the impact at locations other than the RCC leading edge. This may have been due, at least in part, to the transport analysis presentation and the long-standing belief that foam was not a threat to RCC panels. The assumptions and uncertainty embedded in this analysis were never fully presented to the Mission Evaluation Room or the Mission Management Team.

Missed Opportunity 1

On Sunday, Rodney Rocha e-mailed a Johnson Space Center Engineering Directorate manager to ask if a Mission Action Request was in progress for *Columbia*'s crew to visually inspect the left wing for damage. Rocha never received an answer.

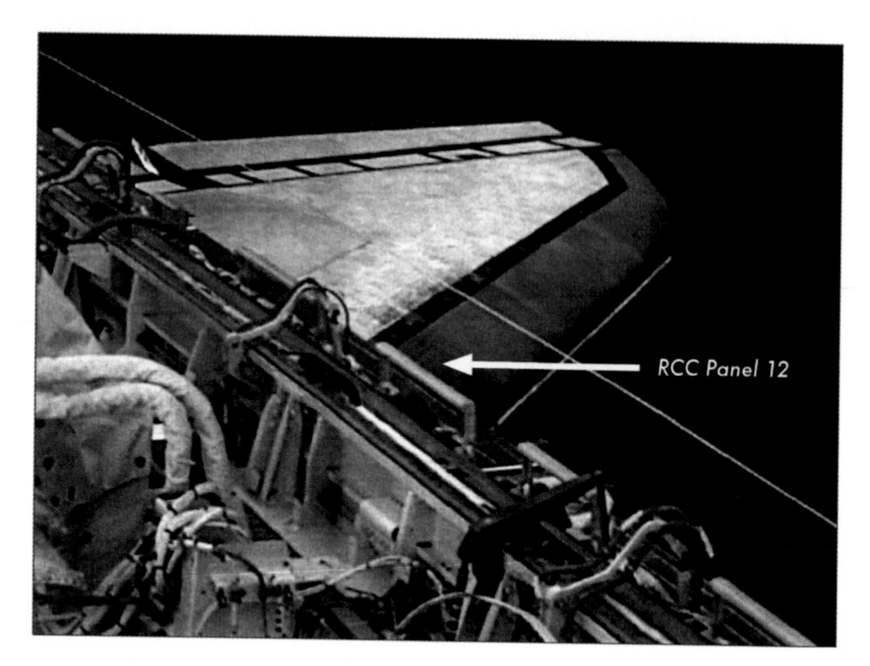

This photo from the aft flight deck window of an Orbiter shows that RCC panels 1 – 11 are not visible from inside the Orbiter. Since Columbia did not have a manipulator arm for STS-107, it would have been necessary for an astronaut to take a spacewalk to visibly inspect the inboard leading edge of the wing.

Flight Day Five, Monday, January 20, 2003

On Monday morning, the Martin Luther King Jr. holiday, the Debris Assessment Team held an informal meeting before its first formal meeting, which was scheduled for Tuesday afternoon. The team expanded to include NASA and Boeing transport analysts expert in the movement of debris in airflows, tile and RCC experts from Boeing and NASA, aerothermal and thermal engineers from NASA, United Space Alliance, and Boeing, and a safety representative from the NASA contractor Science Applications International Corporation.

Engineers emerged from that informal meeting with a goal of obtaining images from ground-based assets. Uncertainty as to precisely where the debris had struck *Columbia* generated concerns about the possibility of a breach in the left main landing gear door seal. They conducted further analysis using angle and thickness variables and thermal data obtained by personnel at Boeing's Huntington Beach facility for STS-87 and STS-50, the two missions that had incurred Thermal Protection System damage. (See Section 6.1.)

Debris Assessment Team Co-Chair Pam Madera distributed an e-mail summarizing the day's events and outlined the agenda for Tuesday's first formal Debris Assessment Team meeting. Included on the agenda was the desire to obtain on-orbit images of *Columbia*'s left wing.

According to an 11:39 a.m. entry in the Mission Evaluation Room Manager's log:

> *"...the debris 'blob' is estimated at 20" +/-10" in some direction, using the Orbiter hatch as a basis. It appears to be similar size as that seen in STS-112. There will be more comparison work done, and more info and details in tomorrow's report."*

This entry illustrates, in NASA language, an initial attempt by managers to classify this bipod ramp foam strike as close to being within the experience base and therefore, being almost an "in-family" event, not necessarily a safety concern. While the size and source of STS-107 debris was somewhat similar to what STS-112 had experienced, the impact sites (the wing versus the Solid Rocket Booster) differed – a distinction not examined by mission managers.

Flight Day Six, Tuesday, January 21, 2003

At 7:00 a.m., the Debris Assessment Team briefed Don McCormack, the chief Mission Evaluation Room manager, that the foam's source and size was similar to what struck STS-112, and that an analysis of measured versus predicted tile damage from STS-87 was being scrutinized by Boeing. An hour later, McCormack related this information to the Mission Management Team at its first post-holiday meeting. Although Space Shuttle Program requirements state that the Mission Management Team will convene daily during a mission, the STS-107 Mission Management Team met only on January 17, 21, 24, 27, and 31. The transcript below is the first record of an official discussion of the debris impact at a Mission Management Team meeting. Before even referring to the debris strike, the Mission Management Team focused on end-of-mission "downweight" (the Orbiter was 150 pounds over the limit), a leaking water separator, a jammed Hasselblad camera, payload and experiment status, and a communications downlink problem. McCormack then stated that engineers planned to determine what could be done if *Columbia* had sustained damage. STS-107 Mission Management Team Chair Linda Ham suggested the team learn what rationale had been used to fly after External Tank foam losses on STS-87 and STS-112.

Transcript Excerpts from the January 21, Mission Management Team Meeting

Ham: *"Alright, I know you guys are looking at the debris."*

McCormack: *"Yeah, as everybody knows, we took a hit on the, somewhere on the left wing leading edge and the photo TV guys have completed I think, pretty much their work although I'm sure they are reviewing their stuff and they've given us an approximate size for the debris and approximate area for where it came from and approximately where it hit, so we are talking about doing some sort of parametric type of analysis and also we're talking about what you can do in the event we have some damage there."*

Ham: *"That comment, I was thinking that the flight rationale at the FRR from tank and orbiter from STS-112 was.... I'm not sure that the area is exactly the same where the foam came from but the carrier properties and density of the foam wouldn't do any damage. So we ought to pull that along with the 87 data where we had some damage, pull this data from 112 or whatever flight it was and make sure that...you know I hope that we had good flight rationale then."*

McCormack: *"Yeah, and we'll look at that, you mentioned 87, you know we saw some fairly significant damage in the area between RCC panels 8 and 9 and the main landing gear door on the bottom on STS-87 we did some analysis prior to STS-89 so uh..."*

Ham: *"And I'm really I don't think there is much we can do so it's not really a factor during the flight because there is not much we can do about it. But what I'm really interested in is making sure our flight rationale to go was good, and maybe this is foam from a different area and I'm not sure and it may not be co-related, but you can try to see what we have."*

McCormack: *"Okay."*

After the meeting, the rationale for continuing to fly after the STS-112 foam loss was sent to Ham for review. She then exchanged e-mails with her boss, Space Shuttle Program Manager Ron Dittemore:

-----Original Message-----
From: DITTEMORE, RONALD D. (JSC-MA) (NASA)
Sent: Wednesday, January 22, 2003 9:14 AM
To: HAM, LINDA J. (JSC-MA2) (NASA)
Subject: RE: ET Briefing - STS-112 Foam Loss

You remember the briefing! Jerry did it and had to go out and say that the hazard report had not changed and that the risk had not changed...But it is worth looking at again.

[continued on next page]

[continued from previous page]

-----Original Message-----
From: HAM, LINDA J. (JSC-MA2) (NASA)
Sent: Tuesday, January 21, 2003 11:14 AM
To: DITTEMORE, RONALD D. (JSC-MA) (NASA)
Subject: FW: ET Briefing - STS-112 Foam Loss

You probably can't open the attachment. But, the ET rationale for flight for the STS-112 loss of foam was lousy. Rationale states we haven't changed anything, we haven't experienced any 'safety of flight' damage in 112 flights, risk of loss of bi-pod ramp TPS is same as previous flghts...So ET is safe to fly with no added risk

Rationale was lousy then and still is....

-----Original Message-----
From: MCCORMACK, DONALD L. (DON) (JSC-MV6) (NASA)
Sent: Tuesday, January 21, 2003 9:45 AM
To: HAM, LINDA J. (JSC-MA2) (NASA)
Subject: FW: ET Briefing - STS-112 Foam Loss
Importance: High

FYI - it kinda says that it will probably be all right

[ORR=Operational Readiness Review, VAB=Vehicle Assembly Building, IFA=In-Flight Anomaly, TPS=Thermal Protection System, ET=External Tank]

Ham's focus on examining the rationale for continuing to fly after the foam problems with STS-87 and STS-112 indicates that her attention had already shifted from the threat the foam posed to STS-107 to the downstream implications of the foam strike. Ham was due to serve, along with Wayne Hale, as the launch integration manager for the next mission, STS-114. If the Shuttle Program's rationale to fly with foam loss was found to be flawed, STS-114, due to be launched in about a month, would have to be delayed per NASA rules that require serious problems to be resolved before the next flight. An STS-114 delay could in turn delay completion of the International Space Station's Node 2, which was a high-priority goal for NASA managers. (See Section 6.2 for a detailed description of schedule pressures.)

During this same Mission Management Team meeting, the Space Shuttle Integration Office's Lambert Austin reported that engineers were reviewing long-range tracking film and that the foam debris that appeared to hit the left wing leading edge may have come from the bipod area of the External Tank. Austin said that the Engineering Directorate would continue to run analyses and compare this foam loss to that of STS-112. Austin also said that after STS-107 landed, engineers were anxious to see the crew-filmed footage of External Tank separation that might show the bipod ramp and therefore could be checked for missing foam.

Missed Opportunity 2

Reviews of flight-deck footage confirm that on Flight Day One, Mission Specialist David Brown filmed parts of the External Tank separation with a Sony PD-100 Camcorder, and Payload Commander Mike Anderson photographed it with a Nikon F-5 camera with a 400-millimeter lens. Brown later downlinked 35 seconds of this video to the ground as part of his Flight Day One mission summary, but the bipod ramp area had rotated out of view, so no evidence of missing foam was seen when this footage was reviewed during the mission. However, after the Intercenter Photo Working Group caught the debris strike on January 17, ground personnel failed to ask Brown if he had additional footage of External Tank separation. Based on how crews are trained to film External Tank separation, the Board concludes Brown did in fact have more film than the 35 seconds he downlinked. Such footage may have confirmed that foam was missing from the bipod ramp area or could have identified other areas of missing foam. Austin's mention of the crew's filming of External Tank separation should have prompted someone at the meeting to ask Brown if he had more External Tank separation film, and if so, to downlink it immediately.

Flight Director Steve Stich discussed the debris strike with Phil Engelauf, a member of the Mission Operations Directorate, after Engelauf returned from the Mission Management Team meeting. As written in a timeline Stich composed after the accident, the conversation included the following.

> *"Phil said the Space Shuttle Program community is not concerned and that Orbiter Project is analyzing ascent debris...relayed that there had been no direction for MOD to ask DOD for any photography of possible damaged tiles"* [MOD=Mission Operations Directorate, or Mission Control, DOD=Department of Defense]

"No direction for DOD photography" seems to refer to either a previous discussion of photography with Mission managers or an expectation of future activity. Since the interagency agreement on imaging support stated that the Flight Dynamics Officer is responsible for initiating such a request, Engelauf's comments demonstrates that an informal chain of command, in which the Mission Operations Directorate figures prominently, was at work.

About an hour later, Calvin Schomburg, a Johnson Space Center engineer with close connections to Shuttle management, sent the following e-mail to other Johnson engineering managers.

-----Original Message-----
From: SCHOMBURG, CALVIN (JSC-EA) (NASA)
Sent: Tuesday, January 21, 2003 9:26 AM
To: SHACK, PAUL E. (JSC-EA42) (NASA); SERIALE-GRUSH, JOYCE M. (JSC-EA) (NASA); HAMILTON, DAVID A. (DAVE) (JSC-EA) (NASA)
Subject: FW: STS-107 Post-Launch Film Review - Day 1

FYI-TPS took a hit-should not be a problem-status by end of week.

[FYI=For Your Information, TPS=Thermal Protection System]

Shuttle Program managers regarded Schomburg as an expert on the Thermal Protection System. His message downplays the possibility that foam damaged the Thermal Protection System. However, the Board notes that Schomburg was not an expert on Reinforced Carbon-Carbon (RCC), which initial debris analysis indicated the foam may have struck. Because neither Schomburg nor Shuttle management rigorously differentiated between tiles and RCC panels, the bounds of Schomburg's expertise were never properly qualified or questioned.

Seven minutes later, Paul Shack, Manager of the Shuttle Engineering Office, Johnson Engineering Directorate, e-mailed to Rocha and other Johnson engineering managers information on how previous bipod ramp foam losses were handled.

-----Original Message-----
From: SHACK, PAUL E. (JSC-EA42) (NASA)
Sent: Tuesday, January 21, 2003 9:33 AM
To: ROCHA, ALAN R. (RODNEY) (JSC-ES2) (NASA); SERIALE-GRUSH, JOYCE M. (JSC-EA) (NASA)
Cc: KRAMER, JULIE A. (JSC-EA4) (NASA); MILLER, GLENN J. (JSC-EA) (NASA); RICKMAN, STEVEN L. (JSC-ES3) (NASA); MADDEN, CHRISTOPHER B. (CHRIS) (JSC-ES3) (NASA)
Subject: RE: STS-107 Debris Analysis Team Plans

This reminded me that at the STS-113 FRR the ET Project reported on foam loss from the Bipod Ramp during STS-112. The foam (estimated 4X5X12 inches) impacted the ET Attach Ring and dented an SRB electronics box cover.

Their charts stated "ET TPS foam loss over the life of the Shuttle program has never been a 'Safety of Flight' issue". They were severely wire brushed over this and Brian O'Conner (Associate Administra-

[continued on next page]

[continued from previous page]

tor for Safety) asked for a hazard assessment for loss of foam.

The suspected cause for foam loss is trapped air pockets which expand due to altitude and aerothermal heating.

[FRR=Flight Readiness Review, ET=External Tank, SRB=Solid Rocket Booster, TPS=Thermal Protection System]

Shack's message informed Rocha that during the STS-113 Flight Readiness Review, foam loss was not considered to be a safety-of-flight issue. The "wirebrushing" that the External Tank Project received for stating that foam loss has "never been a 'Safety of Flight' issue" refers to the wording used to justify continuing to fly. Officials at the Flight Readiness Review insisted on classifying the foam loss as an "accepted risk" rather than "not a safety-of-flight problem" to indicate that although the Shuttle would continue to fly, the threat posed by foam is not zero but rather a known and acceptable risk.

It is here that the decision to fly before resolving the foam problem at the STS-113 Flight Readiness Review influences decisions made during STS-107. Having at hand a previously accepted rationale – reached just one mission ago – that foam strikes are not a safety-of-flight issue provides a strong incentive for Mission managers and working engineers to use that same judgment for STS-107. If managers and engineers were to argue that foam strikes are a safety-of-flight issue, they would contradict an established consensus that was a product of the Shuttle Program's most rigorous review – a review in which many of them were active participants.

An entry in a Mission Evaluation Room console log included a 10:30 a.m. report that compared the STS-107 foam loss to previous foam losses and subsequent tile damage, which reinforced management acceptance about foam strikes by indicating that the foam strike appeared to be more of an "in-family" event.

> *"...STS-107 debris measured at 22" long +/- 10". On STS-112 the debris spray pattern was a lot smaller than that of STS-107. On STS-50 debris that was determined to be the Bipod ramp which measured 26" x 10" caused damage to the left wing...to 1 tile and 20% of the adjacent tile. Same event occurred on STS-7 (no data available)."*

MISSED OPPORTUNITY 3

The foam strike to STS-107 was mentioned by a speaker at an unrelated meeting of NASA Headquarters and National Imagery and Mapping Agency personnel, who then discussed a possible NASA request for Department of Defense imagery support. However, no action was taken.

IMAGERY REQUEST 2

Responding to concerns from his employees who were participating in the Debris Assessment Team, United Space Alliance manager Bob White called Lambert Austin on Flight Day Six to ask what it would take to get imagery of *Columbia* on orbit. They discussed the analytical debris damage work plan, as well as the belief of some integration team members that such imaging might be beneficial.

Austin subsequently telephoned the Department of Defense Manned Space Flight Support Office representative to ask about actions necessary to get imagery of *Columbia* on orbit. Austin emphasized that this was merely information gathering, not a request for action. This call indicates that Austin was unfamiliar with NASA/National Imagery and Mapping Agency imagery request procedures.

An e-mail that Lieutenant Colonel Timothy Lee sent to Don McCormack the following day shows that the Defense Department had begun to implement Austin's request.

-----Original Message-----
From: LEE, TIMOTHY F., LTCOL. (JSC-MT) (USAF)
Sent: Wednesday, January 22, 2003 9:01 AM
To: MCCORMACK, DONALD L. (DON) (JSC-MV6) (NASA)
Subject: NASA request for DOD

Don,

FYI: Lambert Austin called me yesterday requesting DOD photo support for STS-107. Specifically, he is asking us if we have a ground or satellite asset that can take a high resolution photo of the shuttle while on-orbit--to see if there is any FOD damage on the wing. We are working his request.

Tim

[DOD=Department of Defense, FOD=Foreign Object Debris]

At the same time, managers Ralph Roe, Lambert Austin, and Linda Ham referred to conversations with Calvin Schomburg, whom they referred to as a Thermal Protection System "expert." They indicated that Schomburg had advised that any tile damage should be considered a turn-around maintenance concern and not a safety-of-flight issue, and that imagery of *Columbia*'s left wing was not necessary. There was no discussion of potential RCC damage.

First Debris Assessment Team Meeting

On Flight Day Six, the Debris Assessment Team held its first formal meeting to finalize Orbiter damage estimates and their potential consequences. Some participants joined the proceedings via conference call.

IMAGERY REQUEST 3

After two hours of discussing the Crater results and the need to learn precisely where the debris had hit *Columbia*, the Debris Assessment Team assigned its NASA Co-Chair, Rodney Rocha, to pursue a request for imagery of the vehicle on-orbit. Each team member supported the idea to seek imagery from an outside source. Rather than working the request up the usual mission chain of command through the Mission Evaluation Room to the Mission Management Team to the Flight Dynamics Officer, the Debris Assessment Team agreed, largely due to a lack of participation by Mission Management Team and Mission Evaluation Room managers, that Rocha would pursue the request through his division, the Engineering Directorate at Johnson Space Center. Rocha sent the following e-mail to Paul Shack shortly after the meeting adjourned.

-----Original Message-----
From: ROCHA, ALAN R. (RODNEY) (JSC-ES2) (NASA)
Sent: Tuesday, January 21, 2003 4:41 PM
To: SHACK, PAUL E. (JSC-EA42) (NASA); HAMILTON, DAVID A. (DAVE) (JSC-EA) (NASA); MILLER, GLENN J. (JSC-EA) (NASA)
Cc: SERIALE-GRUSH, JOYCE M. (JSC-EA) (NASA); ROGERS, JOSEPH E. (JOE) (JSC-ES2) (NASA); GALBREATH, GREGORY F. (GREG) (JSC-ES2) (NASA)
Subject: STS-107 Wing Debris Impact, Request for Outside Photo-Imaging Help

Paul and Dave,
The meeting participants (Boeing, USA, NASA ES2 and ES3, KSC) all agreed we will always have big uncertainties in any transport/trajectory analyses and applicability/extrapolation of the old Arc-Jet test data until we get definitive, better, clearer photos of the wing and body underside. Without better images it will be very difficult to even bound the problem and initialize thermal, trajectory, and structural analyses. Their answers may have a wide spread ranging from acceptable to not-acceptable to horrible, and no way to reduce uncertainty. Thus, giving MOD options for entry will be very difficult.

[continued on next page]

[continued from previous page]

Can we petition (beg) for outside agency assistance? We are asking for Frank Benz with Ralph Roe or Ron Dittemore to ask for such. Some of the old timers here remember we got such help in the early 1980's when we had missing tile concerns.

Despite some nay-sayers, there are some options for the team to talk about: On-orbit thermal conditioning for the major structure (but is in contradiction with tire pressure temp. cold limits), limiting high cross-range de-orbit entries, constraining right or left had turns during the Heading Alignment Circle (only if there is struc. damage to the RCC panels to the extent it affects flight control.

Rodney Rocha
Structural Engineering Division (ES-SED)
- **ES Div. Chief Engineer (Space Shuttle DCE)**
- **Chair, Space Shuttle Loads & Dynamics Panel**

Mail Code ES2

[USA=United Space Alliance, NASA ES2, ES3=separate divisions of the Johnson Space Center Engineering Directorate, KSC=Kennedy Space Center, MOD=Missions Operations Directorate, or Mission Control]

Routing the request through the Engineering department led in part to it being viewed by Shuttle Program managers as a non-critical engineering desire rather than a critical operational need.

Flight Day Seven, Wednesday, January 22, 2003

Conversations and log entries on Flight Day Seven document how three requests for images (Bob Page to Wayne Hale, Bob White to Lambert Austin, and Rodney Rocha to Paul Shack) were ultimately dismissed by the Mission Management Team, and how the order to halt those requests was then interpreted by the Debris Assessment Team as a direct and final denial of their request for imagery.

MISSED OPPORTUNITY 4

On the morning of Flight Day Seven, Wayne Hale responded to the earlier Flight Day Two request from Bob Page and a call from Lambert Austin on Flight Day Five, during which Austin mentioned that "some analysts" from the Debris Assessment Team were interested in getting imagery. Hale called a Department of Defense representative at Kennedy Space Center (who was not the designated Department of Defense official for coordinating imagery requests) and asked that the military start the planning process for imaging *Columbia* on orbit.

Within an hour, the Defense Department representative at NASA contacted U.S. Strategic Command (USSTRATCOM) at Colorado's Cheyenne Mountain Air Force Station and asked what it would take to get imagery of *Columbia* on orbit. (This call was similar to Austin's call to the Department of Defense Manned Space Flight Support Office in that the caller characterized it as "information gathering" rather than a request for action.) A representative from the USSTRATCOM Plans Office initiated actions to identify ground-based and other imaging assets that could execute the request.

Hale's earlier call to the Defense Department representative at Kennedy Space Center was placed without authorization from Mission Management Team Chair Linda Ham. Also, the call was made to a Department of Defense Representative who was not the designated liaison for handling such requests. In order to initiate the imagery request through official channels, Hale also called Phil Engelauf at the Mission Operations Directorate, told him he had started Defense Department action, and asked if Engelauf could have the Flight Dynamics Officer at Johnson Space Center make an official request to the Cheyenne Mountain Operations Center. Engelauf started to comply with Hale's request.

After the Department of Defense representatives were called, Lambert Austin telephoned Linda Ham to inform her about the imagery requests that he and Hale had initiated. Austin also told Wayne Hale that he had asked Lieutenant Colonel Lee at the Department of Defense Manned Space Flight Support Office about what actions were necessary to get on-orbit imagery.

Missed Opportunities 5 and 6

Mike Card, a NASA Headquarters manager from the Safety and Mission Assurance Office, called Mark Erminger at the Johnson Space Center Safety and Mission Assurance for Shuttle Safety Program and Bryan O'Connor, Associate Administrator for Safety and Mission Assurance, to discuss a potential Department of Defense imaging request. Erminger said that he was told this was an "in-family" event. O'Connor stated he would defer to Shuttle management in handling such a request. Despite two safety officials being contacted, one of whom was NASA's highest-ranking safety official, safety personnel took no actions to obtain imagery.

The following is an 8:09 a.m. entry in the Mission Evaluation Room Console log.

> *"We received a visit from Mission Manager/Vanessa Ellerbe and FD Office/Phil Engelauf regarding two items: (1) the MMT's action item to the MER to determine the impacts to the vehicle's 150 lbs of additional weight...and (2) Mr. Engelauf wants to know who is requesting the Air Force to look at the vehicle."* [FD=Flight Director, MMT=Mission Management Team, MER=Mission Evaluation Room]

Cancellation of the Request for Imagery

At 8:30 a.m., the NASA Department of Defense liaison officer called USSTRATCOM and cancelled the request for imagery. The reason given for the cancellation was that NASA had identified its own in-house resources and no longer needed the military's help. The NASA request to the Department of Defense to prepare to image *Columbia* on-orbit was both made and rescinded within 90 minutes.

The Board has determined that the following sequence of events likely occurred within that 90-minute period. Linda Ham asked Lambert Austin if he knew who was requesting the imagery. After admitting his participation in helping to make the imagery request outside the official chain of command and without first gaining Ham's permission, Austin referred to his conversation with United Space Alliance Shuttle Integration manager Bob White on Flight Day Six, in which White had asked Austin, in response to White's Debris Assessment Team employee concerns, what it would take to get Orbiter imagery.

Even though Austin had already informed Ham of the request for imagery, Ham later called Mission Management Team members Ralph Roe, Manager of the Space Shuttle Vehicle Engineering Office, Loren Shriver, United Space Alliance Deputy Program Manager for Shuttle, and David Moyer, the on-duty Mission Evaluation Room manager, to determine the origin of the request and to confirm that there was a "requirement" for a request. Ham also asked Flight Director Phil Engelauf if he had a "requirement" for imagery of *Columbia's* left wing. These individuals all stated that they had not requested imagery, were not aware of any "official" requests for imagery, and could not identify a "requirement" for imagery. Linda Ham later told several individuals that nobody had a requirement for imagery.

What started as a request by the Intercenter Photo Working Group to seek outside help in obtaining images on Flight Day Two in anticipation of analysts' needs had become by Flight Day Six an actual engineering request by members of the Debris Assessment Team, made informally through Bob White to Lambert Austin, and formally in Rodney Rocha's e-mail to Paul Shack. These requests had then caused Lambert Austin and Wayne Hale to contact Department of Defense representatives. When Ham officially terminated the actions that the Department of Defense had begun, she effectively terminated both the Intercenter Photo Working Group request and the Debris Assessment Team request. While Ham has publicly stated she did not know of the Debris Assessment Team members' desire for imagery, she never asked them directly if the request was theirs, even though they were the team analyzing the foam strike.

Also on Flight Day Seven, Ham raised concerns that the extra time spent maneuvering *Columbia* to make the left wing visible for imaging would unduly impact the mission schedule; for example, science experiments would have to stop while the imagery was taken. According to personal notes obtained by the Board:

> *"Linda Ham said it was no longer being pursued since even if we saw something, we couldn't do anything about it. The Program didn't want to spend the resources."*

Shuttle managers, including Ham, also said they were looking for very small areas on the Orbiter and that past imagery resolution was not very good. The Board notes that no individuals in the STS-107 operational chain of command had the security clearance necessary to know about National imaging capabilities. Additionally, no evidence has been uncovered that anyone from NASA, United Space Alliance, or Boeing sought to determine the expected quality of images and the difficulty and costs of obtaining Department of Defense assistance. Therefore, members of the Mission Management Team were making critical decisions about imagery capabilities based on little or no knowledge.

The following is an entry in the Flight Director Handover Log.

> *"NASA Resident Office, Peterson AFB called and SOI at USSPACECOM was officially turned off. This went all the way up to 4 star General. Post flight we will write a memo to USSPACECOM telling them whom they should take SOI requests from."*[50] *[AFB=Air Force Base, SOI=Spacecraft Object Identification, USSPACECOM=U.S. Space Command]*

After canceling the Department of Defense imagery request, Linda Ham continued to explore whether foam strikes posed a safety of flight issue. She sent an e-mail to Lambert Austin and Ralph Roe.

-----Original Message---
From: HAM, LINDA J. (JSC-MA2) (NASA)
Sent: Wednesday, January 22, 2003 9:33 AM
To: AUSTIN, LAMBERT D. (JSC-MS) (NASA); ROE, RALPH R. (JSC-MV) (NASA)
Subject: ET Foam Loss

Can we say that for any ET foam lost, no 'safety of flight' damage can occur to the Orbiter because of the density?

[ET=External Tank]

Responses included the following.

-----Original Message-----
From: ROE, RALPH R. (JSC-MV) (NASA)
Sent: Wednesday, January 22, 2003 9:38 AM
To: SCHOMBURG, CALVIN (JSC-EA) (NASA)
Subject: FW: ET Foam Loss

Calvin,

I wouldn't think we could make such a generic statement but can we bound it some how by size or acreage?

[Acreage=larger areas of foam coverage]

Ron Dittermore e-mailed Linda Ham the following.

-----Original Message-----
From: DITTEMORE, RONALD D. (JSC-MA) (NASA)
Sent: Wednesday, January 22, 2003 10:15 AM
To: HAM, LINDA J. (JSC-MA2) (NASA)
Subject: RE: ET Briefing - STS-112 Foam Loss

Another thought, we need to make sure that the density of the ET foam cannot damage the tile to where it is an impact to the orbiter...Lambert and Ralph need to get some folks working with ET.

The following is an e-mail from Calvin Schomburg to Ralph Roe.

-----Original Message-----
From: SCHOMBURG, CALVIN (JSC-EA) (NASA)
Sent: Wednesday, January 22, 2003 10:53 AM
To: ROE, RALPH R. (JSC-MV) (NASA)
Subject: RE: ET Foam Loss

No-the amount of damage ET foam can cause to the TPS material-tiles is based on the amount of impact energy-the size of the piece and its velocity(from just after pad clear until about 120 seconds-after that it will not hit or it will not enough energy to cause any damage)-it is a pure kinetic problem-there is a size that can cause enough damage to a tile that enough of the material is lost that we could burn a hole through the skin and have a bad day-(loss of vehicle and crew -about 200-400 tile locations(out of the 23,000 on the lower surface)-the foam usually fails in small popcorn pieces-that is why it is vented-to make small hits-the two or three times we have been hit with a piece as large as the one this flight-we got a gouge about 8-10 inches long about 2 inches wide and 3/4 to an 1 inch deep across two or three tiles. That is what I expect this time-nothing worst. If that is all we get we have have no problem-will have to replace a couple of tiles but nothing else.

[ET=External Tank, TPS=Thermal Protection System]

The following is a response from Lambert Austin to Linda Ham.

-----Original Message-----
From: AUSTIN, LAMBERT D. (JSC-MS) (NASA)
Sent: Wednesday, January 22, 2003 3:22 PM
To: HAM, LINDA J. (JSC-MA2) (NASA)
Cc: WALLACE, RODNEY O. (ROD) (JSC-MS2) (NASA); NOAH, DONALD S. (DON) (JSC-MS) (NASA)
Subject: RE: ET Foam Loss

NO. I will cover some of the pertinent rationale....there could be more if I spent more time thinking about it. Recall this issue has been discussed from time to time since the inception of the basic "no debris" requirement in Vol. X and at each review the SSP has concluded that it is not possible to PRECLUDE a potential catastrophic event as a result of debris impact damage to the flight elements. As regards the Orbiter, both windows and tiles are areas of concern.

You can talk to Cal Schomberg and he will verify the many times we have covered this in SSP reviews. While there is much tolerance to window and tile damage, ET foam loss can result in impact damage that under subsequent entry environments can lead to loss of structural integrity of the Orbiter area impacted or a penetration in a critical function area that results in loss of that function. My recollection of the most critical Orbiter bottom acreage areas are the wing spar, main landing gear door seal and RCC panels...of course Cal can give you a much better rundown.

We can and have generated parametric impact zone characterizations for many areas of the Orbiter for a few of our more typical ET foam loss areas. Of course, the impact/damage significance is always a function of debris size and density, impact velocity, and impact angle--these latter 2 being a function of the flight time at which the ET foam becomes debris. For STS-107 specifically, we have generated

[continued on next page]

[continued from previous page]

> this info and provided it to Orbiter. Of course, even this is based on the ASSUMPTION that the location and size of the debris is the same as occurred on STS-112------this cannot be verified until we receive the on-board ET separation photo evidence post Orbiter landing. We are requesting that this be expedited. I have the STS-107 Orbiter impact map based on the assumptions noted herein being sent down to you. Rod is in a review with Orbiter on this info right now.
>
> [SSP=Space Shuttle Program, ET=External Tank]

The Board notes that these e-mail exchanges indicate that senior Mission Management Team managers, including the Shuttle Program Manager, Mission Management Team Chair, head of Space Shuttle Systems Integration, and a Shuttle tile expert, correctly identified the technical bounds of the foam strike problem and its potential seriousness. Mission managers understood that the relevant question was not whether foam posed a safety-of-flight issue – it did – but rather whether the observed foam strike contained sufficient kinetic energy to cause damage that could lead to a burn-through. Here, all the key managers were asking the right question and admitting the danger. They even identified RCC as a critical impact zone. Yet little follow-through occurred with either the request for imagery or the Debris Assessment Team analysis. (See Section 3.4 and 3.6 for details on the kinetics of foam strikes.)

A Mission Evaluation Room log entry at 10:37 a.m. records the decision not to seek imaging of *Columbia's* left wing.

> *"USA Program Manager/Loren Shriver, NASA Manager, Program Integration/Linda Ham, & NASA SSVEO/Ralph Roe have stated that there is no need for the Air Force to take a look at the vehicle."* [USA=United Space Alliance, SSVEO=Space Shuttle Vehicle Engineering Office]

At 11:22 a.m., Debris Assessment Team Co-Chair Pam Madera sent an e-mail to team members setting the agenda for the team's second formal meeting that afternoon that included:

> *"... Discussion on Need/Rationale for Mandatory Viewing of damage site (All)..."*

Earlier e-mail agenda wording did not include "Need/Rationale for Mandatory" wording as listed here, which indicates that Madera knew of management's decision to not seek images of *Columbia's* left wing and anticipated having to articulate a "mandatory" rationale to reverse that decision. In fact, a United Space Alliance manager had informed Madera that imagery would be sought only if the request was a "mandatory need." Twenty-three minutes later, an e-mail from Paul Shack to Rodney Rocha, who the day before had carried forward the Debris Assessment Team's request for imaging, stated the following.

> *"... FYI, According to the MER, Ralph Roe has told program that Orbiter is not requesting any outside imaging help ..."* [MER=Mission Evaluation Room]

Earlier that morning, Ralph Roe's deputy manager, Trish Petite, had separate conversations with Paul Shack and tile expert Calvin Schomburg. In those conversations, Petite noted that an analysis of potential damage was in progress, and they should wait to see what the analysis showed before asking for imagery. Schomburg, though aware of the Debris Assessment Team's request for imaging, told Shack and Petite that he believed on-orbit imaging of potentially damaged areas was not necessary.

As the morning wore on, Debris Assessment Team engineers, Shuttle Program management, and other NASA personnel exchanged e-mail. Most messages centered on technical matters to be discussed at the Debris Assessment Team's afternoon meeting, including debris density, computer-aided design models, and the highest angle of incidence to use for a particular material property. One e-mail from Rocha to his managers and other Johnson engineers at 11:19 a.m., included the following.

> *"... there are good scenarios (acceptable and minimal damage) to horrible ones, depending on the extent of the damage incurred by the wing and location. The most critical loca-*

tions seem to be the 1191 wing spar region, the main landing gear door seal, and the RCC panels. We do not know yet the exact extent or nature of the damage without being provided better images, and without such all the high powered analyses and assessments in work will retain significant uncertainties …"

Second Debris Assessment Team Meeting

Some but not all of the engineers attending the Debris Assessment Team's second meeting had learned that the Shuttle Program was not pursuing imaging of potentially damaged areas. What team members did not realize was the Shuttle Program's decision not to seek on-orbit imagery was not necessarily a direct and final response to their request. Rather, the "no" was partly in response to the Kennedy Space Center action initiated by United Space Alliance engineers and managers and finally by Wayne Hale.

Not knowing that this was the case, Debris Assessment Team members speculated as to why their request was rejected and whether their analysis was worth pursuing without new imagery. Discussion then moved on to whether the Debris Assessment Team had a "mandatory need" for Department of Defense imaging. Most team members, when asked by the Board what "mandatory need" meant, replied with a shrug of their shoulders. They believed the need for imagery was obvious: without better pictures, engineers would be unable to make reliable predictions of the depth and area of damage caused by a foam strike that was outside of the experience base. However, team members concluded that although their need was important, they could not cite a "mandatory" requirement for the request. *Analysts on the Debris Assessment Team were in the unenviable position of wanting images to more accurately assess damage while simultaneously needing to prove to Program managers, as a result of their assessment, that there was a need for images in the first place.*

After the meeting adjourned, Rocha read the 11:45 a.m. e-mail from Paul Shack, which said that the Orbiter Project was not requesting any outside imaging help. Rocha called Shack to ask if Shack's boss, Johnson Space Center engineering director Frank Benz, knew about the request. Rocha then sent several e-mails consisting of questions about the ongoing analyses and details on the Shuttle Program's cancellation of the imaging request. An e-mail that he did not send but instead printed out and shared with a colleague follows.

> *"In my humble technical opinion, this is the wrong (and bordering on irresponsible) answer from the SSP and Orbiter not to request additional imaging help from any outside source. I must emphasize (again) that severe enough damage (3 or 4 multiple tiles knocked out down to the densification layer) combined with the heating and resulting damage to the underlying structure at the most critical location (viz., MLG door/wheels/tires/hydraulics or the X1191 spar cap) could present potentially grave hazards. The engineering team will admit it might not achieve definitive high confidence answers without additional images, but, without action to request help to clarify the damage visually, we will guarantee it will not. Can we talk to Frank Benz before Friday's MMT? Remember the NASA safety posters everywhere around stating, 'If it's not safe, say so'? Yes, it's that serious."* [SSP=Space Shuttle Program, MLG=Main Landing Gear, MMT=Mission Management Team]

When asked why he did not send this e-mail, Rocha replied that he did not want to jump the chain of command. Having already raised the need to have the Orbiter imaged with Shack, he would defer to management's judgment on obtaining imagery.

Even after the imagery request had been cancelled by Program management, engineers in the Debris Assessment Team and Mission Control continued to analyze the foam strike. A structural engineer in the Mechanical, Maintenance, Arm and Crew Systems sent an e-mail to a flight dynamics engineer that stated:

> *"There is lots of speculation as to extent of the damage, and we could get a burn through into the wheel well upon entry."*

Less than an hour later, at 6:09 p.m., a Mission Evaluation Room Console log entry stated the following.

> *"MMACS is trying to view a Quicktime movie on the debris impact but doesn't have Quick-*

time software on his console. He needs either an avi, mpeg file or a vhs tape. He is asking us for help." [MMACS=Mechanical, Maintenance, Arm and Crew Systems]

The controller at the Mechanical, Maintenance, Arm and Crew Systems console would be among the first in Mission Control to see indications of burn-through during *Columbia*'s re-entry on the morning of February 1. This log entry also indicates that Mission Control personnel were aware of the strike.

Flight Day Eight, Thursday, January 23, 2003

The morning after Shuttle Program Management decided not to pursue on-orbit imagery, Rodney Rocha received a return call from Mission Operations Directorate representative Barbara Conte to discuss what kinds of imaging capabilities were available for STS-107.

MISSED OPPORTUNITY 7

Conte explained to Rocha that the Mission Operations Directorate at Johnson did have U.S. Air Force standard services for imaging the Shuttle during Solid Rocket Booster separation and External Tank separation. Conte explained that the Orbiter would probably have to fly over Hawaii to be imaged. The Board notes that this statement illustrates an unfamiliarity with National imaging assets. Hawaii is only one of many sites where relevant assets are based. Conte asked Rocha if he wanted her to pursue such a request through Missions Operations Directorate channels. Rocha said no, because he believed Program managers would still have to support such a request. Since they had already decided that imaging of potentially damaged areas was not necessary, Rocha thought it unlikely that the Debris Assessment Team could convince them otherwise without definitive data.

Later that day, Conte and another Mission Operations Directorate representative were attending an unrelated meeting with Leroy Cain, the STS-107 ascent/entry Flight Director. At that meeting, they conveyed Rocha's concern to Cain and offered to help with obtaining imaging. After checking with Phil Engelauf, Cain distributed the following e-mail.

-----Original Message-----
From: CAIN, LEROY E. (JSC-DA8) (NASA)
Sent: Thursday, January 23, 2003 12:07 PM
To: JONES, RICHARD S. (JSC-DM) (NASA); OLIVER, GREGORY T. (GREG) (JSC-DM4) (NASA); CONTE, BARBARA A. (JSC-DM) (NASA)
Cc: ENGELAUF, PHILIP L. (JSC-DA8) (NASA); AUSTIN, BRYAN P. (JSC-DA8) (NASA); BECK, KELLY B. (JSC-DA8) (NASA); HANLEY, JEFFREY M. (JEFF) (JSC-DA8) (NASA); STICH, J. S. (STEVE) (JSC-DA8) (NASA)
Subject: Help with debris hit

The SSP was asked directly if they had any interest/desire in requesting resources outside of NASA to view the Orbiter (ref. the wing leading edge debris concern).

They said, No.

After talking to Phil, I consider it to be a dead issue.

[SSP=Space Shuttle Program]

Also on Flight Day Eight, Debris Assessment Team engineers presented their final debris trajectory estimates to their NASA, United Space Alliance, and Boeing managers. These estimates formed the basis for predicting the Orbiter's damaged areas as well as the extent of damage, which in turn determined the ultimate threat to the Orbiter during re-entry.

Mission Control personnel thought they should tell Commander Rick Husband and Pilot William McCool about the debris strike, not because they thought that it was worthy of the crew's attention but because the crew might be asked about it in an upcoming media interview. Flight Director Steve Stitch sent the following e-mail to Husband and McCool and copied other Flight Directors.

-----Original Message-----
From: STICH, J. S. (STEVE) (JSC-DA8) (NASA)
Sent: Thursday, January 23, 2003 11:13 PM
To: CDR; PLT
Cc: BECK, KELLY B. (JSC-DA8) (NASA); ENGELAUF, PHILIP L. (JSC-DA8) (NASA); CAIN, LEROY E. (JSC-DA8) (NASA); HANLEY, JEFFREY M. (JEFF) (JSC-DA8) (NASA); AUSTIN, BRYAN P. (JSC-DA8) (NASA)
Subject: INFO: Possible PAO Event Question

Rick and Willie,

You guys are doing a fantastic job staying on the timeline and accomplishing great science. Keep up the good work and let us know if there is anything that we can do better from an MCC/POCC standpoint.

There is one item that I would like to make you aware of for the upcoming PAO event on Blue FD 10 and for future PAO events later in the mission. This item is not even worth mentioning other than wanting to make sure that you are not surprised by it in a question from a reporter.

During ascent at approximately 80 seconds, photo analysis shows that some debris from the area of the -Y ET Bipod Attach Point came loose and subsequently impacted the orbiter left wing, in the area of transition from Chine to Main Wing, creating a shower of smaller particles. The impact appears to be totally on the lower surface and no particles are seen to traverse over the upper surface of the wing. Experts have reviewed the high speed photography and there is no concern for RCC or tile damage. We have seen this same phenomenon on several other flights and there is absolutely no concern for entry.

That is all for now. It's a pleasure working with you every day.

[MCC/POCC=Mission Control Center/Payload Operations Control Center, PAO=Public Affairs Officer, FD 10=Flight Day Ten, -Y=left, ET=External Tank]

This e-mail was followed by another to the crew with an attachment of the video showing the debris impact. Husband acknowledged receipt of these messages.

Later, a NASA liaison to USSTRATCOM sent an e-mail thanking personnel for the prompt response to the imagery request. The e-mail asked that they help NASA observe "official channels" for this type of support in the future. Excerpts from this message follow.

> *"Let me assure you that, as of yesterday afternoon, the Shuttle was in excellent shape, mission objectives were being performed, and that there were no major debris system problems identified. The request that you received was based on a piece of debris, most likely ice or insulation from the ET, that came off shortly after launch and hit the underside of the vehicle. Even though this is not a common occurrence it is something that has happened before and is not considered to be a major problem. The one problem that this has identified is the need for some additional coordination within NASA to assure that when a request is made it is done through the official channels. The NASA/ USSTRAT (USSPACE) MOA identifies the need for this type of support and that it will be provided by USSTRAT. Procedures have been long established that identifies the Flight Dynamics Officer (for the Shuttle) and the Trajectory Operations Officer (for the International Space Station) as the POCs to work these issues with the personnel in Cheyenne Mountain. One of the primary purposes for this chain is to make sure that requests like this one does not slip through the system and spin the community up about potential problems that have not been fully vetted through the proper channels. Two things that you can help us with is to make sure that future requests of this sort are confirmed through the proper channels. For the Shuttle it is via CMOC to the Flight Dynamics Officer. For the International Space Station it is via CMOC to the Trajectory Operations Officer. The second request is that no resources are spent unless the request has been confirmed. These requests are not meant to diminish the responsibilities of the DDMS office or to change any previous agreements but to eliminate the confusion that can be caused by a lack of proper coordination."* [ET=External Tank,

MOA=Memorandum of Agreement, POC=Point of Contact, CMOC=Cheyenne Mountain Operations Center, DDMS=Department of Defense Manned Space Flight Support Office]

Third Debris Assessment Team Meeting

The Debris Assessment Team met for the third time Thursday afternoon to review updated impact analyses. Engineers noted that there were no alternate re-entry trajectories that the Orbiter could fly to substantially reduce heating in the general area of the foam strike. Engineers also presented final debris trajectory data that included three debris size estimates to cover the continuing uncertainty about the size of the debris. Team members were told that imaging would not be forthcoming. In the face of this denial, the team discussed whether to include a presentation slide supporting their desire for images of the potentially damaged area. Many still felt it was a valid request and wanted their concerns aired at the upcoming Mission Evaluation Room brief and then at the Mission Management Team level. Eventually, the idea of including a presentation slide about the imaging request was dropped.

Just prior to attending the third assessment meeting, tile expert Calvin Schomburg and Rodney Rocha met to discuss foam impacts from other missions. Schomburg implied that the STS-107 foam impact was in the Orbiter's experience base and represented only a maintenance issue. Rocha disagreed and argued about the potential for burn-through on re-entry. Calvin Schomburg stated a belief that if there was severe damage to the tiles, *"nothing could be done."* (See Section 6.4.) Both then joined the meeting already in progress.

According to Boeing analysts who were members of the Debris Assessment Team, Schomburg called to ask about their rationale for pursuing imagery. The Boeing analysts told him that something the size of a large cooler had hit the Orbiter at 500 miles per hour. Pressed for additional reasons and not fully understanding why their original justification was insufficient, the analysts said that at least they would know what happened if something were to go terribly wrong. The Boeing analysts next asked why they were working so hard analyzing potential damage areas if Shuttle Program management believed the damage was minor and that no safety-of-flight issues existed. Schomburg replied that the analysts were new and would learn from this exercise.

Flight Day Nine, Friday, January 24, 2003

At 7:00 a.m., Boeing and United Space Alliance contract personnel presented the Debris Assessment Team's findings to Don McCormack, the Mission Evaluation Room manager. In yet another signal that working engineers and mission personnel shared a high level of concern for *Columbia*'s condition, so many engineers crowded the briefing room that it was standing room only, with people lining the hallway.

The presentation included viewgraphs that discussed the team's analytical methodology and five scenarios for debris damage, each based on different estimates of debris size and impact point. A sixth scenario had not yet been completed, but early indications suggested that it would not differ significantly from the other five. Each case was presented with a general overview of transport mechanics, results from the Crater modeling, aerothermal considerations, and predicted thermal and structural effects for *Columbia*'s re-entry. The briefing focused primarily on potential damage to the tiles, not the RCC panels. (An analysis of how the poor construction of these viewgraphs effectively minimized key assumptions and uncertainties is presented in Chapter 7.)

While the team members were confident that they had conducted the analysis properly – within the limitations of the information they had – they stressed that many uncertainties remained. First, there was great uncertainty about where the debris had struck. Second, Crater, the analytical tool they used to predict the penetration depth of debris impact, was being used on a piece of debris that was 400 times larger than the standard in Boeing's database. (At the time, the team believed that the debris was 640 times larger.) Engineers ultimately concluded that their analysis, limited as it was, did not show that a safety-of-flight issue existed. Engineers who attended this briefing indicated a belief that management focused on the answer – that analysis proved there was no safety-of-flight issue – rather than concerns about the large uncertainties that may have undermined the analysis that provided that answer.

At the Mission Management Team's 8:00 a.m. meeting, Mission Evaluation Room manager Don McCormack verbally summarized the Debris Assessment Team's 7:00 a.m. brief. It was the third topic discussed. Unlike the earlier briefing, McCormack's presentation did not include the Debris Assessment Team's presentation charts. The Board notes that no supporting analysis or examination of minority engineering views was asked for or offered, that neither Mission Evaluation Room nor Mission Management Team members requested a technical paper of the Debris Assessment Team analysis, and that no technical questions were asked.

January 24, 2003, Mission Management Team Meeting Transcript

The following is a transcript of McCormack's verbal briefing to the Mission Management Team, which Linda Ham Chaired. Early in the meeting, Phil Engelauf, Chief of the Flight Director's office, reported that he had made clear in an e-mail to *Columbia's* crew that there were "no concerns" that the debris strike had caused serious damage. The Board notes that this conclusion about whether the debris strike posed a safety-of-flight issue was presented to Mission Management Team members before they discussed the debris strike damage assessment.

Engelauf: "*I will say that crew did send down a note last night asking if anybody is talking about extension days or going to go with that and we sent up to the crew about a 15 second video clip of the strike just so they are armed if they get any questions at the press conferences or that sort of thing, but we made it very clear to them no, no concerns.*"

Linda Ham: "*When is the press conference? Is it today?*"

Engelauf: "*It's later today.*"

Ham: "*They may get asked because the press is aware of it.*"

Engelauf: "*The press is aware of it I know folks have asked me because the press corps at the cape have been asking...wanted to make sure they were properly...*"

Ham: "*Okay, back on the temperature...*"

The meeting went on for another 25 minutes. Other mission-related subjects were discussed before team members returned to the debris strike.

Ham: "*Go ahead, Don.*"

Don McCormack: "*Okay. And also we've received the data from the systems integration guys of the potential ranges of sizes and impact angles and where it might have hit. And the guys have gone off and done an analysis, they use a tool they refer to as Crater which is their official evaluation tool to determine the potential size of the damage. So they went off and done all that work and they've done thermal analysis to the areas where there may be damaged tiles. The analysis is not complete. There is one case yet that they wish to run, but kind of just jumping to the conclusion of all that, they do show that, obviously, a potential for significant tile damage here, but thermal analysis does not indicate that there is potential for a burn-through. I mean there could be localized heating damage. There is... obviously there is a lot of uncertainty in all this in terms of the size of the debris and where it hit and the angle of incidence.*"

Ham: "*No burn through, means no catastrophic damage and the localized heating damage would mean a tile replacement?*"

McCormack: "*Right, it would mean possible impacts to turnaround repairs and that sort of thing, but we do not see any kind of safety of flight issue here yet in anything that we've looked at.*"

Ham: "*And no safety of flight, no issue for this mission, nothing that we're going to do different, there may be a turnaround.*"

McCormack: "*Right, it could potentially hit the RCC and we don't indicate any other possible coating damage or something, we don't see any issue if it hit the RCC. Although we could have some significant tile damage if we don't see a safety-of-flight issue.*"

Ham: "*What do you mean by that?*"

McCormack: "*Well it could be down through the ... we could lose an entire tile and then the ramp into and out of that, I mean it could be a significant area of tile damage down to the SIP perhaps, so it could be a significant piece missing, but...*" [SIP refers to the denser lower layers of tile to which the debris may have penetrated.]

Ham.: "*It would be a turnaround issue only?*"

McCormack: "*Right.*"

(Unintelligible speaker)

At this point, tile expert Calvin Schomburg states his belief that no safety-of-flight issue exists. However, some participants listening via teleconference to the meeting are unable to hear his comments.

Ham: "*Okay. Same thing you told me about the other day in my office. We've seen pieces of this size before haven't we?*"

Unknown speaker. "*Hey Linda, we're missing part of that conversation.*"

Ham: "*Right.*"

Unknown speaker: "*Linda, we can't hear the speaker.*"

Ham: "*He was just reiterating with Calvin that he doesn't believe that there is any burn-through so no safety of flight kind of issue, it's more of a turnaround issue similar to what we've had on other flights. That's it? Alright, any questions on that?*"

The Board notes that when the official minutes of the January 24 Mission Management Team were produced and distributed, there was no mention of the debris strike. These minutes were approved and signed by Frank Moreno, STS-107 Lead Payload Integration Manager, and Linda Ham. For anyone not present at the January 24 Mission Management Team who was relying on the minutes to update them on key issues, they would have read nothing about the debris-strike discussions between Don McCormack and Linda Ham.

A subsequent 8:59 a.m. Mission Evaluation Room console log entry follows.

> "*MMT Summary...McCormack also summarized the debris assessment. Bottom line is that there appears to be no safety of flight issue, but good chance of turnaround impact to repair tile damage.*" [MMT=Mission Management Team]

Flight Day 10 through 16, Saturday through Friday, January 25 through 31, 2003

Although "no safety-of-flight issue" had officially been noted in the Mission Evaluation Room log, the Debris Assessment Team was still working on parts of its analysis of potential damage to the wing and main landing gear door. On Sunday, January 26, Rodney Rocha spoke with a Boeing thermal analyst and a Boeing stress analyst by telephone to express his concern about the Debris Assessment Team's overall analysis, as well as the remaining work on the main landing gear door analysis. After the Boeing engineers stated their confidence with their analyses, Rocha became more comfortable with the damage assessment and sent the following e-mail to his management.

-----Original Message-----
From: ROCHA, ALAN R. (RODNEY) (JSC-ES2) (NASA)
Sent: Sunday, January 26, 2003 7:45 PM
To: SHACK, PAUL E. (JSC-EA42) (NASA); MCCORMACK, DONALD L. (DON) (JSC-MV6) (NASA); OUELLETTE, FRED A. (JSC-MV6) (NASA)
Cc: ROGERS, JOSEPH E. (JOE) (JSC-ES2) (NASA); GALBREATH, GREGORY F. (GREG) (JSC-ES2) (NASA); JACOBS, JEREMY B. (JSC-ES4) (NASA); SERIALE-GRUSH, JOYCE M. (JSC-EA) (NASA); KRAMER, JULIE A. (JSC-EA4) (NASA); CURRY, DONALD M. (JSC-ES3) (NASA); KOWAL, T. J. (JOHN) (JSC-ES3) (NASA); RICKMAN, STEVEN L. (JSC-ES3) (NASA); SCHOMBURG, CALVIN (JSC-EA) (NASA); CAMPBELL, CARLISLE C., JR (JSC-ES2) (NASA)
Subject: STS-107 Wing Debris Impact on Ascent: Final analysis case completed

As you recall from Friday's briefing to the MER, there remained open work to assess analytically predicted impact damage to the wing underside in the region of the main landing gear door. This area was considered a low probability hit area by the image analysis teams, but they admitted a debris strike here could not be ruled out.

As with the other analyses performed and reported on Friday, this assessment by the Boeing multi-technical discipline engineering teams also employed the system integration's dispersed trajectories followed by serial results from the *Crater* damage prediction tool, thermal analysis, and stress analysis. It was reviewed and accepted by the ES-DCE (R. Rocha) by Sunday morning, Jan. 26. The case is defined by a large area gouge about 7 inch wide and about 30 inch long with sloped sides like a crater, and reaching down to the densified layer of the TPS.

SUMMARY: Though this case predicted some higher temperatures at the outer layer of the honeycomb aluminum face sheet and subsequent debonding of the sheet, there is no predicted burn-through of the door, no breeching of the thermal and gas seals, nor is there door structural deformation or thermal warpage to open the seal to hot plasma intrusion. Though degradation of the TPS and door structure is likely (if the impact occurred here), there is no safety of flight (entry, descent, landing) issue.

Note to Don M. and Fred O.: On Friday I believe the MER was thoroughly briefed and it was clear that open work remained (viz., the case summarized above), the message of open work was not clearly given, in my opinion, to Linda Ham at the MMT. I believe we left her the impression that engineering assessments and cases were all finished and we could state with finality no safety of flight issues or questions remaining. This very serious case could not be ruled out and it was a very good thing we carried it through to a finish.

Rodney Rocha (ES2)
- Division Shuttle Chief Engineer (DCE), ES-Structural Engineering Division
- Chair, Space Shuttle Loads & Dynamics Panel

[MER=Mission Evaluation Room, ES-DCE=Structural Engineering-Division Shuttle Chief Engineer, TPS=Thermal Protection System]

In response to this e-mail, Don McCormack told Rocha that he would make sure to correct Linda Ham's possible misconception that the Debris Assessment Team's analysis was finished as of the briefing to the Mission Management Team. McCormack informed Ham at the next Mission Management Team meeting on January 27, that the damage assessment had in fact been ongoing and that their final conclusion was that no safety-of-flight issue existed. The debris strike, in the official estimation of the Debris Assessment Team, amounted to only a post-landing turn-around maintenance issue.

On Monday morning, January 27, Doug Drewry, a structural engineering manager from Johnson Space Center, summoned several Johnson engineers and Rocha to his office and asked them if they all agreed with the completed analyses and with the conclusion that no safety-of-flight issues existed. Although all participants agreed with that conclusion, they also knew that the Debris Assessment Team members and most structural engineers at Johnson still wanted images of *Columbia*'s left wing but had given up trying to make that desire fit the "mandatory" requirement that Shuttle management had set.

Langley Research Center

Although the Debris Analysis Team had completed its analysis and rendered a "no safety-of-flight" verdict, concern persisted among engineers elsewhere at NASA as they learned about the debris strike and potential damage. On Monday, January 27, Carlisle Campbell, the design engineer responsible for landing gear/tires/brakes at Johnson Space Center forwarded Rodney Rocha's January 26, e-mail to Bob Daugherty, an engineer at Langley Research Center who specialized in landing gear design. Engineers at Langley and Ames Research Center and Johnson Space Center did not entertain the possibility of *Columbia* breaking up during re-entry, but rather focused on the idea that landing might not be safe, and that the crew might need to "ditch" the vehicle (crash land in water) or be prepared to land with damaged landing gear.

Campbell initially contacted Daugherty to ask his opinion of the arguments used to declare the debris strike "not a safety-of-flight issue." Campbell commented that someone had brought up worst-case scenarios in which a breach in the main landing gear door causes two tires to go flat. To help Daugherty understand the problem, Campbell forwarded him e-mails, briefing slides, and film clips from the debris damage analysis.

Both engineers felt that the potential ramifications of landing with two flat tires had not been sufficiently explored. They discussed using Shuttle simulator facilities at Ames Research Center to simulate a landing with two flat tires, but initially ruled it out because there was no formal request from the Mission Management Team to work the problem. Because astronauts were training in the Ames simulation facility, the two engineers looked into conducting the simulations after hours. Daugherty contacted his management on Tuesday, January 28, to update them on the plan for after-hours simulations. He reviewed previous data runs, current simulation results, and prepared scenarios that could result from main landing gear problems.

The simulated landings with two flat tires that Daugherty eventually conducted indicated that it was a survivable but very serious malfunction. Of the various scenarios he prepared, Daugherty shared the most unfavorable only with his management and selected Johnson Space Center engineers. In contrast, his favorable simulation results were forwarded to a wider Johnson audience for review, including Rodney Rocha and other Debris Assessment Team members. The Board is disappointed that Daugherty's favorable scenarios received a wider distribution than his discovery of a potentially serious malfunction, and also does not approve of the reticence that he and his managers displayed in not notifying the Mission Management Team of their concerns or his assumption that they could not displace astronauts who were training in the Ames simulator.

At 4:36 p.m. on Monday, January 27, Daugherty sent the following to Campbell.

-----Original Message-----
From: Robert H. Daugherty
Sent: Monday, January 27, 2003 3:35 PM
To: CAMPBELL, CARLISLE C., JR (JSC-ES2) (NASA)
Subject: Video you sent

WOW!!!
I bet there are a few pucker strings pulled tight around there!
Thinking about a belly landing versus bailout...... (I would say that if there is a question about main gear well burn thru that its crazy to even hit the deploy gear button...the reason being that you might have failed the wheels since they are aluminum..they will fail before the tire heating/pressure makes them fail..and you will send debris all over the wheel well making it a possibility that the gear would not even deploy due to ancillary damage...300 feet is the wrong altitude to find out you have one gear down and the other not down...you're dead in that case)
Think about the pitch-down moment for a belly landing when hitting not the main gear but the trailing edge of the wing or body flap when landing gear up...even if you come in fast and at slightly less pitch attitude...the nose slapdown with that pitching moment arm seems to me to be pretty scary...so much so that I would bail out before I would let a loved one land like that.
My two cents.
See ya,
Bob

The following reply from Campbell to Daugherty was sent at 4:49 p.m.

-----Original Message-----
From: "CAMPBELL, CARLISLE C., JR (JSC-ES2) (NASA)"
To: "'Bob Daugherty'"
Subject: FW: Video you sent
Date: Mon, 27 Jan 2003 15:59:53 -0600
X-Mailer: ßInternet Mail Service (5.5.2653.19)

Thanks. That's why they need to get all the facts in early on--such as look at impact damage from the spy telescope. Even then, we may not know the real effect of the damage.

The LaRC ditching model tests 20 some years ago showed that the Orbiter was the best ditching shape that they had ever tested, of many. But, our structures people have said that if we ditch we would blow such big holes in the lower panels that the orbiter might break up. Anyway, they refuse to even consider water ditching any more--I still have the test results[Bailout seems best.

[LaRC=Langley Research Center]

On the next day, Tuesday, Daugherty sent the following to Campbell.

-----Original Message-----
From: Robert H. Daugherty
Sent: Tuesday, January 28, 2003 12:39 PM
To: CAMPBELL, CARLISLE C., JR (JSC-ES2) (NASA)
Subject: Tile Damage

Any more activity today on the tile damage or are people just relegated to crossing their fingers and hoping for the best?
See ya,
Bob

Campbell's reply:

-----Original Message-----
From: "CAMPBELL, CARLISLE C., JR (JSC-ES2) (NASA)"
To: "'Robert H. Daugherty'"
Subject: RE: Tile Damage
Date: Tue, 28 Jan 2003 13:29:58 -0600
X-Mailer: Internet Mail Service (5.5.2653.19)

I have not heard anything new. I'll let you know if I do.

CCC

Carlisle Campbell sent the following e-mail to Johnson Space Center engineering managers on January 31.

> *"In order to alleviate concerns regarding the worst case scenario which could potentially be caused by the debris impact under the Orbiter's left wing during launch, EG conducted some landing simulations on the Ames Vertical Motion Simulator which tested the ability of the crew and vehicle to survive a condition where two main gear tires are deflated before landing. The results, although limited, showed that this condition is controllable, including the nose slap down rates. These results may give MOD a different decision path should this scenario become a reality. Previous opinions were that bailout was the only answer."*
> **[EG=Aeroscience and Flight Mechanics Division, MOD=Mission Operations Directorate]**

In the Mission Evaluation Room, a safety representative from Science Applications International Corporation, NASA's contract safety company, made a log entry at the Safety and Quality Assurance console on January 28, at 12:15 p.m. It was only the second mention of the debris strike in the safety console log during the mission (the first was also minor).

> *"[MCC SAIC] called asking if any SR&QA people were involved in the decision to say that the ascent debris hit (left wing) is safe. [SAIC engineer] has indeed been involved in the analysis and stated that he concurs with the analysis. Details about the debris hit are found in the Flight Day 12 MER Manager and our Daily Report."* [MCC=Mission Control Center, SAIC=Science Applications International Corporation, SR&QA=Safety, Reliability, and Quality Assurance, MER=Mission Evaluation Room]

MISSED OPPORTUNITY 8

According to a Memorandum for the Record written by William Readdy, Associate Administrator for Space Flight, Readdy and Michael Card, from NASA's Safety and Mission Assurance Office, discussed an offer of Department of Defense imagery support for *Columbia*. This January 29, conversation ended with Readdy telling Card that NASA would accept the offer but because the Mission Management Team had concluded that this was not a safety-of-flight issue, the imagery should be gathered only on a low priority "not-to-interfere" basis. Ultimately, no imagery was taken.

The Board notes that at the January 31, Mission Management Team meeting, there was only a minor mention of the debris strike. Other issues discussed included onboard crew consumables, the status of the leaking water separator, an intercom anomaly, SPACEHAB water flow rates, an update of the status of onboard experiments, end-of-mission weight concerns, landing day weather forecasts, and landing opportunities. The only mention of the debris strike was a brief comment by Bob Page, representing Kennedy Space Center's Launch Integration Office, who stated that the crew's hand-held cameras and External Tank films would be expedited to Marshall Space Flight Center via the Shuttle Training Aircraft for post-flight foam/debris imagery analysis, per Linda Ham's request.

Summary: Mission Management Decision Making

Discovery and Initial Analysis of Debris Strike

In the course of examining film and video images of *Columbia's* ascent, the Intercenter Photo Working Group identified, on the day after launch, a large debris strike to the leading edge of *Columbia*'s left wing. Alarmed at seeing so severe a hit so late in ascent, and at not having a clear view of damage the strike might have caused, Intercenter Photo Working Group members alerted senior Program managers by phone and sent a digitized clip of the strike to hundreds of NASA personnel via e-mail. These actions initiated a contingency plan that brought together an interdisciplinary group of experts from NASA, Boeing, and the United Space Alliance to analyze the strike. So concerned were Intercenter Photo Working Group personnel that on the day they discovered the debris strike, they tapped their Chair, Bob Page, to see through a request to image the left wing with Department of Defense assets in anticipation of analysts needing these images to better determine potential damage. By the Board's count, this would be the first of three requests to secure imagery of *Columbia* on-orbit during the 16-day mission.

IMAGERY REQUESTS

1. Flight Day 2. Bob Page, Chair, Intercenter Photo Working Group to Wayne Hale, Shuttle Program Manager for Launch Integration at Kennedy Space Center (in person).
2. Flight Day 6. Bob White, United Space Alliance manager, to Lambert Austin, head of the Space Shuttle Systems Integration at Johnson Space Center (by phone).
3. Flight Day 6. Rodney Rocha, Co-Chair of Debris Assessment Team to Paul Shack, Manager, Shuttle Engineering Office (by e-mail).

Missed Opportunities

1. Flight Day 4. Rodney Rocha inquires if crew has been asked to inspect for damage. No response.
2. Flight Day 6. Mission Control fails to ask crew member David Brown to downlink video he took of External Tank separation, which may have revealed missing bipod foam.
3. Flight Day 6. NASA and National Imagery and Mapping Agency personnel discuss possible request for imagery. No action taken.
4. Flight Day 7. Wayne Hale phones Department of Defense representative, who begins identifying imaging assets, only to be stopped per Linda Ham's orders.
5. Flight Day 7. Mike Card, a NASA Headquarters manager from the Safety and Mission Assurance Office, discusses imagery request with Mark Erminger, Johnson Space Center Safety and Mission Assurance. No action taken.
6. Flight Day 7. Mike Card discusses imagery request with Bryan O'Connor, Associate Administrator for Safety and Mission Assurance. No action taken.
7. Flight Day 8. Barbara Conte, after discussing imagery request with Rodney Rocha, calls LeRoy Cain, the STS-107 ascent/entry Flight Director. Cain checks with Phil Engelauf, and then delivers a "no" answer.
8. Flight Day 14. Michael Card, from NASA's Safety and Mission Assurance Office, discusses the imaging request with William Readdy, Associate Administrator for Space Flight. Readdy directs that imagery should only be gathered on a "not-to-interfere" basis. None was forthcoming.

Upon learning of the debris strike on Flight Day Two, the responsible system area manager from United Space Alliance and her NASA counterpart formed a team to analyze the debris strike in accordance with mission rules requiring the careful examination of any "out-of-family" event. Using film from the Intercenter Photo Working Group, Boeing systems integration analysts prepared a preliminary analysis that afternoon. (Initial estimates of debris size and speed, origin of debris, and point of impact would later prove remarkably accurate.)

As Flight Day Three and Four unfolded over the Martin Luther King Jr. holiday weekend, engineers began their analysis. One Boeing analyst used Crater, a mathematical prediction tool, to assess possible damage to the Thermal Protection System. Analysis predicted tile damage deeper than the actual tile depth, and penetration of the RCC coating at impact angles above 15 degrees. This suggested the potential for a burn-through during re-entry. Debris Assessment Team members judged that the actual damage would not be as severe as predicted because of the inherent conservatism in the Crater model and because, in the case of tile, Crater does not take into account the tile's stronger and more impact-resistant "densified" layer, and in the case of RCC, the lower density of foam would preclude penetration at impact angles under 21 degrees.

On Flight Day Five, impact assessment results for tile and RCC were presented at an informal meeting of the Debris Assessment Team, which was operating without direct Shuttle Program or Mission Management leadership. Mission Control's engineering support, the Mission Evaluation Room, provided no direction for team activities other than to request the team's results by January 24. As the problem was being worked, Shuttle managers did not formally direct the actions of or consult with Debris Assessment Team leaders about the team's assumptions, uncertainties, progress, or interim results, an unusual circumstance given that NASA managers are normally engaged in analyzing what they view as problems. At this meeting, participants agreed that an image of the area of the wing in question was essential to refine their analysis and reduce the uncertainties in their damage assessment.

Each member supported the idea to seek imagery from an outside source. Due in part to a lack of guidance from the Mission Management Team or Mission Evaluation Room managers, the Debris Assessment Team chose an unconventional route for its request. Rather than working the request up the normal chain of command – through the Mission Evaluation Room to the Mission Management Team for action to Mission Control – team members nominated Rodney Rocha, the team's Co-Chair, to pursue the request through the Engineering Directorate at Johnson Space Center. As a result, even after the accident the Debris Assessment Team's request was viewed by Shuttle Program managers as a non-critical engineering desire rather than a critical operational need.

When the team learned that the Mission Management Team was not pursuing on-orbit imaging, members were concerned. What Debris Assessment Team members did not realize was the negative response from the Program was not necessarily a direct and final response to their official request. Rather, the "no" was in part a response to requests for imagery initiated by the Intercenter Photo Working Group at Kennedy on Flight Day 2 in anticipation of analysts' needs that had become by Flight Day 6 an actual engineering request by the Debris Assessment Team, made informally through Bob White to Lambert Austin, and formally through Rodney Rocha's e-mail to Paul Shack. Even after learning that the Shuttle Program was not going to provide the team with imagery, some members sought information on how to obtain it anyway.

Debris Assessment Team members believed that imaging of potentially damaged areas was necessary even after the January 24, Mission Management Team meeting, where they had reported their results. Why they did not directly approach Shuttle Program managers and share their concern and uncertainty, and why Shuttle Program managers claimed to be isolated from engineers, are points that the Board labored to understand. Several reasons for this communications failure relate to NASA's internal culture and the climate established by Shuttle Program management, which are discussed in more detail in Chapters 7 and 8.

A Flawed Analysis

An inexperienced team, using a mathematical tool that was not designed to assess an impact of this estimated size, performed the analysis of the potential effect of the debris impact. Crater was designed for "in-family" impact events and was intended for day-of-launch analysis of debris impacts. It was not intended for large projectiles like those observed on STS-107. Crater initially predicted possible damage, but the Debris Assessment Team assumed, without theoretical or experimental validation, that because Crater is a conservative tool – that is, it predicts more damage than will actually occur – the debris would stop at the tile's densified layer, even though their experience did not involve debris strikes as large as STS-107's. Crater-like equations were also used as part of the analysis to assess potential impact damage to the wing leading edge RCC. Again, the tool was used for something other than that for which it was designed; again, it predicted possible penetration; and again, the Debris Assessment Team used engineering arguments and their experience to discount the results.

As a result of a transition of responsibility for Crater analysis from the Boeing Huntington Beach facility to the Houston-based Boeing office, the team that conducted the Crater analyses had been formed fairly recently, and therefore could be considered less experienced when compared with the more senior Huntington Beach analysts. In fact, STS-107 was the first mission for which they were solely responsible for providing analysis with the Crater tool. Though post-accident interviews suggested that the training for the Houston Boeing analysts was of high quality and adequate in substance and duration, communications and theoretical understandings of the Crater model among the Houston-based team members had not yet developed to the standard of a more senior team. Due in part to contractual arrangements related to the transition, the Houston-based team did not take full advantage of the Huntington Beach engineers' experience.

At the January 24, Mission Management Team meeting at which the "no safety-of-flight" conclusion was presented, there was little engineering discussion about the assumptions made, and how the results would differ if other assumptions were used.

Engineering solutions presented to management should have included a quantifiable range of uncertainty and risk analysis. Those types of tools were readily available, routinely used, and would have helped management understand the risk involved in the decision. Management, in turn, should have demanded such information. The very absence of a clear and open discussion of uncertainties and assumptions in the analysis presented should have caused management to probe further.

Shuttle Program Management's Low Level of Concern

While the debris strike was well outside the activities covered by normal mission flight rules, Mission Management Team members and Shuttle Program managers did not treat the debris strike as an issue that required operational action by Mission Control. Program managers, from Ron Dittemore to individual Mission Management Team members, had, over the course of the Space Shuttle Program, gradually become inured to External Tank foam losses and on a fundamental level did not believe foam striking the vehicle posed a critical threat to the Orbiter. In particular, Shuttle managers exhibited a belief that RCC panels are impervious to foam impacts. Even after seeing the video of *Columbia*'s debris impact, learning estimates of the size and location of the strike, and noting that a foam strike with sufficient kinetic energy could cause Thermal Protection System damage, management's level of concern did not change.

The opinions of Shuttle Program managers and debris and photo analysts on the potential severity of the debris strike diverged early in the mission and continued to diverge as the mission progressed, making it increasingly difficult for the Debris Assessment Team to have their concerns heard by those in a decision-making capacity. In the face of Mission managers' low level of concern and desire to get on with the mission, Debris Assessment Team members had to prove unequivocally that a safety-of-flight issue existed before Shuttle Program management would move to obtain images of the left wing. The engineers found themselves in the unusual position of having to prove that the situation was *unsafe* – a reversal of the usual requirement to prove that a situation *is safe*.

Other factors contributed to Mission management's ability to resist the Debris Assessment Team's concerns. A tile expert told managers during frequent consultations that strike damage was only a maintenance-level concern and that on-orbit imaging of potential wing damage was not necessary. Mission management welcomed this opinion and sought no others. This constant reinforcement of managers' pre-existing beliefs added another block to the wall between decision makers and concerned engineers.

Another factor that enabled Mission management's detachment from the concerns of their own engineers is rooted in the culture of NASA itself. The Board observed an unofficial hierarchy among NASA programs and directorates that hindered the flow of communications. The effects of this unofficial hierarchy are seen in the attitude that members of the Debris Assessment Team held. Part of the reason they chose the institutional route for their imagery request was that without direction from the Mission Evaluation Room and Mission Management Team, they felt more comfortable with their own chain of command, which was outside the Shuttle Program. Further, when asked by investigators why they were not more vocal about their concerns, Debris Assessment Team members opined that by raising contrary points of view about Shuttle mission safety, they would be singled out for possible ridicule by their peers and managers.

A Lack of Clear Communication

Communication did not flow effectively up to or down from Program managers. As it became clear during the mission that managers were not as concerned as others about the danger of the foam strike, the ability of engineers to challenge those beliefs greatly diminished. Managers' tendency to accept opinions that agree with their own dams the flow of effective communications.

After the accident, Program managers stated privately and publicly that if engineers had a safety concern, they were obligated to communicate their concerns to management. Managers did not seem to understand that as leaders they had a corresponding and perhaps greater obligation to create viable routes for the engineering community to express their views and receive information. This barrier to communications not only blocked the flow of information to managers, but it also prevented the downstream flow of information from managers to engineers, leaving Debris Assessment Team members no basis for understanding the reasoning behind Mission Management Team decisions.

The January 27 to January 31, phone and e-mail exchanges, primarily between NASA engineers at Langley and Johnson, illustrate another symptom of the "cultural fence" that impairs open communications between mission managers and working engineers. These exchanges and the reaction to them indicated that during the evaluation of a mission contingency, the Mission Management Team failed to disseminate information to all system and technology experts who could be consulted. Issues raised by two Langley and Johnson engineers led to the development of "what-if" landing scenarios of the potential outcome if the main landing gear door sustained damaged. This led to behind-the-scenes networking by these engineers to use NASA facilities to make simulation runs of a compromised landing configuration. These engineers – who understood their systems and related technology – saw the potential for a problem on landing and ran it down in case the unthinkable occurred. But their concerns never reached the managers on the Mission Management Team that had operational control over *Columbia*.

A Lack of Effective Leadership

The Shuttle Program, the Mission Management Team, and through it the Mission Evaluation Room, were not actively directing the efforts of the Debris Assessment Team. These management teams were not engaged in scenario selection or discussions of assumptions and did not actively seek status, inputs, or even preliminary results from the individuals charged with analyzing the debris strike. They did not investigate the value of imagery, did not intervene to consult the more experienced Crater analysts at Boeing's Huntington Beach facility, did not probe the assumptions of the Debris Assessment Team's analysis, and did not consider actions to mitigate the effects of the damage on re-entry. Managers' claims that they didn't hear the engineers' concerns were due in part to their not asking or listening.

The Failure of Safety's Role

As will be discussed in Chapter 7, safety personnel were present but passive and did not serve as a channel for the voicing of concerns or dissenting views. Safety representatives attended meetings of the Debris Assessment Team, Mission Evaluation Room, and Mission Management Team, but were merely party to the analysis process and conclusions instead of an independent source of questions and challenges. Safety contractors in the Mission Evaluation Room were only marginally aware of the debris strike analysis. One contractor did question the Debris Assessment Team safety representative about the analysis and was told that it was adequate. No additional inquiries were made. The highest-ranking safety representative at NASA headquarters deferred to Program managers when asked for an opinion on imaging of *Columbia*. The safety manager he spoke to also failed to follow up.

Summary

Management decisions made during *Columbia*'s final flight reflect missed opportunities, blocked or ineffective communications channels, flawed analysis, and ineffective leadership. Perhaps most striking is the fact that management – including Shuttle Program, Mission Management Team, Mission Evaluation Room, and Flight Director and Mission Control – displayed no interest in understanding a problem and its implications. Because managers failed to avail themselves of the wide range of expertise and opinion necessary to achieve the best answer to the debris strike question – *"Was this a safety-of-flight concern?"* – some Space Shuttle Program managers failed to fulfill the implicit contract to do whatever is possible to ensure the safety of the crew. In fact, their management techniques unknowingly imposed barriers that kept at bay both engineering concerns and dissenting views, and ultimately helped create "blind spots" that prevented them from seeing the danger the foam strike posed.

Because this chapter has focused on key personnel who participated in STS-107 bipod foam debris strike decisions, it is tempting to conclude that replacing them will solve all NASA's problems. However, solving NASA's problems is not quite so easily achieved. Peoples' actions are influenced by the organizations in which they work, shaping their choices in directions that even they may not realize. The Board explores the organizational context of decision making more fully in Chapters 7 and 8.

Findings

Intercenter Photo Working Group

F6.3-1 The foam strike was first seen by the Intercenter Photo Working Group on the morning of Flight Day Two during the standard review of launch video and high-speed photography. The strike was larger than any seen in the past, and the group was concerned about possible damage to the Orbiter. No conclusive images of the strike existed. One camera that may have provided an additional view was out of focus because of an improperly maintained lens.

F6.3-2 The Chair of the Intercenter Photo Working Group asked management to begin the process of getting outside imagery to help in damage assessment. This request, the first of three, began its journey through the management hierarchy on Flight Day Two.

F6.3-3 The Intercenter Photo Working Group distributed its first report, including a digitized video clip and initial assessment of the strike, on Flight Day Two. This information was widely disseminated to NASA and contractor engineers, Shuttle Program managers, and Mission Operations Directorate personnel.

F6.3-4 Initial estimates of debris size, speed, and origin were remarkably accurate. Initial information available to managers stated that the debris originated in the left bipod area of the External Tank, was quite large, had a high velocity, and struck the underside of the left wing near its leading edge. The report stated that the debris could have hit the RCC or tile.

The Debris Assessment Team

F6.3-5 A Debris Assessment Team began forming on Flight Day two to analyze the impact. Once the debris strike was categorized as "out of family" by United Space Alliance, contractual obligations led to the Team being Co-Chaired by the cognizant contractor sub-system manager and her NASA counterpart. The team was not designated a Tiger Team by the Mission Evaluation Room or Mission Management Team.

F6.3-6 Though the Team was clearly reporting its plans (and final results) through the Mission Evaluation Room to the Mission Management Team, no Mission manager appeared to "own" the Team's actions. The Mission Management Team, through the Mission Evaluation Room, provided no direction for team activities, and Shuttle managers did not formally consult the Team's leaders about their progress or interim results.

F6.3-7 During an organizational meeting, the Team discussed the uncertainty of the data and the value of on-orbit imagery to "bound" their analysis. In its first official meeting the next day, the Team gave its NASA Co-Chair the action to request imagery of *Columbia* on-orbit.

F6.3-8 The Team routed its request for imagery through Johnson Space Center's Engineering Directorate rather than through the Mission Evaluation Room to the Mission Management Team to the Flight Dynamics Officer, the channel used during a mission. This routing diluted the urgency of their request. Managers viewed it as a non-critical engineering desire rather than a critical operational need.

F6.3-9 Team members never realized that management's decision against seeking imagery was not intended as a direct or final response to their request.

F6.3-10 The Team's assessment of possible tile damage was performed using an impact simulation that was well outside Crater's test database. The Boeing analyst was inexperienced in the use of Crater and the interpretation of its results. Engineers with extensive Thermal Protection System expertise at Huntington Beach were not actively involved in determining if the Crater results were properly interpreted.

F6.3-11 Crater initially predicted tile damage deeper than the actual tile depth, but engineers used their judgment to conclude that damage would not penetrate the densified layer of tile. Similarly, RCC damage conclusions were based primarily on judgment and experience rather than analysis.

F6.3-12 For a variety of reasons, including management failures, communication breakdowns, inadequate imagery, inappropriate use of assessment tools, and flawed engineering judgments, the damage assessments contained substantial uncertainties.

F6.3-13 The assumptions (and their uncertainties) used in the analysis were never presented or discussed in full to either the Mission Evaluation Room or the Mission Management Team.

F6.3-14 While engineers and managers knew the foam could have struck RCC panels; the briefings on the analysis to the Mission Evaluation Room and Mission Management Team did not address RCC damage, and neither Mission Evaluation Room nor Mission Management Team managers asked about it.

Space Shuttle Program Management

F6.3-15 There were lapses in leadership and communication that made it difficult for engineers to raise concerns or understand decisions. Management failed to actively engage in the analysis of potential damage caused by the foam strike.

F6.3-16 Mission Management Team meetings occurred infrequently (five times during a 16 day mission), not every day, as specified in Shuttle Program management rules.

F6.3-17 Shuttle Program Managers entered the mission with the belief, recently reinforced by the STS-113 Flight Readiness Review, that a foam strike is not a safety-of-flight issue.

F6.3-18 After Program managers learned about the foam strike, their belief that it would not be a problem was confirmed (early, and without analysis) by a trusted expert who was readily accessible and spoke from "experience." No one in management questioned this conclusion.

F6.3-19 Managers asked *"Who's requesting the photos?"* instead of assessing the merits of the request. Management seemed more concerned about the staff following proper channels (even while they were themselves taking informal advice) than they were about the analysis.

F6.3-20 No one in the operational chain of command for STS-107 held a security clearance that would enable them to understand the capabilities and limitations of National imagery resources.

F6.3-21 Managers associated with STS-107 began investigating the implications of the foam strike on the launch schedule, and took steps to expedite post-flight analysis.

F6.3-22 Program managers required engineers to prove that the debris strike created a safety-of-flight issue: that is, engineers had to produce evidence that the system was unsafe rather than prove that it was safe.

F6.3-23 In both the Mission Evaluation Room and Mission Management Team meetings over the Debris Assessment Team's results, the focus was on the bottom line – was there a safety-of-flight issue, or not? There was little discussion of analysis, assumptions, issues, or ramifications.

Communication

F6.3-24 Communication did not flow effectively up to or down from Program managers.

F6.3-25 Three independent requests for imagery were initiated.

F6.3-26 Much of Program managers' information came through informal channels, which prevented relevant opinion and analysis from reaching decision makers.

F6.3-27 Program Managers did not actively communicate with the Debris Assessment Team. Partly as a result of this, the Team went through institutional, not mission-related, channels with its request for imagery, and confusion surrounded the origin of imagery requests and their subsequent denial.

F6.3-28 Communication was stifled by the Shuttle Program attempts to find out who had a "mandatory requirement" for imagery.

Safety Representative's Role

F6.3-29 Safety representatives from the appropriate organizations attended meetings of the Debris Assessment Team, Mission Evaluation Room, and Mission Management Team, but were passive, and therefore were not a channel through which to voice concerns or dissenting views.

Recommendation:

R6.3-1 Implement an expanded training program in which the Mission Management Team faces potential crew and vehicle safety contingences beyond launch and ascent. These contingences should involve potential loss of Shuttle or crew, contain numerous uncertainties and unknowns, and require the Mission Management Team to assemble and interact with support organizations across NASA/Contractor lines and in various locations.

R6.3-2 Modify the Memorandum of Agreement with the National Imagery and Mapping Agency (NIMA) to make the imaging of each Shuttle flight while on orbit a standard requirement.

6.4 Possibility of Rescue or Repair

To put the decisions made during the flight of STS-107 into perspective, the Board asked NASA to determine if there were options for the safe return of the STS-107 crew. In this study, NASA was to assume that the extent of damage to the leading edge of the left wing was determined by national imaging assets or by a spacewalk. NASA was then asked to evaluate the possibility of:

1. Rescuing the STS-107 crew by launching *Atlantis*. *Atlantis* would be hurried to the pad, launched, rendezvous with *Columbia*, and take on *Columbia*'s crew for a return. It was assumed that NASA would be willing to expose *Atlantis* and its crew to the same possibility of External Tank bipod foam loss that damaged *Columbia*.
2. Repairing damage to *Columbia*'s wing on orbit. In the repair scenario, astronauts would use onboard materials to rig a temporary fix. Some of *Columbia*'s cargo might be jettisoned and a different re-entry profile would be flown to lessen heating on the left wing leading edge. The crew would be prepared to bail out if the wing structure was predicted to fail on landing.

In its study of these two options, NASA assumed the following timeline. Following the debris strike discovery on Flight Day Two, Mission Managers requested imagery by Flight Day Three. That imagery was inconclusive, leading to a decision on Flight Day Four to perform a spacewalk on Flight Day Five. That spacewalk revealed potentially catastrophic damage. The crew was directed to begin conserving consumables, such as oxygen and water, and Shuttle managers began around-the-clock processing of *Atlantis* to prepare it for launch. Shuttle managers pursued both the rescue and the repair options from Flight Day Six to Flight Day 26, and on that day (February 10) decided which one to abandon.

The NASA team deemed this timeline realistic for several reasons. First, the team determined that a spacewalk to inspect the left wing could be easily accomplished. The team then assessed how the crew could limit its use of consumables to determine how long *Columbia* could stay in orbit. The limiting consumable was the lithium hydroxide canisters, which scrub from the cabin atmosphere the carbon dioxide the crew exhales. After consulting with flight surgeons, the team concluded that by modifying crew activity and sleep time carbon dioxide could be kept to acceptable levels until Flight Day 30 (the morning of February 15). All other consumables would last longer. Oxygen, the next most critical, would require the crew to return on Flight Day 31.

Repairing Damage On Orbit

The repair option (see Figure 6.4-1), while logistically viable using existing materials onboard *Columbia*, relied on so many uncertainties that NASA rated this option "high risk." To complete a repair, the crew would perform a spacewalk to fill an assumed 6-inch hole in an RCC panel with heavy metal tools, small pieces of titanium, or other metal scavenged from the crew cabin. These heavy metals, which would help protect the wing structure, would be held in place during re-entry by a water-filled bag that had turned into ice in the cold of space. The ice and metal would help restore wing leading edge geometry, preventing a turbulent airflow over the wing and therefore keeping heating and burn-through levels low enough for the crew to survive re-entry and bail out before landing. Because the NASA team could not verify that the repairs would survive even a modified re-entry, the rescue option had a considerably higher chance of bringing *Columbia*'s crew back alive.

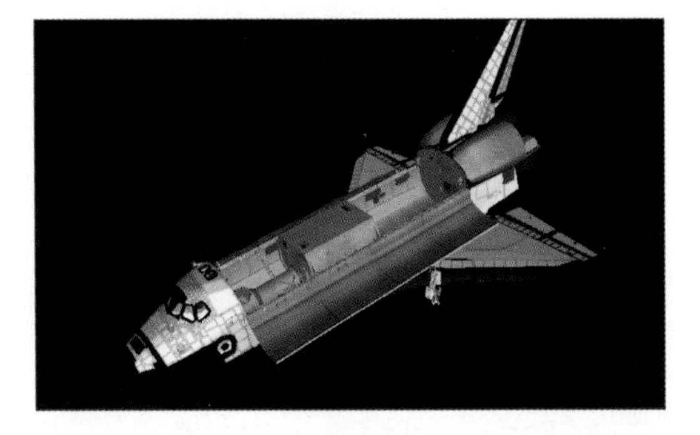

Figure 6.4-1. The speculative repair option would have sent astronauts hanging over the payload bay door to reach the left wing RCC panels using a ladder scavenged from the crew module.

Rescuing the STS-107 Crew with Atlantis

Accelerating the processing of *Atlantis* for early launch and rendezvous with *Columbia* was by far the most complex task in the rescue scenario. On *Columbia's* Flight Day Four, *Atlantis* was in the Orbiter Processing Facility at Kennedy Space Center with its main engines installed and only 41 days from its scheduled March 1 launch. The Solid Rocket Boosters were already mated with the External Tank in the Vehicle Assembly Building. By working three around-the-clock shifts seven days a week, *Atlantis* could be readied for launch, with no necessary testing skipped, by February 10. If launch processing and countdown proceeded smoothly, this would provide a five-day window, from February 10 to February 15, in which *Atlantis* could rendezvous with *Columbia* before *Columbia*'s consumables ran out. According to records, the weather on these days allowed a launch. *Atlantis* would be launched with a crew of four: a command-

Figure 6.4-2. The rescue option had Atlantis (lower vehicle) rendezvousing with Columbia and the STS-107 crew transferring via ropes. Note that the payload bay of Atlantis is empty except for the external airlock/docking adapter.

er, pilot, and two astronauts trained for spacewalks. In January, seven commanders, seven pilots, and nine spacewalk-trained astronauts were available. During the rendezvous on *Atlantis*'s first day in orbit, the two Orbiters would maneuver to face each other with their payload bay doors open (see Figure 6.4-2). Suited *Columbia* crew members would then be transferred to *Atlantis* via spacewalks. *Atlantis* would return with four crew members on the flight deck and seven in the mid-deck. Mission Control would then configure *Columbia* for a de-orbit burn that would ditch the Orbiter in the Pacific Ocean, or would have the *Columbia* crew take it to a higher orbit for a possible subsequent repair mission if more thorough repairs could be developed.

This rescue was considered challenging but feasible. To succeed, it required problem-free processing of *Atlantis* and a flawless launch countdown. If Program managers had understood the threat that the bipod foam strike posed and were able to unequivocally determine before Flight Day Seven that there was potentially catastrophic damage to the left wing, these repair and rescue plans would most likely have been developed, and a rescue would have been conceivable. For a detailed discussion of the rescue and repair options, see Appendix D.13.

Findings:

F6.4-1 The repair option, while logistically viable using existing materials onboard *Columbia*, relied on so many uncertainties that NASA rated this option "high risk."

F6.4-2 If Program managers were able to unequivocally determine before Flight Day Seven that there was potentially catastrophic damage to the left wing, accelerated processing of *Atlantis* might have provided a window in which *Atlantis* could rendezvous with *Columbia* before *Columbia*'s limited consumables ran out.

Recommendation:

R6.4-1 For missions to the International Space Station, develop a practicable capability to inspect and effect emergency repairs to the widest possible range of damage to the Thermal Protection System, including both tile and Reinforced Carbon-Carbon, taking advantage of the additional capabilities available when near to or docked at the International Space Station.

For non-Station missions, develop a comprehensive autonomous (independent of Station) inspection and repair capability to cover the widest possible range of damage scenarios.

Accomplish an on-orbit Thermal Protection System inspection, using appropriate assets and capabilities, early in all missions.

The ultimate objective should be a fully autonomous capability for all missions to address the possibility that an International Space Station mission fails to achieve the correct orbit, fails to dock successfully, or is damaged during or after undocking.

The crew cabin access arm in position against Columbia on Launch Complex 39-A.

ENDNOTES FOR CHAPTER 6

The citations that contain a reference to "CAIB document" with CAB or CTF followed by seven to eleven digits, such as CAB001-0010, refer to a document in the Columbia Accident Investigation Board database maintained by the Department of Justice and archived at the National Archives.

[1] "Space Shuttle Program Description and Requirements Baseline," NSTS-07700, Volume X, Book 1. CAIB document CTF028-32643667.

[2] "External Tank End Item (CEI) Specification – Part 1," CPT01M09A, contract NAS8 –30300, April 9, 1980, WBS 1.6.1.2 and 1.6.2.2.

[3] "STS-1 Orbiter Final Mission Report," JSC-17378, August 1981, p. 85.

[4] Discussed in Craig Covault, "Investigators Studying Shuttle Tiles, *Aviation Week & Space Technology*, May 11, 1981, pg. 40.

[5] *Report of the Presidential Commission on the Space Shuttle Challenger Accident*, Volume V, 1986, pp. 1028-9, hearing section pp. 1845-1849.

[6] "Orbiter Vehicle End Item Specification for the Space Shuttle System, Part 1, Performance and Design Requirements," contract NAS9-20000, November 7, 2002. CAIB documents CAB006-06440645 and CAB033-20242971.

[7] "Problem Reporting and Corrective Action System Requirements," NSTS-08126, Revision H, November 22, 2000, Appendix C, Definitions, In Family. CAIB document CTF044-28652894.

[8] Ibid.

[9] Ibid.

[10] The umbilical wells are compartments on the underside of the Orbiter where External Tank liquid oxygen and hydrogen lines connect. After the Orbiters land, the umbilical well camera film is retrieved and developed.

[11] NSTS-08126, Paragraph 3.4, Additional Requirements for In-Flight Anomaly (IFA) Reporting.

[12] Integrated Hazard Analysis INTG 037, "Degraded Functioning of Orbiter TPS or Damage to the Windows Caused by SRB/ET Ablatives or Debonded ET or SRB TPS."

[13] Ibid.

[14] Ibid.

[15] During the flight of STS-112, the Intercenter Photo Working Group speculated that a second debris strike occurred at 72 seconds, possibly to the right wing. Although post-flight analysis showed that this did not occur, the Board notes that the Intercenter Photo Working Group failed to properly inform the Mission Management Team of this strike, and that the Mission Management Team subsequently failed to aggressively address the event during flight.

[16] "Safety and Mission Assurance Report for the STS-113 Mission, Pre-Launch Mission Management Team Edition," Enterprise Safety and Mission Assurance Division, November 7, 2002. CAIB Document CTF024-00430061.

[17] Orbiter TPS damage numbers come from the Shuttle Flight Data and In-Flight Anomaly List (JSC-19413).

[18] CAIB Meeting Minutes, presentation and discussion on IFAs for STS-27 and STS-28, March 28, 2003, Houston, Texas.

[19] "STS-27R National Space Transportation System Mission Report," NSTS-23370, February 1989, p. 2.

[20] CAIB Meeting Minutes, presentation and discussion on IFAs for STS-27 and STS-28, March 28, 2003, Houston, Texas.

[21] Corrective Action Record, 27RF13, Closeout Report (no date). CAIB document CTF010-20822107.

[22] "STS-27R OV-104 Orbiter TPS Damage Review Team Summary Report," Volume I, February 1989, TM-100355, p. 64. CAIB document CAB035-02290303.

[23] Ibid.

[24] "In-Flight Anomaly: STS-35/ET-35," External Tank Flight Readiness Report 3500.2.3/91. CAIB document CAB057-51185119.

[25] STS-36 PRCB, IFA Closure Rationale for STS-35. CAIB document CAB029-03620433.

[26] Identified by MSFC in PRACA database as "not a safety of flight" concern. Briefed at post-STS-42 PRCB and STS-45 Flight Readiness Review.

[27] "STS-45 Space Shuttle Mission Report," NSTS-08275, May 1992, pg. 17. CAIB document CTF003-00030006.

[28] "STS-45 Space Shuttle Mission Report," NSTS-08275, May 1992. CAIB document CTF003-00030006.

[29] Both STS-56 and STS-58 post mission PRCBs discussed the debris events and IFAs. Closeout rationale was based upon the events being considered "in family" and "within experience base."

[30] "Problem Reporting and Corrective Action System Requirements," NSTS-08126, Revision H, November 22, 2000, Appendix C, Definitions, Out of Family. CAIB document CTF044-28652894.

[31] Post STS-87 PRCBD, S 062127, 18 Dec 1997.

[32] M. Elisabeth Paté-Cornell and Paul S. Fischbeck, "Risk Management for the Tiles of the Space Shuttle," pp. 64-86, *Interfaces* 24, January-February 1994. CAIB document CAB005-0141.

[33] Letter to M. Elisabeth Paté-Cornell, Stanford University, from Benjamin Buchbinder, Risk Management Program Manager, NASA, 10 May 1993. CAIB document CAB038-36973698.

[34] M. Elisabeth Paté-Cornell, "Follow-up on the Standard 1990 Study of the Risk of Loss of Vehicle and Crew of the NASA Space Shuttle Due to Tile Failure," Report to the Columbia Accident Investigation Board, 18 June 2003. CAIB document CAB006-00970104.

[35] M. Litwinsk and G. Wilson, et al., "End-to-End TPS Upgrades Plan for Space Shuttle Orbiter," February 1997; K. Hinkle and G. Wilson, "Advancements in TPS," *M&P Engineering*, 22 October 1998.

[36] Daniel B. Leiser, et al., "Toughened Uni-piece Fibrous Insulation (TUFI)" Patent #5,079,082, 7 January 1992.

[37] Karrie Hinkle, "High Density Tile for Enhanced Dimensional Stability," Briefing to Space Shuttle Program, October 19, 1998. CAIB document CAB033-32663280.

[38] Daniel B. Leiser, "Present/Future Tile Thermal Protection Systems," A presentation to the CAIB (Group 1), 16 May 2003.

[39] John Kowal, "Orbiter Thermal Protection System (TPS) Upgrades." Space Shuttle Upgrades Safety Panel Review, 10 February 2003.

[40] "Problem Reporting and Corrective Action System Requirements," NSTS-08126, Revision H, November 22, 2000. CAIB document CTF044-28652894.

[41] Diane Vaughan, *The Challenger Launch Decision: Risky Technology, Culture, and Deviance at NASA* (Chicago: University of Chicago Press, 1996).

[42] Richard Feynman, Minority Report on Challenger, *The Pleasure of Finding Things Out*, (New York: Perseus Publishing, 2002).

[43] See Appendix D.17 Tiger Team Checklists.

[44] Allen J. Richardson and A. H. McHugh, "Hypervelocity Impact Penetration Equation for Metal By Multiple Regression Analysis," STR153, North American Aviation, Inc., March 1966.

[45] Allen J. Richardson and J. C. Chou, "Correlation of TPS Tile Penetration Equation & Impact Test Data," 3 March 1985.

[46] "Review of Crater Program for Evaluating Impact Damage to Orbiter TPS Tiles," presented at Boeing-Huntington Beach, 29 Apr 2003. CAIB document CTF070-29492999.

[47] J. L. Rand, "Impact Testing of Orbiter HRSI Tiles," Texas Engineering Experiment Station Report (Texas A&M), 1979; Tests conducted by NASA (D. Arabian) ca. 1979.

[48] Drew L. Goodlin, "Orbiter Tile Impact Testing, Final Report", SwRI Project # 18-7503-005, March 5, 1999.

[49] Allen J. Richardson, "Evaluation of Flight Experience & Test Results for Ice Impaction on Orbiter RCC & ACC Surfaces," Rockwell International, November 26, 1984.

[50] Though this entry indicates that NASA contacted USSPACECOM, the correct entity is USSTRATCOM. USSPACECOM ceased to exist in October 2002.

CHAPTER 7

The Accident's Organizational Causes

Many accident investigations make the same mistake in defining causes. They identify the widget that broke or malfunctioned, then locate the person most closely connected with the technical failure: the engineer who miscalculated an analysis, the operator who missed signals or pulled the wrong switches, the supervisor who failed to listen, or the manager who made bad decisions. When causal chains are limited to technical flaws and individual failures, the ensuing responses aimed at preventing a similar event in the future are equally limited: they aim to fix the technical problem and replace or retrain the individual responsible. Such corrections lead to a misguided and potentially disastrous belief that the underlying problem has been solved. The Board did not want to make these errors. A central piece of our expanded cause model involves NASA as an organizational whole.

ORGANIZATIONAL CAUSE STATEMENT

The organizational causes of this accident are rooted in the Space Shuttle Program's history and culture, including the original compromises that were required to gain approval for the Shuttle Program, subsequent years of resource constraints, fluctuating priorities, schedule pressures, mischaracterizations of the Shuttle as operational rather than developmental, and lack of an agreed national vision. Cultural traits and organizational practices detrimental to safety and reliability were allowed to develop, including: reliance on past success as a substitute for sound engineering practices (such as testing to understand why systems were not performing in accordance with requirements/specifications); organizational barriers which prevented effective communication of critical safety information and stifled professional differences of opinion; lack of integrated management across program elements; and the evolution of an informal chain of command and decision-making processes that operated outside the organization's rules.

UNDERSTANDING CAUSES

In the Board's view, NASA's organizational culture and structure had as much to do with this accident as the External Tank foam. Organizational culture refers to the values, norms, beliefs, and practices that govern how an institution functions. At the most basic level, organizational culture defines the assumptions that employees make as they carry out their work. It is a powerful force that can persist through reorganizations and the reassignment of key personnel.

Given that today's risks in human space flight are as high and the safety margins as razor thin as they have ever been, there is little room for overconfidence. Yet the attitudes and decision-making of Shuttle Program managers and engineers during the events leading up to this accident were clearly overconfident and often bureaucratic in nature. They deferred to layered and cumbersome regulations rather than the fundamentals of safety. The Shuttle Program's safety culture is straining to hold together the vestiges of a once robust systems safety program.

As the Board investigated the *Columbia* accident, it expected to find a vigorous safety organization, process, and culture at NASA, bearing little resemblance to what the Rogers Commission identified as the ineffective "silent safety" system in which budget cuts resulted in a lack of resources, personnel, independence, and authority. NASA's initial briefings to the Board on its safety programs espoused a risk-averse philosophy that empowered any employee to stop an operation at the mere glimmer of a problem. Unfortunately, NASA's views of its safety culture in those briefings did not reflect reality. Shuttle Program safety personnel failed to adequately assess anomalies and frequently accepted critical risks without qualitative or quantitative support, even when the tools to provide more comprehensive assessments were available.

Similarly, the Board expected to find NASA's Safety and Mission Assurance organization deeply engaged at every level of Shuttle management: the Flight Readiness Review, the Mission Management Team, the Debris Assessment Team, the Mission Evaluation Room, and so forth. This was not the case. In briefing after briefing, interview after interview, NASA remained in denial: in the agency's eyes, "there were no safety-of-flight issues," and no safety compromises in the long history of debris strikes on the Thermal Protection System. The silence of Program-level safety processes undermined oversight; when they did not speak up, safety personnel could not fulfill their stated mission to provide "checks and balances." A pattern of acceptance prevailed throughout the organization that tolerated foam problems without sufficient engineering justification for doing so.

This chapter presents an organizational context for understanding the *Columbia* accident. Section 7.1 outlines a short history of safety at NASA, beginning in the pre-Apollo era when the agency reputedly had the finest system safety-engineering programs in the world. Section 7.2 discusses organizational theory and its importance to the Board's investigation, and Section 7.3 examines the practices of three organizations that successfully manage high risk. Sections 7.4 and 7.5 look at NASA today and answer the question, "How could NASA have missed the foam signal?" by highlighting the blind spots that rendered the Shuttle Program's risk perspective myopic. The Board's conclusion and recommendations are presented in 7.6. (See Chapter 10 for a discussion of the differences between industrial safety and mission assurance/quality assurance.)

7.1 Organizational Causes: Insights from History

NASA's organizational culture is rooted in history and tradition. From NASA's inception in 1958 to the *Challenger* accident in 1986, the agency's Safety, Reliability, and Quality Assurance (SRQA) activities, "although distinct disciplines," were "typically treated as one function in the design, development, and operations of NASA's manned space flight programs."[1] Contractors and NASA engineers collaborated closely to assure the safety of human space flight. Solid engineering practices emphasized defining goals and relating system performance to them; establishing and using decision criteria; developing alternatives; modeling systems for analysis; and managing operations.[2] Although a NASA Office of Reliability and Quality Assurance existed for a short time during the early 1960s, it was funded by the human space flight program. By 1963, the office disappeared from the agency's organization charts. For the next few years, the only type of safety program that existed at NASA was a decentralized "loose federation" of risk assessment oversight run by each program's contractors and the project offices at each of the three Human Space Flight Centers.

Fallout from Apollo – 1967

In January 1967, months before the scheduled launch of *Apollo 1*, three astronauts died when a fire erupted in a ground-test capsule. In response, Congress, seeking to establish an independent safety organization to oversee space flight, created the Aerospace Safety Advisory Panel (ASAP). The ASAP was intended to be a senior advisory committee to NASA, reviewing space flight safety studies and operations plans, and evaluating "systems procedures and management policies that contribute to risk." The panel's main priority was human space flight missions.[3] Although four of the panel's nine members can be NASA employees, in recent years few have served as members. While the panel's support staff generally consists of full-time NASA employees, the group technically remains an independent oversight body.

Congress simultaneously mandated that NASA create separate safety and reliability offices at the agency's headquarters and at each of its Human Space Flight Centers and Programs. Overall safety oversight became the responsibility of NASA's Chief Engineer. Although these offices were not totally independent – their funding was linked with the very programs they were supposed to oversee – their existence allowed NASA to treat safety as a unique function. Until the *Challenger* accident in 1986, NASA safety remained linked organizationally and financially to the agency's Human Space Flight Program.

Challenger – 1986

In the aftermath of the *Challenger* accident, the Rogers Commission issued recommendations intended to remedy what it considered to be basic deficiencies in NASA's safety system. These recommendations centered on an underlying theme: the lack of independent safety oversight at NASA. Without independence, the Commission believed, the slate of safety failures that contributed to the *Challenger* accident – such as the undue influence of schedule pressures and the flawed Flight Readiness process – would not be corrected. "NASA should establish an Office of Safety, Reliability, and Quality Assurance to be headed by an Associate Administrator, reporting directly to the NASA Administrator," concluded the Commission. "It would have *direct authority* for safety, reliability, and quality assurance throughout the Agency. The office should be assigned the workforce to ensure adequate oversight of its functions and should be independent of other NASA functional and program responsibilities" [emphasis added].

In July 1986, NASA Administrator James Fletcher created a Headquarters Office of Safety, Reliability, and Quality Assurance, which was given responsibility for all agency-wide safety-related policy functions. In the process, the position of Chief Engineer was abolished.[4] The new office's Associate Administrator promptly initiated studies on Shuttle in-flight anomalies, overtime levels, the lack of spare parts, and landing and crew safety systems, among other issues.[5] Yet NASA's response to the Rogers Commission recommendation did not meet the Commission's intent: the Associate Administrator did not have direct authority, and safety, reliability, and mission assurance activities across the agency remained dependent on other programs and Centers for funding.

General Accounting Office Review – 1990

A 1990 review by the U.S. General Accounting Office questioned the effectiveness of NASA's new safety organizations in a report titled "Space Program Safety: Funding for NASA's Safety Organizations Should Be Centralized."[6] The report concluded *"NASA did not have an independent and effective safety organization"* [emphasis added]. Although the safety organizational structure may have "appeared adequate," in the late 1980s the space agency had concentrated most of its efforts on creating an independent safety office at NASA Headquarters. In contrast, the safety offices at NASA's field centers "were not entirely independent because they obtained most of their funds from activities whose safety-related performance they were responsible for overseeing." The General Accounting Office worried that "the lack of centralized independent funding may also restrict the flexibility of center safety managers." It also suggested "most NASA safety managers believe that centralized SRM&QA [Safety, Reliability, Maintainability and Quality Assurance] funding would ensure independence." NASA did not institute centralized funding in response to the General Accounting Office report, nor has it since. The problems outlined in 1990 persist to this day.

Space Flight Operations Contract – 1996

The Space Flight Operations Contract was intended to streamline and modernize NASA's cumbersome contracting practices, thereby freeing the agency to focus on research and development (see Chapter 5). Yet its implementation complicated issues of safety independence. A single contractor would, in principle, provide "oversight" on production, safety, and mission assurance, as well as cost management, while NASA maintained "insight" into safety and quality assurance through reviews and metrics. Indeed, the reduction to a single primary contract simplified some aspects of the NASA/contractor interface. However, as a result, experienced engineers changed jobs, NASA grew dependent on contractors for technical support, contract monitoring requirements increased, and positions were subsequently staffed by less experienced engineers who were placed in management roles.

Collectively, this eroded NASA's in-house engineering and technical capabilities and increased the agency's reliance on the United Space Alliance and its subcontractors to identify, track, and resolve problems. The contract also involved substantial transfers of safety responsibility from the government to the private sector; rollbacks of tens of thousands of Government Mandated Inspection Points; and vast reductions in NASA's in-house safety-related technical expertise (see Chapter 10). In the aggregate, these mid-1990s transformations rendered NASA's already problematic safety system simultaneously weaker and more complex.

The effects of transitioning Shuttle operations to the Space Flight Operations Contract were not immediately apparent in the years following implementation. In November 1996, as the contract was being implemented, the Aerospace Safety Advisory Panel published a comprehensive contract review, which concluded that the effort "to streamline the Space Shuttle program has not inadvertently created unacceptable flight or ground risks."[7] The Aerospace Safety Advisory Panel's passing grades proved temporary.

Shuttle Independent Assessment Team – 1999

Just three years later, after a number of close calls, NASA chartered the Shuttle Independent Assessment Team to examine Shuttle sub-systems and maintenance practices (see Chapter 5). The Shuttle Independent Assessment Team Report sounded a stern warning about the quality of NASA's Safety and Mission Assurance efforts and noted that the Space Shuttle Program had undergone a massive change in structure and was transitioning to "a slimmed down, contractor-run operation."

The team produced several pointed conclusions: the Shuttle Program was inappropriately *using previous success as a justification* for accepting increased risk; the Shuttle Program's *ability to manage risk was being eroded* "by the desire to reduce costs;" the size and complexity of the Shuttle Program and NASA/contractor relationships *demanded better communication practices*; NASA's safety and mission assurance organization was *not sufficiently independent*; and "the workforce has received a conflicting message due to the emphasis on achieving cost and staff reductions, and the *pressures placed on increasing scheduled flights* as a result of the Space Station" [emphasis added].[8] The Shuttle Independent Assessment Team found failures of communication to flow up from the "shop floor" and down from supervisors to workers, deficiencies in problem and waiver-tracking systems, potential conflicts of interest between Program and contractor goals, and a general failure to communicate requirements and changes across organizations. In general, the Program's organizational culture was deemed "too insular."[9]

NASA subsequently formed an Integrated Action Team to develop a plan to address the recommendations from previous Program-specific assessments, including the Shuttle Independent Assessment Team, and to formulate improvements.[10] In part this effort was also a response to program missteps in the drive for efficiency seen in the "faster, better, cheaper" NASA of the 1990s. The NASA Integrated Action Team observed: *"NASA should continue to remove communication barriers and foster an inclusive environment where open communication is the norm."* The intent was to establish an initiative where *"the importance of communication and a culture of trust and openness permeate all facets of the organization."* The report indicated that *"multiple processes to get the messages across the organizational structure"* would need to be explored and fostered [emphasis added]. The report recommended that NASA solicit expert advice in identifying and removing barriers, providing tools, training, and education, and facilitating communication processes.

The Shuttle Independent Assessment Team and NASA Integrated Action Team findings mirror those presented by the Rogers Commission. The same communication problems persisted in the Space Shuttle Program at the time of the *Columbia* accident.

Space Shuttle Competitive Source Task Force – 2002

In 2002, a 14-member Space Shuttle Competitive Task Force supported by the RAND Corporation examined competitive sourcing options for the Shuttle Program. In its final report to NASA, the team highlighted several safety-related concerns, which the Board shares:

- Flight and ground hardware and software are obsolete, and safety upgrades and aging infrastructure repairs have been deferred.
- Budget constraints have impacted personnel and resources required for maintenance and upgrades.
- International Space Station schedules exert significant pressures on the Shuttle Program.
- Certain mechanisms may impede worker anonymity in reporting safety concerns.
- NASA does not have a truly independent safety function with the authority to halt the progress of a critical mission element. [11]

Based on these findings, the task force suggested that an Independent Safety Assurance function should be created that would hold one of "three keys" in the Certification of Flight Readiness process (NASA and the operating contractor would hold the other two), effectively giving this function the ability to stop any launch. Although in the Board's view the "third key" Certification of Flight Readiness process is not a perfect solution, independent safety and verification functions are vital to continued Shuttle operations. This independent function should possess the authority to shut down the flight preparation processes or intervene post-launch when an anomaly occurs.

7.2 Organizational Causes: Insights from Theory

To develop a thorough understanding of accident causes and risk, and to better interpret the chain of events that led to the *Columbia* accident, the Board turned to the contemporary social science literature on accidents and risk and sought insight from experts in High Reliability, Normal Accident, and Organizational Theory.[12] Additionally, the Board held a forum, organized by the National Safety Council, to define the essential characteristics of a sound safety program.[13]

High Reliability Theory argues that organizations operating high-risk technologies, if properly designed and managed, can compensate for inevitable human shortcomings, and therefore avoid mistakes that under other circumstances would lead to catastrophic failures.[14] Normal Accident Theory, on the other hand, has a more pessimistic view of the ability of organizations and their members to manage high-risk technology. Normal Accident Theory holds that organizational and technological complexity contributes to failures. Organizations that aspire to failure-free performance are inevitably doomed to fail because of the inherent risks in the technology they operate.[15] Normal Accident models also emphasize systems approaches and systems thinking, while the High Reliability model works from the bottom up: if each component is highly reliable, then the system will be highly reliable and safe.

Though neither High Reliability Theory nor Normal Accident Theory is entirely appropriate for understanding this accident, insights from each figured prominently in the Board's deliberation. Fundamental to each theory is the importance of strong organizational culture and commitment to building successful safety strategies.

The Board selected certain well-known traits from these models to use as a yardstick to assess the Space Shuttle Program, and found them particularly useful in shaping its views on whether NASA's current organization of its Human Space Flight Program is appropriate for the remaining years of Shuttle operation and beyond. Additionally, organizational theory, which encompasses organizational culture, structure, history, and hierarchy, is used to explain the *Columbia* accident, and, ultimately, combines with Chapters 5 and 6 to produce an expanded explanation of the accident's causes.[16] The Board believes the following considerations are critical to understand what went wrong during STS-107. They will become the central motifs of the Board's analysis later in this chapter.

- **Commitment to a Safety Culture:** NASA's safety culture has become reactive, complacent, and dominated by unjustified optimism. Over time, slowly and unintentionally, independent checks and balances intended to increase safety have been eroded in favor of detailed processes that produce massive amounts of data and unwarranted consensus, but little effective communication. Organizations that successfully deal with high-risk technologies create and sustain a disciplined safety system capable of identifying, analyzing, and controlling hazards throughout a technology's life cycle.

- **Ability to Operate in Both a Centralized and Decentralized Manner:** The ability to operate in a centralized manner when appropriate, and to operate in a decentralized manner when appropriate, is the hallmark of a high-reliability organization. On the operational side, the Space Shuttle Program has a highly centralized structure. Launch commit criteria and flight rules govern every imaginable contingency. The Mission Control Center and the Mission Management Team have very capable decentralized processes to solve problems that are not covered by such rules. The process is so highly regarded that it is considered one of the best problem-solving organizations of its type.[17] In these situations, mature processes anchor rules, procedures, and routines to make the Shuttle Program's matrixed workforce seamless, at least on the surface.

  Nevertheless, it is evident that the position one occupies in this structure makes a difference. When supporting organizations try to "push back" against centralized Program direction – like the Debris Assessment Team did during STS-107 – independent analysis generated by a decentralized decision-making process can be stifled. The Debris Assessment Team, working in an essentially decentralized format, was well-led and had the right expertise to work the problem, but their charter was "fuzzy," and the team had little direct connection to the Mission Management Team. This lack of connection to the Mission Management Team and the Mission Evaluation Room is the single most compelling reason why communications were so poor during the debris assessment. In this case, the Shuttle Program was unable to simultaneously manage both the centralized and decentralized systems.

- **Importance of Communication:** At every juncture of STS-107, the Shuttle Program's structure and processes, and therefore the managers in charge, resisted new information. Early in the mission, it became clear that the Program was not going to authorize imaging of the Orbiter because, in the Program's opinion, images were not needed. Overwhelming evidence indicates that Program leaders decided the foam strike was merely a maintenance problem long before any analysis had begun. Every manager knew the party line: "we'll wait for the analysis – no safety-of-flight issue expected." Program leaders spent at least as much time making sure hierarchical rules and processes were followed as they did trying to establish why anyone would want a picture of the Orbiter. These attitudes are incompatible with an organization that deals with high-risk technology.

- **Avoiding Oversimplification:** The *Columbia* accident is an unfortunate illustration of how NASA's strong cultural bias and its optimistic organizational thinking undermined effective decision-making. Over the course of 22 years, foam strikes were normalized to the point where they were simply a "maintenance" issue – a concern that did not threaten a mission's success. This oversimplification of the threat posed by foam debris rendered the issue a low-level concern in the minds of Shuttle managers. Ascent risk, so evident in *Challenger*, biased leaders to focus on strong signals from the Shuttle System Main Engine and the Solid Rocket Boosters. Foam strikes, by comparison, were a weak and consequently overlooked signal, although they turned out to be no less dangerous.

- **Conditioned by Success:** Even after it was clear from the launch videos that foam had struck the Orbiter in a manner never before seen, Space Shuttle Program managers were not unduly alarmed. They could not imagine why anyone would want a photo of something that could be fixed after landing. More importantly, learned attitudes about foam strikes diminished management's wariness of their danger. The Shuttle Program turned "the experience of failure into the memory of success."[18] Managers also failed to develop simple contingency plans for a re-entry emergency. They were convinced, without study, that nothing could be done about such an emergency. The intellectual curiosity and skepticism that a solid safety culture requires was almost entirely absent. Shuttle managers did not embrace safety-conscious attitudes. Instead, their attitudes were shaped and reinforced by an organization that, in this instance, was incapable of stepping back and gauging its biases. Bureaucracy and process trumped thoroughness and reason.

- **Significance of Redundancy:** The Human Space Flight Program has compromised the many redundant processes, checks, and balances that should identify and correct small errors. Redundant systems essential to every high-risk enterprise have fallen victim to bureaucratic efficiency. Years of workforce reductions and outsourcing have culled from NASA's workforce the layers of experience and hands-on systems knowledge that once provided a capacity for safety oversight. Safety and Mission Assurance personnel have been eliminated, careers in safety have lost organizational prestige, and the Program now decides on its own how much safety and engineering oversight it needs. Aiming to align its inspection regime with the International Organization for Standardization 9000/9001 protocol, commonly used in industrial environments – environments very different than the Shuttle Program – the Human Space Flight Program shifted from a comprehensive "oversight" inspection process to a more limited "insight" process, cutting mandatory inspection points by more than half and leaving even fewer workers to make "second" or "third" Shuttle systems checks (see Chapter 10).

Implications for the Shuttle Program Organization

The Board's investigation into the *Columbia* accident revealed two major causes with which NASA has to contend: one technical, the other organizational. As mentioned earlier, the Board studied the two dominant theories on complex organizations and accidents involving high-risk technologies. These schools of thought were influential in shaping the Board's organizational recommendations, primarily because each takes a different approach to understanding accidents and risk.

The Board determined that high-reliability theory is extremely useful in describing the culture that should exist in the human space flight organization. NASA and the Space Shuttle Program must be committed to a strong safety culture, a view that serious accidents can be prevented, a willingness to learn from mistakes, from technology, and from others, and a realistic training program that empowers employees to know when to decentralize or centralize problem-solving. The Shuttle Program cannot afford the mindset that accidents are inevitable because it may lead to unnecessarily accepting known and preventable risks.

The Board believes normal accident theory has a key role in human spaceflight as well. Complex organizations need specific mechanisms to maintain their commitment to safety and assist their understanding of how complex interactions can make organizations accident-prone. Organizations cannot put blind faith into redundant warning systems because they inherently create more complexity, and this complexity in turn often produces unintended system interactions that can lead to failure. The Human Space Flight Program must realize that additional protective layers are not always the best choice. The Program must also remain sensitive to the fact that despite its best intentions, managers, engineers, safety professionals, and other employees, can, when confronted with extraordinary demands, act in counterproductive ways.

The challenges to failure-free performance highlighted by these two theoretical approaches will always be present in an organization that aims to send humans into space. What can the Program do about these difficulties? The Board considered three alternatives. First, the Board could recommend that NASA follow traditional paths to improving safety by making changes to policy, procedures, and processes. These initiatives could improve organizational culture. The analysis provided by experts and the literature leads the Board to conclude that although reforming management practices has certain merits, it also has critical limitations. Second, the Board could recommend that the Shuttle is simply too risky and should be grounded. As will be discussed in Chapter 9, the Board is committed to continuing human space exploration, and believes the Shuttle Program can and should continue to operate. Finally, the Board could recommend a significant change to the organizational structure that controls the Space Shuttle Program's technology. As will be discussed at length in this chapter's conclusion, the Board believes this option has the best chance to successfully manage the complexities and risks of human space flight.

7.3 Organizational Causes: Evaluating Best Safety Practices

Many of the principles of solid safety practice identified as crucial by independent reviews of NASA and in accident and risk literature are exhibited by organizations that, like NASA, operate risky technologies with little or no margin for error. While the Board appreciates that organizations dealing with high-risk technology cannot sustain accident-free performance indefinitely, evidence suggests that there are effective ways to minimize risk and limit the number of accidents.

In this section, the Board compares NASA to three specific examples of independent safety programs that have strived for accident-free performance and have, by and large, achieved it: the U.S. Navy Submarine Flooding Prevention and Recovery (SUBSAFE), Naval Nuclear Propulsion (Naval Reactors) programs, and the Aerospace Corporation's Launch Verification Process, which supports U.S. Air Force space launches.[19] The safety cultures and organizational structure of all three make them highly adept in dealing with inordinately high risk by designing hardware and management systems that prevent seemingly inconsequential failures from leading to major accidents. Although size, complexity, and missions in these organizations and NASA differ, the following comparisons yield valuable lessons for the space agency to consider when re-designing its organization to increase safety.

Navy Submarine and Reactor Safety Programs

Human space flight and submarine programs share notable similarities. Spacecraft and submarines both operate in hazardous environments, use complex and dangerous systems, and perform missions of critical national significance. Both NASA and Navy operational experience include failures (for example, USS *Thresher*, USS *Scorpion*, *Apollo 1* capsule fire, *Challenger*, and *Columbia*). Prior to the *Columbia* mishap, Administrator Sean O'Keefe initiated the NASA/Navy Benchmarking Exchange to compare and contrast the programs, specifically in safety and mission assurance.[20]

The Navy SUBSAFE and Naval Reactor programs exercise a high degree of engineering discipline, emphasize total responsibility of individuals and organizations, and provide redundant and rapid means of communicating problems to decision-makers. The Navy's nuclear safety program emerged with its first nuclear-powered warship (USS *Nautilus*), while non-nuclear SUBSAFE practices evolved from from past flooding mishaps and philosophies first introduced by Naval Reactors. The Navy lost two nuclear-powered submarines in the 1960s – the USS *Thresher* in 1963 and the *Scorpion* 1968 – which resulted in a renewed effort to prevent accidents.[21] The SUBSAFE program was initiated just two months after the *Thresher* mishap to identify critical changes to submarine certification requirements. Until a ship was independently recertified, its operating depth and maneuvers were limited. SUBSAFE proved its value as a means of verifying the readiness and safety of submarines, and continues to do so today.[22]

The Naval Reactor Program is a joint Navy/Department of Energy organization responsible for all aspects of Navy nuclear propulsion, including research, design, construction, testing, training, operation, maintenance, and the disposition of the nuclear propulsion plants onboard many Naval ships and submarines, as well as their radioactive materials. Although the naval fleet is ultimately responsible for day-to-day operations and maintenance, those operations occur within parameters established by an entirely independent division of Naval Reactors.

The U.S. nuclear Navy has more than 5,500 reactor years of experience without a reactor accident. Put another way, nuclear-powered warships have steamed a cumulative total of over 127 million miles, which is roughly equivalent to over 265 lunar roundtrips. In contrast, the Space Shuttle Program has spent about three years on-orbit, although its spacecraft have traveled some 420 million miles.

Naval Reactor success depends on several key elements:

- Concise and timely communication of problems using redundant paths
- Insistence on airing minority opinions
- Formal written reports based on independent peer-reviewed recommendations from prime contractors
- Facing facts objectively and with attention to detail
- Ability to manage change and deal with obsolescence of classes of warships over their lifetime

These elements can be grouped into several thematic categories:

- **Communication and Action:** Formal and informal practices ensure that relevant personnel at all levels are informed of technical decisions and actions that affect their area of responsibility. Contractor technical recommendations and government actions are documented in peer-reviewed formal written correspondence. Unlike NASA, PowerPoint briefings and papers for technical seminars are not substitutes for completed staff work. In addition, contractors strive to provide recommendations based on a technical need, uninfluenced by headquarters or its representatives. Accordingly, division of responsibilities between the contractor and the Government remain clear, and a system of checks and balances is therefore inherent.

- **Recurring Training and Learning From Mistakes:** The Naval Reactor Program has yet to experience a reactor accident. This success is partially a testament to design, but also due to relentless and innovative training, grounded on lessons learned both inside and outside the program. For example, since 1996, Naval Reactors has educated more than 5,000 Naval Nuclear Propulsion Program personnel on the lessons learned from the *Challenger* accident.[23] Senior NASA managers recently attended the 143rd presentation of the Naval Reactors seminar entitled "The Challenger Accident Re-examined." The Board credits NASA's interest in the Navy nuclear community, and encourages the agency to continue to learn from the mistakes of other organizations as well as from its own.

- **Encouraging Minority Opinions:** The Naval Reactor Program encourages minority opinions and "bad news." Leaders continually emphasize that when no minority opinions are present, the responsibility for a thorough and critical examination falls to management. Alternate perspectives and critical questions are always encouraged. In practice, NASA does not appear to embrace these attitudes. Board interviews revealed that it is difficult for minority and dissenting opinions to percolate up through the agency's hierarchy, despite processes like the anonymous NASA Safety Reporting System that supposedly encourages the airing of opinions.

- **Retaining Knowledge:** Naval Reactors uses many mechanisms to ensure knowledge is retained. The Director serves a minimum eight-year term, and the program documents the history of the rationale for every technical requirement. Key personnel in Headquarters routinely rotate into field positions to remain familiar with every aspect of operations, training, maintenance, development and the workforce. Current and past issues are discussed in open forum with the Director and immediate staff at "all-hands" informational meetings under an in-house professional development program. NASA lacks such a program.

- **Worst-Case Event Failures:** Naval Reactors hazard analyses evaluate potential damage to the reactor plant, potential impact on people, and potential environmental impact. The Board identified NASA's failure to adequately prepare for a range of worst-case scenarios as a weakness in the agency's safety and mission assurance training programs.

SUBSAFE

The Board observed the following during its study of the Navy's SUBSAFE Program.

- SUBSAFE requirements are clearly documented and achievable, with minimal "tailoring" or granting of waivers. NASA requirements are clearly documented but are also more easily waived.

- A separate compliance verification organization independently assesses program management.[24] NASA's Flight Preparation Process, which leads to Certification of Flight Readiness, is supposed to be an independent check-and-balance process. However, the Shuttle Program's control of both engineering and safety compromises the independence of the Flight Preparation Process.

- The submarine Navy has a strong safety culture that emphasizes understanding and learning from past failures. NASA emphasizes safety as well, but training programs are not robust and methods of learning from past failures are informal.

- The Navy implements extensive safety training based on the *Thresher* and *Scorpion* accidents. NASA has not focused on any of its past accidents as a means of mentoring new engineers or those destined for management positions.

- The SUBSAFE structure is enhanced by the clarity, uniformity, and consistency of submarine safety requirements and responsibilities. Program managers are not permitted to "tailor" requirements without approval from the organization with final authority for technical requirements and the organization that verifies SUBSAFE's compliance with critical design and process requirements.[25]

- The SUBSAFE Program and implementing organization are relatively immune to budget pressures. NASA's program structure requires the Program Manager position to consider such issues, which forces the manager to juggle cost, schedule, and safety considerations. Independent advice on these issues is therefore inevitably subject to political and administrative pressure.

- Compliance with critical SUBSAFE design and process requirements is *independently verified* by a highly capable centralized organization that also "owns" the processes and monitors the program for compliance.

- Quantitative safety assessments in the Navy submarine program are deterministic rather than probabilistic. NASA does not have a quantitative, program-wide risk and safety database to support future design capabilities and assist risk assessment teams.

Comparing Navy Programs with NASA

Significant differences exist between NASA and Navy submarine programs.

- **Requirements Ownership (Technical Authority):** Both the SUBSAFE and Naval Reactors' organizational approach separates the technical and funding authority from program management in safety matters. The Board believes this separation of authority of program managers – who, by nature, must be sensitive to costs and schedules – and "owners" of technical requirements and waiver capabilities – who, by nature, are more sensitive to safety and technical rigor – is crucial. In the Naval Reactors Program, safety matters are the responsibility of the technical authority. They are not merely relegated to an independent safety organization with oversight responsibilities. This creates valuable checks and balances for safety matters in the Naval Reactors Program technical "requirements owner" community.

- **Emphasis on Lessons Learned:** Both Naval Reactors and the SUBSAFE have "institutionalized" their "lessons learned" approaches to ensure that knowledge gained from both good and bad experience is maintained in corporate memory. This has been accomplished by designating a central technical authority responsible for establishing and maintaining functional technical requirements as well as providing an organizational and institutional focus for capturing, documenting, and using operational lessons to improve future designs. NASA has an impressive history of scientific discovery, but can learn much from the application of lessons learned, especially those that relate to future vehicle design and training for contingencies. NASA has a broad Lessons Learned Information System that is strictly voluntary for program/project managers and management teams. Ideally, the Lessons Learned Information System should support overall program management and engineering functions and provide a historical experience base to aid conceptual developments and preliminary design.

The Aerospace Corporation

The Aerospace Corporation, created in 1960, operates as a Federally Funded Research and Development Center that supports the government in science and technology that is critical to national security. It is the equivalent of a $500 million enterprise that supports U.S. Air Force planning, development, and acquisition of space launch systems. The Aerospace Corporation employs approximately 3,200 people including 2,200 technical staff (29 percent Doctors of Philosophy, 41 percent Masters of Science) who conduct advanced planning, system design and integration, verify readiness, and provide technical oversight of contractors.[26]

The Aerospace Corporation's independent launch verification process offers another relevant benchmark for NASA's safety and mission assurance program. Several aspects of the Aerospace Corporation launch verification process and independent mission assurance structure could be tailored to the Shuttle Program.

Aerospace's primary product is a formal verification letter to the Air Force Systems Program Office stating a vehicle has been *independently* verified as ready for launch. The verification includes an independent General Systems Engineering and Integration review of launch preparations by Aerospace staff, a review of launch system design and payload integration, and a review of the adequacy of flight and ground hardware, software, and interfaces. This "concept-to-orbit" process begins in the design requirements phase, continues through the formal verification to countdown and launch, and concludes with a post-flight evaluation of events with findings for subsequent missions. Aerospace Corporation personnel cover the depth and breadth of space disciplines, and the organization has its own integrated engineering analysis, laboratory, and test matrix capability. This enables the Aerospace Corporation to rapidly transfer lessons learned and respond to program anomalies. Most importantly, Aerospace is uniquely independent and is not subject to any schedule or cost pressures.

The Aerospace Corporation and the Air Force have found the independent launch verification process extremely valuable. Aerospace Corporation involvement in Air Force launch verification has significantly reduced engineering errors, resulting in a 2.9 percent "probability-of-failure" rate for expendable launch vehicles, compared to 14.6 percent in the commercial sector.[27]

Conclusion

The practices noted here suggest that responsibility and authority for decisions involving technical requirements and safety should rest with an independent technical authority. Organizations that successfully operate high-risk technologies have a major characteristic in common: they place a premium on safety and reliability by structuring their programs so that technical and safety engineering organizations own the process of determining, maintaining, and waiving technical requirements with a voice that is equal to yet independent of Program Managers, who are governed by cost, schedule and mission-accomplishment goals. The Naval Reactors Program, SUBSAFE program, and the Aerospace Corporation are examples of organizations that have invested in redundant technical authorities and processes to become highly reliable.

7.4 ORGANIZATIONAL CAUSES: A BROKEN SAFETY CULTURE

Perhaps the most perplexing question the Board faced during its seven-month investigation into the *Columbia* accident was "How could NASA have missed the signals the foam was sending?" Answering this question was a challenge. The investigation revealed that in most cases, the Human Space Flight Program is extremely aggressive in reducing threats to safety. But we also know – in hindsight – that detection of the dangers posed by foam was impeded by "blind spots" in NASA's safety culture.

From the beginning, the Board witnessed a consistent lack of concern about the debris strike on *Columbia*. NASA managers told the Board "there was no safety-of-flight issue" and "we couldn't have done anything about it anyway." The investigation uncovered a troubling pattern in which Shuttle Program management made erroneous assumptions about the robustness of a system based on prior success rather than on dependable engineering data and rigorous testing.

The Shuttle Program's complex structure erected barriers to effective communication and its safety culture no longer asks enough hard questions about risk. (Safety culture refers to an organization's characteristics and attitudes – promoted by its leaders and internalized by its members – that serve to make safety the top priority.) In this context, the Board believes the mistakes that were made on STS-107 are not isolated failures, but are indicative of systemic flaws that existed prior to the accident. Had the Shuttle Program observed the principles discussed in the previous two sections, the threat that foam posed to the Orbiter, particularly after the STS-112 and STS-107 foam strikes, might have been more fully appreciated by Shuttle Program management.

In this section, the Board examines the NASA's safety policy, structure, and process, communication barriers, the risk assessment systems that govern decision-making and risk management, and the Shuttle Program's penchant for substituting analysis for testing.

NASA's Safety: Policy, Structure, and Process

Safety Policy

NASA's current philosophy for safety and mission assurance calls for centralized policy and oversight at Headquarters and decentralized execution of safety programs at the enterprise, program, and project levels. Headquarters dictates what must be done, not how it should be done. The operational premise that logically follows is that safety is the responsibility of program and project managers. Managers are subsequently given flexibility to organize safety efforts as they see fit, while NASA Headquarters is charged with maintaining oversight through independent surveillance and assessment.[28] NASA policy dictates that safety programs should be placed high enough in the organization, and be vested with enough authority and seniority, to "maintain independence." Signals of potential danger, anomalies, and critical information should, in principle, surface in the hazard identification process and be tracked with risk assessments supported by engineering analyses. In reality, such a process demands a more independent status than NASA has ever been willing to give its safety organizations, despite the recommendations of numerous outside experts over nearly two decades, including the Rogers Commission (1986), General Accounting Office (1990), and the Shuttle Independent Assessment Team (2000).

Safety Organization Structure

Center safety organizations that support the Shuttle Program are tailored to the missions they perform. Johnson and

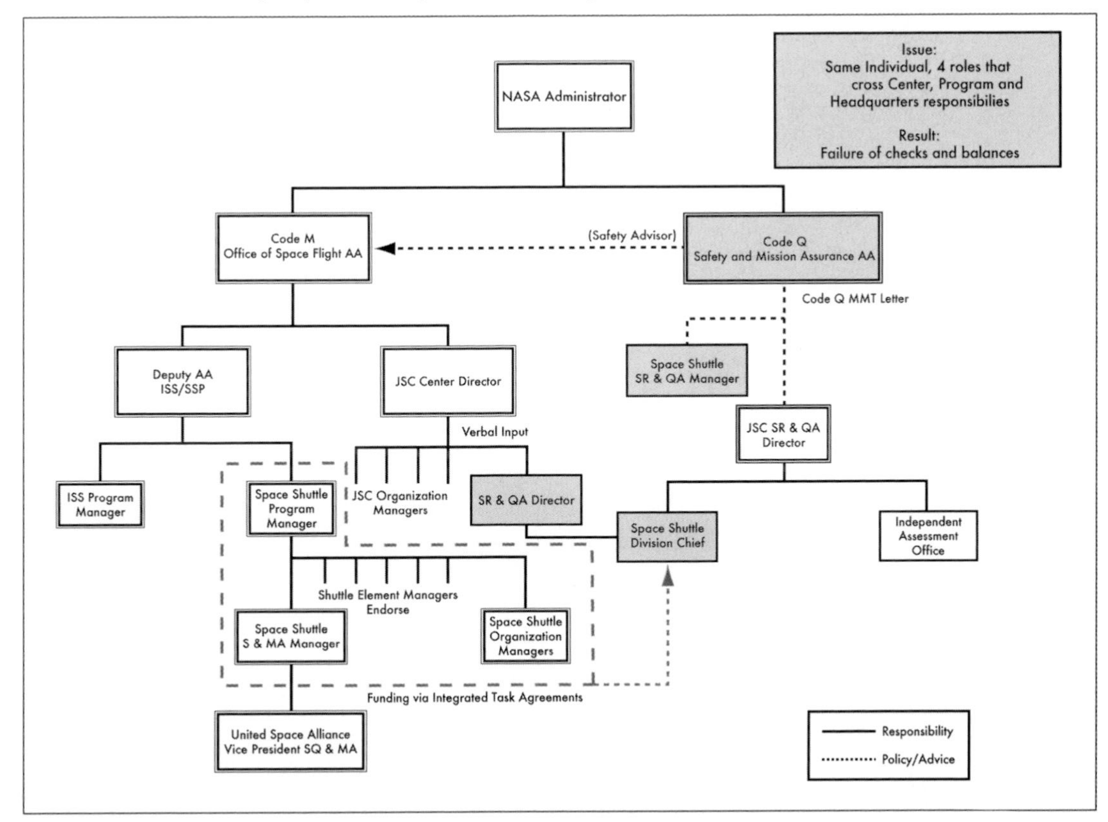

Figure 7.4-1. Independent safety checks and balance failure.

Marshall Safety and Mission Assurance organizations are organized similarly. In contrast, Kennedy has decentralized its Safety and Mission Assurance components and assigned them to the Shuttle Processing Directorate. This management change renders Kennedy's Safety and Mission Assurance structure even more dependent on the Shuttle Program, which reduces effective oversight.

At Johnson, safety programs are centralized under a Director who oversees five divisions and an Independent Assessment Office. Each division has clearly-defined roles and responsibilities, with the exception of the Space Shuttle Division Chief, whose job description does not reflect the full scope of authority and responsibility ostensibly vested in the position. Yet the Space Shuttle Division Chief is empowered to represent the Center, the Shuttle Program, and NASA Headquarters Safety and Mission Assurance at critical junctures in the safety process. The position therefore represents a critical node in NASA's Safety and Mission Assurance architecture that seems to the Board to be plagued by conflict of interest. It is a single point of failure without any checks or balances.

Johnson also has a Shuttle Program Safety and Mission Assurance Manager who oversees United Space Alliance's safety organization. The Shuttle Program further receives program safety support from the Center's Safety, Reliability, and Quality Assurance Space Shuttle Division. Johnson's Space Shuttle Division Chief has the additional role of Shuttle Program Safety, Reliability, and Quality Assurance Manager (see Figure 7.4-1). Over the years, this dual designation has resulted in a general acceptance of the fact that the Johnson Space Shuttle Division Chief performs duties on both the Center's and Program's behalf. The detached nature of the support provided by the Space Shuttle Division Chief, and the wide band of the position's responsibilities throughout multiple layers of NASA's hierarchy, confuses lines of authority, responsibility, and accountability in a manner that almost defies explanation.

A March 2001 NASA Office of Inspector General Audit Report on Space Shuttle Program Management Safety Observations made the same point:

> *The job descriptions and responsibilities of the Space Shuttle Program Manager and Chief, Johnson Safety Office Space Shuttle Division, are nearly identical with each official reporting to a different manager. This overlap in responsibilities conflicts with the SFOC* [Space Flight Operations Contract] *and NSTS 07700, which requires the Chief, Johnson Safety Office Space Shuttle Division, to provide matrixed personnel support to the Space Shuttle Program Safety Manager in fulfilling requirements applicable to the safety, reliability, and quality assurance aspects of the Space Shuttle Program.*

The fact that Headquarters, Center, and Program functions are rolled-up into one position is an example of how a carefully designed oversight process has been circumvented and made susceptible to conflicts of interest. This organizational construct is unnecessarily bureaucratic and defeats NASA's stated objective of providing an independent safety function. A similar argument can be made about the placement of quality assurance in the Shuttle Processing Divisions at Kennedy, which increases the risk that quality assurance personnel will become too "familiar" with programs they are charged to oversee, which hinders oversight and judgment.

The Board believes that although the Space Shuttle Program has effective safety practices at the "shop floor" level, its operational and systems safety program is flawed by its dependence on the Shuttle Program. Hindered by a cumbersome organizational structure, chronic understaffing, and poor management principles, the safety apparatus is not currently capable of fulfilling its mission. An independent safety structure would provide the Shuttle Program a more effective operational safety process. Crucial components of this structure include a comprehensive integration of safety across all the Shuttle programs and elements, and a more independent system of checks and balances.

Safety Process

In response to the Rogers Commission Report, NASA established what is now known as the Office of Safety and Mission Assurance at Headquarters to independently monitor safety and ensure communication and accountability agency-wide. The Office of Safety and Mission Assurance monitors unusual events like "out of family" anomalies and establishes agency-wide Safety and Mission Assurance policy. (An out-of-family event is an operation or performance outside the expected performance range for a given parameter or which has not previously been experienced.) The Office of Safety and Mission Assurance also screens the Shuttle Program's Flight Readiness Process and signs the Certificate of Flight Readiness. The Shuttle Program Manager, in turn, is responsible for overall Shuttle safety and is supported by a one-person safety staff.

The Shuttle Program has been permitted to organize its safety program as it sees fit, which has resulted in a lack of standardized structure throughout NASA's various Centers, enterprises, programs, and projects. The level of funding a program is granted impacts how much safety the Program can "buy" from a Center's safety organization. In turn, Safety and Mission Assurance organizations struggle to anticipate program requirements and guarantee adequate support for the many programs for which they are responsible.

It is the Board's view, shared by previous assessments, that the current safety system structure leaves the Office of Safety and Mission Assurance ill-equipped to hold a strong and central role in integrating safety functions. NASA Headquarters has not effectively integrated safety efforts across its culturally and technically distinct Centers. In addition, the practice of "buying" safety services establishes a relationship in which programs sustain the very livelihoods of the safety experts hired to oversee them. These idiosyncrasies of structure and funding preclude the safety organization from effectively providing independent safety analysis.

The commit-to-flight review process, as described in Chapters 2 and 6, consists of program reviews and readiness polls that are structured to allow NASA's senior leaders to assess mission readiness. In like fashion, safety organizations affiliated with various projects, programs, and Centers at NASA, conduct a Pre-launch Assessment Review of safety preparations and mission concerns. The Shuttle Program does not officially sanction the Pre-launch Assessment Review, which updates the Associate Administrator for Safety and Mission Assurance on safety concerns during the Flight Readiness Review/Certification of Flight Readiness process.

The Johnson Space Shuttle Safety, Reliability, and Quality Assurance Division Chief orchestrates this review on behalf of Headquarters. Note that this division chief also advises the Shuttle Program Manager of Safety. Because it lacks independent analytical rigor, the Pre-launch Assessment Review is only marginally effective. In this arrangement, the Johnson Shuttle Safety, Reliability, and Quality Assurance Division Chief is expected to *render an independent assessment of his own activities*. Therefore, the Board is concerned that the Pre-Launch Assessment Review is not an effective check and balance in the Flight Readiness Review.

Given that the entire Safety and Mission Assurance organization depends on the Shuttle Program for resources and simultaneously lacks the independent ability to conduct detailed analyses, cost and schedule pressures can easily and unintentionally influence safety deliberations. Structure and process places Shuttle safety programs in the unenviable position of having to choose between rubber-stamping engineering analyses, technical efforts, and Shuttle program decisions, or trying to carry the day during a committee meeting in which the other side almost always has more information and analytic capability.

NASA Barriers to Communication: Integration, Information Systems, and Databases

By their very nature, high-risk technologies are exceptionally difficult to manage. Complex and intricate, they consist of numerous interrelated parts. Standing alone, components may function adequately, and failure modes may be anticipated. Yet when components are integrated into a total system and work in concert, unanticipated interactions can occur that can lead to catastrophic outcomes.[29] The risks inherent in these technical systems are heightened when they are produced and operated by complex organizations that can also break down in unanticipated ways. The Shuttle Program is such an organization. All of these factors make effective communication – between individuals and between programs – absolutely critical. However, the structure and complexity of the Shuttle Program hinders communication.

The Shuttle Program consists of government and contract personnel who cover an array of scientific and technical disciplines and are affiliated with various dispersed space, research, and test centers. NASA derives its organizational complexity from its origins as much as its widely varied missions. NASA Centers naturally evolved with different points of focus, a "divergence" that the Rogers Commission found evident in the propensity of Marshall personnel to resolve problems without including program managers outside their Center – especially managers at Johnson, to whom they officially reported (see Chapter 5).

Despite periodic attempts to emphasize safety, NASA's frequent reorganizations in the drive to become more efficient reduced the budget for safety, sending employees conflicting messages and creating conditions more conducive to the development of a conventional bureaucracy than to the maintenance of a safety-conscious research-and-development organization. Over time, a pattern of ineffective communication has resulted, leaving risks improperly defined, problems unreported, and concerns unexpressed.[30] The question is, why?

The transition to the Space Flight Operations Contract – and the effects it initiated – provides part of the answer. In the Space Flight Operations Contract, NASA encountered a completely new set of structural constraints that hindered effective communication. New organizational and contractual requirements demanded an even more complex system of shared management reviews, reporting relationships, safety oversight and insight, and program information development, dissemination, and tracking.

The Shuttle Independent Assessment Team's report documented these changes, noting that "the size and complexity of the Shuttle system and of the NASA/contractor relationships place extreme importance on understanding, communication, and information handling."[31] Among other findings, the Shuttle Independent Assessment Team observed that:

- The current Shuttle program culture is too insular
- There is a potential for conflicts between contractual and programmatic goals
- There are deficiencies in problem and waiver-tracking systems
- The exchange of communication across the Shuttle program hierarchy is structurally limited, both upward and downward.[32]

The Board believes that deficiencies in communication, including those spelled out by the Shuttle Independent Assessment Team, were a foundation for the *Columbia* accident. These deficiencies are byproducts of a cumbersome, bureaucratic, and highly complex Shuttle Program structure and the absence of authority in two key program areas that are responsible for integrating information across all programs and elements in the Shuttle program.

Integration Structures

NASA did not adequately prepare for the consequences of adding organizational structure and process complexity in the transition to the Space Flight Operations Contract. The agency's lack of a centralized clearinghouse for integration and safety further hindered safe operations. In the Board's opinion, the Shuttle Integration and Shuttle Safety, Reliability, and Quality Assurance Offices do not fully integrate information on behalf of the Shuttle Program. This is due, in part, to an irregular division of responsibilities between the Integration Office and the Orbiter Vehicle Engineering Office and the absence of a truly independent safety organization.

Within the Shuttle Program, the Orbiter Office handles many key integration tasks, even though the Integration Office appears to be the more logical office to conduct them; the Orbiter Office does not actively participate in the Integration Control Board; and Orbiter Office managers are actually ranked above their Integration Office counterparts. These uncoordinated roles result in conflicting and erroneous information, and support the perception that the Orbiter Office is isolated from the Integration Office and has its own priorities.

The Shuttle Program's structure and process for Safety and Mission Assurance activities further confuse authority and responsibility by giving the Program's Safety and Mission Assurance Manager technical oversight of the safety aspects of the Space Flight Operations Contract, while simultaneously making the Johnson Space Shuttle Division Chief responsible for advising the Program on safety performance. As a result, no one office or person in Program management is responsible for developing an integrated risk assessment above the sub-system level that would provide a comprehensive picture of total program risks. The net effect is that many Shuttle Program safety, quality, and mission assurance roles are never clearly defined.

Safety Information Systems

Numerous reviews and independent assessments have noted that NASA's safety system does not effectively manage risk. In particular, these reviews have observed that the processes in which NASA tracks and attempts to mitigate the risks posed by components on its Critical Items List is flawed. The Post Challenger Evaluation of Space Shuttle Risk Assessment and Management Report (1988) concluded that:

> *The committee views NASA critical items list (CIL) waiver decision-making process as being subjective, with little in the way of formal and consistent criteria for approval or rejection of waivers. Waiver decisions appear to be driven almost exclusively by the design based Failure Mode Effects Analysis (FMEA)/CIL retention rationale, rather than being based on an integrated assessment of all inputs to risk management. The retention rationales appear biased toward proving that the design is "safe," sometimes ignoring significant evidence to the contrary.*

The report continues, "… the Committee has not found an independent, detailed analysis or assessment of the CIL retention rationale which considers all inputs to the risk assessment process."[33] Ten years later, the Shuttle Independent Assessment Team reported "Risk Management process erosion created by the desire to reduce costs …" [34] The Shuttle Independent Assessment Team argued strongly that NASA Safety and Mission Assurance should be restored to its previous role of an independent oversight body, and Safety and Mission Assurance not be simply a "safety auditor."

The Board found similar problems with integrated hazard analyses of debris strikes on the Orbiter. In addition, the information systems supporting the Shuttle – intended to be tools for decision-making – are extremely cumbersome and difficult to use at any level.

The following addresses the hazard tracking tools and major databases in the Shuttle Program that promote risk management.

- **Hazard Analysis:** A fundamental element of system safety is managing and controlling hazards. NASA's only guidance on hazard analysis is outlined in the Methodology for Conduct of Space Shuttle Program Hazard Analysis, which merely lists tools available.[35] Therefore, it is not surprising that hazard analysis processes are applied inconsistently across systems, subsystems, assemblies, and components.

  United Space Alliance, which is responsible for both Orbiter integration and Shuttle Safety Reliability and Quality Assurance, delegates hazard analysis to Boeing. However, as of 2001, the Shuttle Program no longer requires Boeing to conduct integrated hazard analyses. Instead, Boeing now performs hazard analysis only at the sub-system level. In other words, Boeing analyzes hazards to components and elements, but is not required to consider the Shuttle as a whole. Since the current Failure Mode Effects Analysis/Critical Item List process is designed for bottom-up analysis at the component level, it cannot effectively support the kind of "top-down" hazard analysis that is needed to inform managers on risk trends and identify potentially harmful interactions between systems.

  The Critical Item List (CIL) tracks 5,396 individual Shuttle hazards, of which 4,222 are termed "Critical-

Space Shuttle Safety Upgrade Program

NASA presented a Space Shuttle Safety Upgrade Initiative to Congress as part of its Fiscal Year 2001 budget in March 2000. This initiative sought to create a "Pro-active upgrade program to keep Shuttle flying safely and efficiently to 2012 and beyond to meet agency commitments and goals for human access to space."

The planned Shuttle safety upgrades included: Electric Auxiliary Power Unit, Improved Main Landing Gear Tire, Orbiter Cockpit/Avionics Upgrades, Space Shuttle Main Engine Advanced Health Management System, Block III Space Shuttle Main Engine, Solid Rocket Booster Thrust Vector Control/Auxiliary Power Unit Upgrades Plan, Redesigned Solid Rocket Motor – Propellant Grain Geometry Modification, and External Tank Upgrades – Friction Stir Weld. The plan called for the upgrades to be completed by 2008.

However, as discussed in Chapter 5, every proposed safety upgrade – with a few exceptions – was either not approved or was deferred.

The irony of the Space Shuttle Safety Upgrade Program was that the strategy placed emphasis on keeping the "Shuttle flying safely and efficiently to 2012 and beyond," yet the Space Flight Leadership Council accepted the upgrades **only as long as they were financially feasible**. *Funding a safety upgrade in order to fly safely, and then canceling it for budgetary reasons, makes the concept of mission safety rather hollow.*

ity 1/1R." Of those, 3,233 have waivers. CRIT 1/1R component failures are defined as those that will result in loss of the Orbiter and crew. Waivers are granted whenever a Critical Item List component cannot be redesigned or replaced. More than 36 percent of these waivers have not been reviewed in 10 years, a sign that NASA is not aggressively monitoring changes in system risk.

It is worth noting that the Shuttle's Thermal Protection System is on the Critical Item List, and an existing hazard analysis and hazard report deals with debris strikes. As discussed in Chapter 6, Hazard Report #37 is ineffectual as a decision aid, yet the Shuttle Program never challenged its validity at the pivotal STS-113 Flight Readiness Review.

Although the Shuttle Program has undoubtedly learned a great deal about the technological limitations inherent in Shuttle operations, it is equally clear that risk – as represented by the number of critical items list and waivers – has grown substantially without a vigorous effort to assess and reduce technical problems that increase risk. An information system bulging with over 5,000 critical items and 3,200 waivers is exceedingly difficult to manage.

- **Hazard Reports:** Hazard reports, written either by the Space Shuttle Program or a contractor, document conditions that threaten the safe operation of the Shuttle. Managers use these reports to evaluate risk and justify flight.[36] During mission preparations, contractors and Centers review all baseline hazard reports to ensure they are current and technically correct.

Board investigators found that a large number of hazard reports contained subjective and qualitative judgments, such as "believed" and "based on experience from previous flights this hazard is an 'Accepted Risk.'" A critical ingredient of a healthy safety program is the rigorous implementation of technical standards. These standards must include more than hazard analysis or low-level technical activities. Standards must integrate project engineering and management activities. Finally, a mechanism for feedback on the effectiveness of system safety engineering and management needs to be built into procedures to learn if safety engineering and management methods are weakening over time.

Dysfunctional Databases

In its investigation, the Board found that the information systems that support the Shuttle program are extremely cumbersome and difficult to use in decision-making at any level. For obvious reasons, these shortcomings imperil the Shuttle Program's ability to disseminate and share critical information among its many layers. This section explores the report databases that are crucial to effective risk management.

- **Problem Reporting and Corrective Action:** The Problem Reporting and Corrective Action database records any non-conformances (instances in which a requirement is not met). Formerly, different Centers and contractors used the Problem Reporting and Corrective Action database differently, which prevented comparisons across the database. NASA recently initiated an effort to integrate these databases to permit anyone in the agency to access information from different Centers. This system, Web Program Compliance Assurance and Status System (WEBPCASS), is supposed to provide easier access to consolidated information and facilitates higher-level searches.

However, NASA safety managers have complained that the system is too time-consuming and cumbersome. Only employees trained on the database seem capable of using WEBPCASS effectively. One particularly frustrating aspect of which the Board is acutely aware is the database's waiver section. It is a critical information source, but only the most expert users can employ it effectively. The database is also incomplete. For instance, in the case of foam strikes on the Thermal Protection System, only strikes that were declared "In-Fight Anomalies" are added to the Problem Reporting and Corrective Action database, which masks the full extent of the foam debris trends.

- **Lessons Learned Information System:** The Lessons Learned Information System database is a much simpler system to use, and it can assist with hazard identification and risk assessment. However, personnel familiar with the Lessons Learned Information System indicate that design engineers and mission assurance personnel use it only on an *ad hoc* basis, thereby limiting its utility. The Board is not the first to note such deficiencies. Numerous reports, including most recently a General Accounting Office 2001 report, highlighted fundamental weaknesses in the collection and sharing of lessons learned by program and project managers.[37]

Conclusions

Throughout the course of this investigation, the Board found that the Shuttle Program's complexity demands highly effective communication. Yet integrated hazard reports and risk analyses are rarely communicated effectively, nor are the many databases used by Shuttle Program engineers and managers capable of translating operational experiences into effective risk management practices. Although the Space Shuttle system has conducted a relatively small number of missions, there is more than enough data to generate performance trends. As it is currently structured, the Shuttle Program does not use data-driven safety methodologies to their fullest advantage.

7.5 Organizational Causes: Impact of a Flawed Safety Culture on STS-107

In this section, the Board examines how and why an array of processes, groups, and individuals in the Shuttle Program failed to appreciate the severity and implications of the foam strike on STS-107. The Board believes that the Shuttle Program should have been able to detect the foam trend and more fully appreciate the danger it represented. Recall that "safety culture" refers to the collection of characteristics and attitudes in an organization – promoted by its leaders and internalized by its members – that makes safety an overriding priority. In the following analysis, the Board outlines shortcomings in the Space Shuttle Program, Debris Assessment Team, and Mission Management Team that resulted from a flawed safety culture.

Shuttle Program Shortcomings

The flight readiness process, which involves every organization affiliated with a Shuttle mission, missed the danger signals in the history of foam loss.

Generally, the higher information is transmitted in a hierarchy, the more it gets "rolled-up," abbreviated, and simplified. Sometimes information gets lost altogether, as weak signals drop from memos, problem identification systems, and formal presentations. The same conclusions, repeated over time, can result in problems eventually being deemed non-problems. An extraordinary example of this phenomenon is how Shuttle Program managers assumed the foam strike on STS-112 was not a warning sign (see Chapter 6).

During the STS-113 Flight Readiness Review, the bipod foam strike to STS-112 was rationalized by simply restating earlier assessments of foam loss. The question of why bipod foam would detach and strike a Solid Rocket Booster spawned no further analysis or heightened curiosity; nor did anyone challenge the weakness of External Tank Project Manager's argument that backed launching the next mission. After STS-113's successful flight, once again the STS-112 foam event was not discussed at the STS-107 Flight Readiness Review. The failure to mention an outstanding technical anomaly, even if not technically a violation of NASA's own procedures, desensitized the Shuttle Program to the dangers of foam striking the Thermal Protection System, and demonstrated just how easily the flight preparation process can be compromised. In short, the dangers of bipod foam got "rolled-up," which resulted in a missed opportunity to make Shuttle managers aware that the Shuttle required, and did not yet have a fix for the problem.

Once the *Columbia* foam strike was discovered, the Mission Management Team Chairperson asked for the rationale the STS-113 Flight Readiness Review used to launch in spite of the STS-112 foam strike. In her e-mail, she admitted that the analysis used to continue flying was, in a word, "lousy" (Chapter 6). This admission – that the rationale to fly was rubber-stamped – is, to say the least, unsettling.

The Flight Readiness process is supposed to be shielded from outside influence, and is viewed as both rigorous and systematic. Yet the Shuttle Program is inevitably influenced by external factors, including, in the case of the STS-107, schedule demands. Collectively, such factors shape how the Program establishes mission schedules and sets budget priorities, which affects safety oversight, workforce levels, facility maintenance, and contractor workloads. Ultimately, external expectations and pressures impact even data collection, trend analysis, information development, and the reporting and disposition of anomalies. These realities contradict NASA's optimistic belief that pre-flight reviews provide true safeguards against unacceptable hazards. The schedule pressure to launch International Space Station Node 2 is a powerful example of this point (Section 6.2).

The premium placed on maintaining an operational schedule, combined with ever-decreasing resources, gradually led Shuttle managers and engineers to miss signals of potential danger. Foam strikes on the Orbiter's Thermal Protection System, no matter what the size of the debris, were "normalized" and accepted as not being a "safety-of-flight risk." Clearly, the risk of Thermal Protection damage due to such a strike needed to be better understood in quantifiable terms. External Tank foam loss should have been eliminated or mitigated with redundant layers of protection. If there was in fact a strong safety culture at NASA, safety experts would have had the authority to test the actual resilience of the leading edge Reinforced Carbon-Carbon panels, as the Board has done.

Debris Assessment Team Shortcomings

Chapter Six details the Debris Assessment Team's efforts to obtain additional imagery of *Columbia*. When managers in the Shuttle Program denied the team's request for imagery, the Debris Assessment Team was put in the untenable position of having to prove that a safety-of-flight issue existed without the very images that would permit such a determination. This is precisely the opposite of how an effective safety culture would act. Organizations that deal with high-risk operations must always have a healthy fear of failure – operations must be proved safe, rather than the other way around. NASA inverted this burden of proof.

Another crucial failure involves the Boeing engineers who conducted the Crater analysis. The Debris Assessment Team relied on the inputs of these engineers along with many others to assess the potential damage caused by the foam strike. Prior to STS-107, Crater analysis was the responsibility of a team at Boeing's Huntington Beach facility in California, but this responsibility had recently been transferred to Boeing's Houston office. In October 2002, the Shuttle Program completed a risk assessment that predicted the move of Boeing functions from Huntington Beach to Houston would increase risk to Shuttle missions through the end of 2003, because of the small number of experienced engineers who were willing to relocate. To mitigate this risk, NASA and United Space Alliance developed a transition plan to run through January 2003.

The Board has discovered that the implementation of the transition plan was incomplete and that training of replacement personnel was not uniform. STS-107 was the first mission during which Johnson-based Boeing engineers conducted analysis without guidance and oversight from engineers at Huntington Beach.

Even though STS-107's debris strike was 400 times larger than the objects Crater is designed to model, neither Johnson engineers nor Program managers appealed for assistance from the more experienced Huntington Beach engineers,

ENGINEERING BY VIEWGRAPHS

The Debris Assessment Team presented its analysis in a formal briefing to the Mission Evaluation Room that relied on PowerPoint slides from Boeing. When engineering analyses and risk assessments are condensed to fit on a standard form or overhead slide, information is inevitably lost. In the process, the priority assigned to information can be easily misrepresented by its placement on a chart and the language that is used. Dr. Edward Tufte of Yale University, an expert in information presentation who also researched communications failures in the *Challenger* accident, studied how the slides used by the Debris Assessment Team in their briefing to the Mission Evaluation Room misrepresented key information.[38]

The slide created six levels of hierarchy, signified by the title and the symbols to the left of each line. These levels prioritized information that was already contained in 11 simple sentences. Tufte also notes that the title is confusing. "Review of Test Data Indicates Conservatism" refers not to the predicted tile damage, *but to the choice of test models used to predict the damage.*

Only at the bottom of the slide do engineers state a key piece of information: that one estimate of the debris that struck *Columbia* was 640 times larger than the data used to calibrate the model on which engineers based their damage assessments. (Later analysis showed that the debris object was actually 400 times larger). This difference led Tufte to suggest that a more appropriate headline would be "Review of Test Data Indicates Irrelevance of Two Models."[39]

Tufte also criticized the sloppy language on the slide. "The vaguely quantitative words 'significant' and 'significantly' are used 5 times on this slide," he notes, "with de facto meanings ranging from 'detectable in largely irrelevant calibration case study' to 'an amount of damage so that everyone dies' to 'a difference of 640-fold.' "[40] Another example of sloppiness is that "cubic inches" is written inconsistently: "3cu. In," "1920cu in," and "3 cu in." While such inconsistencies might seem minor, in highly technical fields like aerospace engineering a misplaced decimal point or mistaken unit of measurement can easily engender inconsistencies and inaccuracies. In another phrase "Test results do show that it is possible at sufficient mass and velocity," the word "it" actually refers to "damage to the protective tiles."

As information gets passed up an organization hierarchy, from people who do analysis to mid-level managers to high-level leadership, key explanations and supporting information is filtered out. In this context, it is easy to understand how a senior manager might read this PowerPoint slide and not realize that it addresses a life-threatening situation.

At many points during its investigation, the Board was surprised to receive similar presentation slides from NASA officials in place of technical reports. The Board views the endemic use of PowerPoint briefing slides instead of technical papers as an illustration of the problematic methods of technical communication at NASA.

The vaguely quantitative words "significant" and "significantly" are used 5 times on this slide, with *de facto* meanings ranging from "detectable in largely irrelevant calibration case study" to "an amount of damage so that everyone dies" to "a difference of 640-fold." None of these 5 usages appears to refer to the technical meaning of "statistical significance."

The low resolution of PowerPoint slides promotes the use of compressed phrases like "Tile Penetration." As is the case here, such phrases may well be ambiguous. (The low resolution and large font generate 3 typographic orphans, lonely words dangling on a seperate line.)

This vague pronoun reference "it" alludes to *damage to the protective tiles*, which caused the destruction of the Columbia. The slide weakens important material with ambiguous language (sentence fragments, passive voice, multiple meanings of "significant"). The 3 reports were created by engineers for high-level NASA officials who were deciding whether the threat of wing damage required further investigation before the Columbia attempted return. The officials were satisfied that the reports indicated that the Columbia was not in danger, and no attempts to further examine the threat were made. The slides were part of an oral presentation and also were circulated as e-mail attachments.

In this slide the same unit of measure for volume (cubic inches) is shown a different way every time

3cu. in **1920cu. in** **3 cu. in**

rather than in clear and tidy exponential form **1920 in³**. Perhaps the available font cannot show exponents. Shakiness in units of measurement provokes concern. Slides that use hierarchical bullet-outlines here do not handle statistical data and scientific notation gracefully. If PowerPoint is a corporate-mandated format for all engineering reports, then some competent scientific typography (rather than the PP market-pitch style) is essential. In this slide, the typography is so choppy and clunky that it impedes understanding.

The analysis by Dr. Edward Tufte of the slide from the Debris Assessment Team briefing. [SOFI=Spray-On Foam Insulation] who might have cautioned against using Crater so far outside its validated limits. Nor did safety personnel provide any additional oversight. NASA failed to connect the dots: the engineers who misinterpreted Crater – a tool already unsuited to the task at hand – were the very ones the Shuttle Program identified as engendering the most risk in their transition from Huntington Beach. The Board views this example as characteristic of the greater turbulence the Shuttle Program experienced in the decade before *Columbia* as a result of workforce reductions and management reforms.

Mission Management Team Shortcomings

In the Board's view, the decision to fly STS-113 without a compelling explanation for why bipod foam had separated on ascent during the preceding mission, combined with the low number of Mission Management Team meetings during STS-107, indicates that the Shuttle Program had become overconfident. Over time, the organization determined it did not need daily meetings during a mission, despite regulations that state otherwise.

Status update meetings should provide an opportunity to raise concerns and hold discussions across structural and technical boundaries. The leader of such meetings must encourage participation and guarantee that problems are assessed and resolved fully. All voices must be heard, which can be difficult when facing a hierarchy. An employee's location in the hierarchy can encourage silence. Organizations interested in safety must take steps to guarantee that all relevant information is presented to decision-makers. This did not happen in the meetings during the *Columbia* mission (see Chapter 6). For instance, e-mails from engineers at Johnson and Langley conveyed the depth of their concern about the foam strike, the questions they had about its implications, and the actions they wanted to take as a follow-up. However, these e-mails did not reach the Mission Management Team.

The failure to convey the urgency of engineering concerns was caused, at least in part, by organizational structure and spheres of authority. The Langley e-mails were circulated among co-workers at Johnson who explored the possible effects of the foam strike and its consequences for landing. Yet, like Debris Assessment Team Co-Chair Rodney Rocha, they kept their concerns within local channels and did not forward them to the Mission Management Team. They were separated from the decision-making process by distance and rank.

Similarly, Mission Management Team participants felt pressured to remain quiet unless discussion turned to their particular area of technological or system expertise, and, even then, to be brief. The initial damage assessment briefing prepared for the Mission Evaluation Room was cut down considerably in order to make it "fit" the schedule. Even so, it took 40 minutes. It was cut down further to a three-minute discussion topic at the Mission Management Team. Tapes of STS-107 Mission Management Team sessions reveal a noticeable "rush" by the meeting's leader to the preconceived bottom line that there was "no safety-of-flight" issue (see Chapter 6). Program managers created huge barriers against dissenting opinions by stating preconceived conclusions based on subjective knowledge and experience, rather than on solid data. Managers demonstrated little concern for mission safety.

Organizations with strong safety cultures generally acknowledge that a leader's best response to unanimous consent is to play devil's advocate and encourage an exhaustive debate. Mission Management Team leaders failed to seek out such minority opinions. Imagine the difference if any Shuttle manager had simply asked, "Prove to me that *Columbia* has *not* been harmed."

Similarly, organizations committed to effective communication seek avenues through which unidentified concerns and dissenting insights can be raised, so that weak signals are not lost in background noise. Common methods of bringing minority opinions to the fore include hazard reports, suggestion programs, and empowering employees to call "time out" (Chapter 10). For these methods to be effective, they must mitigate the fear of retribution, and management and technical staff must pay attention. Shuttle Program hazard reporting is seldom used, safety time outs are at times disregarded, and informal efforts to gain support are squelched. The very fact that engineers felt inclined to conduct simulated blown tire landings at Ames "after hours," indicates their reluctance to bring the concern up in established channels.

Safety Shortcomings

The Board believes that the safety organization, due to a lack of capability and resources independent of the Shuttle Program, was not an effective voice in discussing technical issues or mission operations pertaining to STS-107. The safety personnel present in the Debris Assessment Team, Mission Evaluation Room, and on the Mission Management Team were largely silent during the events leading up to the loss of *Columbia*. That silence was not merely a failure of safety, but a failure of the entire organization.

7.6 FINDINGS AND RECOMMENDATIONS

The evidence that supports the organizational causes also led the Board to conclude that NASA's current organization, which combines in the Shuttle Program all authority and responsibility for schedule, cost, manifest, safety, technical requirements, and waivers to technical requirements, is not an effective check and balance to achieve safety and mission assurance. Further, NASA's Office of Safety and Mission Assurance does not have the independence and authority that the Board and many outside reviews believe is necessary. Consequently, the Space Shuttle Program does not consistently demonstrate the characteristics of organizations that effectively manage high risk. Therefore, the Board offers the following Findings and Recommendations:

Findings:

F7.1-1 Throughout its history, NASA has consistently struggled to achieve viable safety programs and adjust them to the constraints and vagaries of changing budgets. Yet, according to multiple high level independent reviews, NASA's safety system has fallen short of the mark.

F7.4-1 The Associate Administrator for Safety and Mission Assurance is not responsible for safety and mission assurance execution, as intended by the Rogers Commission, but is responsible for Safety and Mission Assurance policy, advice, coordination, and budgets. This view is consistent with NASA's recent philosophy of management at a strategic level at NASA Headquarters but contrary to the Rogers' Commission recommendation.

F7.4-2 Safety and Mission Assurance organizations supporting the Shuttle Program are largely dependent upon the Program for funding, which hampers their status as independent advisors.

F7.4-3 Over the last two decades, little to no progress has been made toward attaining integrated, independent, and detailed analyses of risk to the Space Shuttle system.

F7.4-4 System safety engineering and management is separated from mainstream engineering, is not vigorous enough to have an impact on system design, and is hidden in the other safety disciplines at NASA Headquarters.

F7.4-5 Risk information and data from hazard analyses are not communicated effectively to the risk assessment and mission assurance processes. The Board could not find adequate application of a process, database, or metric analysis tool that took an integrated, systemic view of the entire Space Shuttle system.

F7.4-6 The Space Shuttle Systems Integration Office handles all Shuttle systems except the Orbiter. Therefore, it is not a true integration office.

F7.4-7 When the Integration Office convenes the Integration Control Board, the Orbiter Office usually does not send a representative, and its staff makes verbal inputs only when requested.

F7.4-8 The Integration office did not have continuous responsibility to integrate responses to bipod foam shedding from various offices. Sometimes the Orbiter Office had responsibility, sometimes the External Tank Office at Marshall Space Flight Center had responsibility, and sometime the bipod shedding did not result in any designation of an In-Flight Anomaly. Integration did not occur.

F7.4-9 NASA information databases such as The Problem Reporting and Corrective Action and the Web Program Compliance Assurance and Status System are marginally effective decision tools.

F7.4-10 Senior Safety, Reliability & Quality Assurance and element managers do not use the Lessons Learned Information System when making decisions. NASA subsequently does not have a constructive program to use past lessons to educate engineers, managers, astronauts, or safety personnel.

F7.4-11 The Space Shuttle Program has a wealth of data tucked away in multiple databases without a convenient way to integrate and use the data for management, engineering, or safety decisions.

F7.4-12 The dependence of Safety, Reliability & Quality Assurance personnel on Shuttle Program support limits their ability to oversee operations and communicate potential problems throughout the organization.

F7.4-13 There are conflicting roles, responsibilities, and guidance in the Space Shuttle safety programs. The Safety & Mission Assurance Pre-Launch Assessment Review process is not recognized by the Space Shuttle Program as a requirement that must be followed (NSTS 22778). Failure to consistently apply the Pre-Launch Assessment Review as a requirements document creates confusion about roles and responsibilities in the NASA safety organization.

Recommendations:

R7.5-1 Establish an independent Technical Engineering Authority that is responsible for technical requirements and all waivers to them, and will build a disciplined, systematic approach to identifying, analyzing, and controlling hazards throughout the life cycle of the Shuttle System. The independent technical authority does the following as a minimum:

- Develop and maintain technical standards for all Space Shuttle Program projects and elements
- Be the sole waiver-granting authority for all technical standards
- Conduct trend and risk analysis at the sub-system, system, and enterprise levels
- Own the failure mode, effects analysis and hazard reporting systems
- Conduct integrated hazard analysis
- Decide what is and is not an anomalous event
- Independently verify launch readiness
- Approve the provisions of the recertification program called for in Recommendation R9.1-1

The Technical Engineering Authority should be funded directly from NASA Headquarters, and should have no connection to or responsibility for schedule or program cost.

R7.5-2 NASA Headquarters Office of Safety and Mission Assurance should have direct line authority over the entire Space Shuttle Program safety organization and should be independently resourced.

R7.5-3 Reorganize the Space Shuttle Integration Office to make it capable of integrating all elements of the Space Shuttle Program, including the Orbiter.

ENDNOTES FOR CHAPTER 7

The citations that contain a reference to "CAIB document" with CAB or CTF followed by seven to eleven digits, such as CAB001-0010, refer to a document in the Columbia Accident Investigation Board database maintained by the Department of Justice and archived at the National Archives.

1 Sylvia Kramer, "History of NASA Safety Office from 1958-1980's," NASA History Division Record Collection, 1986, p. 1. CAIB document CAB065-0358.

2 Ralph M. Miles Jr. "Introduction." In Ralph M. Miles Jr., editor, *System Concepts: Lectures on Contemporary Approaches to Systems*, p. 1-12 (New York: John F. Wiley & Sons, 1973).

3 "The Aerospace Safety Advisory Panel, " NASA History Office, July 1, 1987, p. 1.

4 On Rodney's appointment, see *NASA Management Instruction* 1103.39, July 3, 1986, and NASA News July 8, 1986.

5 *NASA Facts*, "Brief Overview, Office of Safety, Reliability, Maintainability and Quality Assurance," circa 1987.

6 "Space Program Safety: Funding for NASA's Safety Organizations Should Be Centralized," General Accounting Office Report, NSIAD-90-187, 1990.

7 "Aerospace Safety Advisory Panel Annual Report," 1996.

8 The quotes are from the Executive Summary of National Aeronautics and Space Administration Space Shuttle Independent Assessment Team, "Report to Associate Administrator, Office of Space Flight," October-December 1999. CAIB document CTF017-0169.

9 Harry McDonald, "SIAT Space Shuttle Independent Assessment Team Report."

10 NASA Chief Engineer and NASA Integrated Action Team, "Enhancing Mission Success – A Framework for the Future," December 21, 2000.

11 The information in this section is derived from a briefing titled, "Draft Final Report of the Space Shuttle Competitive Source Task Force," July 12, 2002. Mr. Liam Sarsfield briefed this report to NASA Headquarters.

12 Dr. Karl Weick, University of Michigan; Dr. Karlene Roberts, University of California-Berkley; Dr. Howard McCurdy, American University; and Dr. Diane Vaughan, Boston College.

13 Dr. David Woods, Ohio State University; Dr. Nancy G. Leveson, Massachusetts Institute of Technology; Mr. James Wick, Intel Corporation; Ms. Deborah L. Grubbe, DuPont Corporation; Dr. M. Sam Mannan, Texas A&M University; Douglas A. Wiegmann, University of Illinois at Urbana-Champaign; and Mr. Alan C. McMillan, President and Chief Executive Officer, National Safety Council.

14 Todd R. La Porte and Paula M. Consolini, "Working in Practice but Not in Theory," *Journal of Public Administration Research and Theory*, 1 (1991) pp. 19-47.

15 Scott Sagan, *The Limits of Safety* (Princeton: Princeton University Press, 1995).

16 Dr. Diane Vaughan, Boston College; Dr. David Woods, Ohio State University; Dr. Howard E. McCurdy, American University; Dr. Karl E. Weick, University of Michigan; Dr. Karlene H. Roberts; Dr. M. Elisabeth Paté-Cornell; Dr. Douglas A. Wiegmann, University of Illinois at Urbana-Champaign; Dr. Nancy G. Leveson, Massachusetts Institute of Technology; Mr. James Wick, Intel Corporation; Ms. Deborah L. Grubbe, Dupont Corporation; Dr. M. Sam Mannan, Texas A&M University; and Mr. Alan C. McMillan, President and Chief Executive Officer, National Safety Council.

17 Dr. David Woods of Ohio State University speaking to the Board on Hind-Sight Bias. April 28, 2003.

18 Sagan, The Limits of Safety, p.258.

19 LaPorte and Consolini, "Working In Practice."

20 Notes from "NASA/Navy Benchmarking Exchange (NNBE), Interim Report, Observations & Opportunities Concerning Navy Submarine Program Safety Assurance," Joint NASA and Naval Sea Systems Command NNBE Interim Report, December 20, 2002.

21 Theodore Rockwell, *The Rickover Effect, How One Man Made a Difference*. (Annapolis, Maryland: Naval Institute Press, 1992), p. 318.

22 Rockwell, *Rickover*, p. 320.

23 For more information, see Dr. Diane Vaughn, *The Challenger Launch Decision, Risky Technology, Culture, and Deviance at NASA* (Chicago: University of Chicago Press, 1996).

24 Presentation to the Board by Admiral Walter Cantrell, Aerospace Advisory Panel member, April 7, 2003.

25 Presentation to the Board by Admiral Walter Cantrell, Aerospace Advisory Panel member, April 7, 2003.

26 Aerospace's Launch Verification Process and its Contribution to Titan Risk Management, Briefing given to Board, May 21, 2003, Mr. Ken Holden, General Manager, Launch Verification Division.

27 Joe Tomei, "ELV Launch Risk Assessment Briefing," 3rd Government/Industry Mission Assurance Forum, Aerospace Corporation, September 24, 2002.

28 NASA Policy Directive 8700.1A, "NASA Policy for Safety and Mission Success", Para 1.b, 5.b(1), 5.e(1), and 5.f(1).

29 Charles B. Perrow. *Normal Accidents* (New York: Basic Books, 1984).

30 A. Shenhar, "Project management style and the space shuttle program (part 2): A retrospective look," *Project Management Journal*, 23 (1), pp. 32-37.

31 Harry McDonald, "SIAT Space Shuttle Independent Assessment Team Report."

32 Ibid.

33 "Post Challenger Evaluation of Space Shuttle Risk Assessment and Management Report, National Academy Press 1988," section 5.1, pg. 40.

34 Harry McDonald, "SIAT Space Shuttle Independent Assessment Team Report."

35 NSTS-22254 Rev B.

36 Ibid.

37 GAO Report, "Survey of NASA Lessons Learned," GAO-01-1015R, September 5, 2001.

38 E. Tufte, *Beautiful Evidence* (Cheshire, CT: Graphics Press). [in press.]

39 Ibid., Edward R. Tufte, "The Cognitive Style of PowerPoint," (Cheshire, CT: Graphics Press, May 2003).

40 Ibid.

CHAPTER 8

History As Cause: *Columbia* and *Challenger*

The Board began its investigation with two central questions about NASA decisions. Why did NASA continue to fly with known foam debris problems in the years preceding the *Columbia* launch, and why did NASA managers conclude that the foam debris strike 81.9 seconds into *Columbia*'s flight was not a threat to the safety of the mission, despite the concerns of their engineers?

8.1 ECHOES OF *CHALLENGER*

As the investigation progressed, Board member Dr. Sally Ride, who also served on the Rogers Commission, observed that there were "echoes" of *Challenger* in *Columbia*. Ironically, the Rogers Commission investigation into *Challenger* started with two remarkably similar central questions: Why did NASA continue to fly with known O-ring erosion problems in the years before the *Challenger* launch, and why, on the eve of the *Challenger* launch, did NASA managers decide that launching the mission in such cold temperatures was an acceptable risk, despite the concerns of their engineers?

The echoes did not stop there. The foam debris hit was not the single cause of the *Columbia* accident, just as the failure of the joint seal that permitted O-ring erosion was not the single cause of *Challenger*. Both *Columbia* and *Challenger* were lost also because of the failure of NASA's organizational system. Part Two of this report cites failures of the three parts of NASA's organizational system. This chapter shows how previous political, budgetary, and policy decisions by leaders at the White House, Congress, and NASA (Chapter 5) impacted the Space Shuttle Program's structure, culture, and safety system (Chapter 7), and how these in turn resulted in flawed decision-making (Chapter 6) for both accidents. The explanation is about system effects: how actions taken in one layer of NASA's organizational system impact other layers. History is not just a backdrop or a scene-setter. History is cause. History set the *Columbia* and *Challenger* accidents in motion. Although Part Two is separated into chapters and sections to make clear what happened in the political environment, the organization, and managers' and engineers' decision-making, the three worked together. Each is a critical link in the causal chain.

This chapter shows that both accidents were "failures of foresight" in which history played a prominent role.[1] First, the history of engineering decisions on foam and O-ring incidents had identical trajectories that "normalized" these anomalies, so that flying with these flaws became routine and acceptable. Second, NASA history had an effect. In response to White House and Congressional mandates, NASA leaders took actions that created systemic organizational flaws at the time of *Challenger* that were also present for *Columbia*. The final section compares the two critical decision sequences immediately before the loss of both Orbiters – the pre-launch teleconference for *Challenger* and the post-launch foam strike discussions for *Columbia*. It shows history again at work: how past definitions of risk combined with systemic problems in the NASA organization caused both accidents.

Connecting the parts of NASA's organizational system and drawing the parallels with *Challenger* demonstrate three things. First, despite all the post-*Challenger* changes at NASA and the agency's notable achievements since, the causes of the institutional failure responsible for *Challenger* have not been fixed. Second, the Board strongly believes that if these persistent, systemic flaws are not resolved, the scene is set for another accident. Therefore, the recommendations for change are not only for fixing the Shuttle's technical system, but also for fixing each part of the organizational system that produced *Columbia*'s failure. Third, the Board's focus on the context in which decision making occurred does not mean that individuals are not responsible and accountable. To the contrary, individuals always must assume responsibility for their actions. What it does mean is that NASA's problems cannot be solved simply by retirements, resignations, or transferring personnel.[2]

The constraints under which the agency has operated throughout the Shuttle Program have contributed to both

Shuttle accidents. Although NASA leaders have played an important role, these constraints were not entirely of NASA's own making. The White House and Congress must recognize the role of their decisions in this accident and take responsibility for safety in the future.

8.2 FAILURES OF FORESIGHT: TWO DECISION HISTORIES AND THE NORMALIZATION OF DEVIANCE

Foam loss may have occurred on all missions, and left bipod ramp foam loss occurred on 10 percent of the flights for which visible evidence exists. The Board had a hard time understanding how, after the bitter lessons of *Challenger*, NASA could have failed to identify a similar trend. Rather than view the foam decision only in hindsight, the Board tried to see the foam incidents as NASA engineers and managers saw them as they made their decisions. This section gives an insider perspective: how NASA defined risk and how those definitions changed over time for both foam debris hits and O-ring erosion. In both cases, engineers and managers conducting risk assessments continually normalized the technical deviations they found.[3] In all official engineering analyses and launch recommendations prior to the accidents, evidence that the design was not performing as expected was reinterpreted as acceptable and non-deviant, which diminished perceptions of risk throughout the agency.

The initial Shuttle design predicted neither foam debris problems nor poor sealing action of the Solid Rocket Booster joints. To experience either on a mission was a violation of design specifications. These anomalies were signals of potential danger, not something to be tolerated, but in both cases after the first incident the engineering analysis concluded that the design could tolerate the damage. These engineers decided to implement a temporary fix and/or accept the risk, and fly. For both O-rings and foam, that first decision was a turning point. It established a precedent for accepting, rather than eliminating, these technical deviations. As a result of this new classification, subsequent incidents of O-ring erosion or foam debris strikes were not defined as signals of danger, but as evidence that the design was now acting as predicted. Engineers and managers incorporated worsening anomalies into the engineering experience base, which functioned as an elastic waistband, expanding to hold larger deviations from the original design. Anomalies that did not lead to catastrophic failure were treated as a source of valid engineering data that justified further flights. These anomalies were translated into a safety margin that was extremely influential, allowing engineers and managers to add incrementally to the amount and seriousness of damage that was acceptable. Both O-ring erosion and foam debris events were repeatedly "addressed" in NASA's Flight Readiness Reviews but never fully resolved. In both cases, the engineering analysis was incomplete and inadequate. Engineers understood what was happening, but they never understood why. NASA continued to implement a series of small corrective actions, living with the problems until it was too late.[4]

NASA documents show how official classifications of risk were downgraded over time.[5] Program managers designated both the foam problems and O-ring erosion as "acceptable risks" in Flight Readiness Reviews. NASA managers also assigned each bipod foam event In-Flight Anomaly status, and then removed the designation as corrective actions were implemented. But when major bipod foam-shedding occurred on STS-112 in October 2002, Program management did not assign an In-Flight Anomaly. Instead, it downgraded the problem to the lower status of an "action" item. Before *Challenger*, the problematic Solid Rocket Booster joint had been elevated to a Criticality 1 item on NASA's Critical Items List, which ranked Shuttle components by failure consequences and noted why each was an acceptable risk. The joint was later demoted to a Criticality 1-R (redundant), and then in the month before *Challenger*'s launch was "closed out" of the problem-reporting system. Prior to both accidents, this demotion from high-risk item to low-risk item was very similar, but with some important differences. Damaging the Orbiter's Thermal Protection System, especially its fragile tiles, was normalized even before Shuttle launches began: it was expected due to forces at launch, orbit, and re-entry.[6] So normal was replacement of Thermal Protection System materials that NASA managers budgeted for tile cost and turnaround maintenance time from the start.

It was a small and logical next step for the discovery of foam debris damage to the tiles to be viewed by NASA as part of an already existing maintenance problem, an assessment based on experience, not on a thorough hazard analysis. Foam debris anomalies came to be categorized by the reassuring term "in-family," a formal classification indicating that new occurrences of an anomaly were within the engineering experience base. "In-family" was a strange term indeed for a violation of system requirements. Although "in-family" was a designation introduced post-*Challenger* to separate problems by seriousness so that "out-of-family" problems got more attention, by definition the problems that were shifted into the lesser "in-family" category got less attention. The Board's investigation uncovered no paper trail showing escalating concern about the foam problem like the one that Solid Rocket Booster engineers left prior to *Challenger*.[7] So ingrained was the agency's belief that foam debris was not a threat to flight safety that in press briefings after the *Columbia* accident, the Space Shuttle Program Manager still discounted the foam as a probable cause, saying that Shuttle managers were "comfortable" with their previous risk assessments.

From the beginning, NASA's belief about both these problems was affected by the fact that engineers were evaluating them in a work environment where technical problems were normal. Although management treated the Shuttle as operational, it was in reality an experimental vehicle. Many anomalies were expected on each mission. Against this backdrop, an anomaly was not in itself a warning sign of impending catastrophe. Another contributing factor was that both foam debris strikes and O-ring erosion events were examined separately, one at a time. Individual incidents were not read by engineers as strong signals of danger. What NASA engineers and managers saw were pieces of ill-structured problems.[8] An incident of O-ring erosion or foam bipod debris would be followed by several launches where the machine behaved properly, so that signals of danger were followed by all-clear signals – in other words, NASA managers and engineers were receiving mixed signals.[9] Some signals defined as weak at the time were, in retrospect, warnings of danger. Foam debris damaged tile was assumed (erroneously) not to pose a danger to the wing. If a primary O-ring failed, the secondary was assumed (erroneously) to provide a backup. Finally, because foam debris strikes were occurring frequently, like O-ring erosion in the years before *Challenger*, foam anomalies became routine signals – a normal part of Shuttle operations, not signals of danger. Other anomalies gave signals that were strong, like wiring malfunctions or the cracked balls in Ball Strut Tie Rod Assemblies, which had a clear relationship to a "loss of mission." On those occasions, NASA stood down from launch, sometimes for months, while the problems were corrected. In contrast, foam debris and eroding O-rings were defined as nagging issues of seemingly little consequence. Their significance became clear only in retrospect, after lives had been lost.

History became cause as the repeating pattern of anomalies was ratified as safe in Flight Readiness Reviews. The official definitions of risk assigned to each anomaly in Flight Readiness Reviews limited the actions taken and the resources spent on these problems. Two examples of the road not taken and the devastating implications for the future occurred close in time to both accidents. On the October 2002 launch of STS-112, a large piece of bipod ramp foam hit and damaged the External Tank Attachment ring on the Solid Rocket Booster skirt, a strong signal of danger 10 years after the last known bipod ramp foam event. Prior to *Challenger,* there was a comparable surprise. After a January 1985 launch, for which the Shuttle sat on the launch pad for three consecutive nights of unprecedented cold temperatures, engineers discovered upon the Orbiter's return that hot gases had eroded the primary and reached the secondary O-ring, blackening the putty in between – an indication that the joint nearly failed.

But accidents are not always preceded by a wake-up call.[10] In 1985, engineers realized they needed data on the relationship between cold temperatures and O-ring erosion. However, the task of getting better temperature data stayed on the back burner because of the definition of risk: the primary erosion was within the experience base; the secondary O-ring (thought to be redundant) was not damaged and, significantly, there was a low probability that such cold Florida temperatures would recur.[11] The scorched putty, initially a strong signal, was redefined after analysis as weak. On the eve of the *Challenger* launch, when cold temperature became a concern, engineers had no test data on the effect of cold temperatures on O-ring erosion. Before *Columbia*, engineers concluded that the damage from the STS-112 foam hit in October 2002 was not a threat to flight safety. The logic was that, yes, the foam piece was large and there was damage, but no serious consequences followed. Further, a hit this size, like cold temperature, was a low-probability event. After analysis, the biggest foam hit to date was redefined as a weak signal. Similar self-defeating actions and inactions followed. Engineers were again dealing with the poor quality of tracking camera images of strikes during ascent. Yet NASA took no steps to improve imagery and took no immediate action to reduce the risk of bipod ramp foam shedding and potential damage to the Orbiter before *Columbia*. Furthermore, NASA performed no tests on what would happen if a wing leading edge were struck by bipod foam, even though foam had repeatedly separated from the External Tank.

During the *Challenger* investigation, Rogers Commission member Dr. Richard Feynman famously compared launching Shuttles with known problems to playing Russian roulette.[12] But that characterization is only possible in hindsight. It is not how NASA personnel perceived the risks as they were being assessed, one launch at a time. Playing Russian roulette implies that the pistol-holder realizes that death might be imminent and still takes the risk. For both foam debris and O-ring erosion, fixes were in the works at the time of the accidents, but there was no rush to complete them because neither problem was defined as a show-stopper. Each time an incident occurred, the Flight Readiness process declared it safe to continue flying. Taken one at a time, each decision seemed correct. The agency allocated attention and resources to these two problems accordingly. The consequences of living with both of these anomalies were, in its view, minor. Not all engineers agreed in the months immediately preceding *Challenger*, but the dominant view at NASA – the managerial view – was, as one manager put it, "we were just eroding rubber O-rings," which was a low-cost problem.[13] The financial consequences of foam debris also were relatively low: replacing tiles extended the turnaround time between launches. In both cases, NASA was comfortable with its analyses. Prior to each accident, the agency saw no greater consequences on the horizon.

8.3 System Effects: The Impact of History and Politics on Risky Work

The series of engineering decisions that normalized technical deviations shows one way that history became cause in both accidents. But NASA's own history encouraged this pattern of flying with known flaws. Seventeen years separated the two accidents. NASA Administrators, Congresses, and political administrations changed. However, NASA's political and budgetary situation remained the same in principle as it had been since the inception of the Shuttle Program. NASA remained a politicized and vulnerable agency, dependent on key political players who accepted NASA's ambitious proposals and then imposed strict budget limits. Post-*Challenger* policy decisions made by the White House, Congress, and NASA leadership resulted in the agency reproducing many of the failings identified by the Rogers Commission. Policy constraints affected the Shuttle Program's organization culture, its structure, and the structure of the safety system. The three combined to keep NASA on its slippery slope toward *Challenger* and *Columbia*. NASA culture allowed flying with flaws when problems were defined as normal and routine; the structure of NASA's Shuttle Program blocked the flow of critical information up the hierarchy, so definitions of risk continued unaltered. Finally, a perennially weakened safety system, unable to critically analyze and intervene, had no choice but to ratify the existing risk assessments on these two problems. The following comparison shows that these system effects persisted through time, and affected engineering decisions in the years leading up to both accidents.

The Board found that dangerous aspects of NASA's 1986 culture, identified by the Rogers Commission, remained unchanged. The Space Shuttle Program had been built on compromises hammered out by the White House and NASA headquarters.[14] As a result, NASA was transformed from a research and development agency to more of a business, with schedules, production pressures, deadlines, and cost efficiency goals elevated to the level of technical innovation and safety goals.[15] The Rogers Commission dedicated an entire chapter of its report to production pressures.[16] Moreover, the Rogers Commission, as well as the 1990 Augustine Committee and the 1999 Shuttle Independent Assessment Team, criticized NASA for treating the Shuttle as if it were an operational vehicle. Launching on a tight schedule, which the agency had pursued as part of its initial bargain with the White House, was not the way to operate what was in fact an experimental vehicle. The Board found that prior to *Columbia*, a budget-limited Space Shuttle Program, forced again and again to refashion itself into an efficiency model because of repeated government cutbacks, was beset by these same ills. The harmful effects of schedule pressure identified in previous reports had returned.

Prior to both accidents, NASA was scrambling to keep up. Not only were schedule pressures impacting the people who worked most closely with the technology – technicians, mission operators, flight crews, and vehicle processors – engineering decisions also were affected.[17] For foam debris and O-ring erosion, the definition of risk established during the Flight Readiness process determined actions taken and not taken, but the schedule and shoestring budget were equally influential. NASA was cutting corners. Launches proceeded with incomplete engineering work on these flaws. *Challenger*-era engineers were working on a permanent fix for the booster joints while launches continued.[18] After the major foam bipod hit on STS-112, management made the deadline for corrective action on the foam problem *after* the next launch, STS-113, and then slipped it again until *after* the flight of STS-107. Delays for flowliner and Ball Strut Tie Rod Assembly problems left no margin in the schedule between February 2003 and the management-imposed February 2004 launch date for the International Space Station Node 2. Available resources – including time out of the schedule for research and hardware modifications – went to the problems that were designated as serious – those most likely to bring down a Shuttle. The NASA culture encouraged flying with flaws because the schedule could not be held up for routine problems that were not defined as a threat to mission safety.[19]

The question the Board had to answer was why, since the foam debris anomalies went on for so long, had no one recognized the trend and intervened? The O-ring history prior to *Challenger* had followed the same pattern. This question pointed the Board's attention toward the NASA organization structure and the structure of its safety system. Safety-oriented organizations often build in checks and balances to identify and monitor signals of potential danger. If these checks and balances were in place in the Shuttle Program, they weren't working. Again, past policy decisions produced system effects with implications for both *Challenger* and *Columbia*.

Prior to *Challenger*, Shuttle Program structure had hindered information flows, leading the Rogers Commission to conclude that critical information about technical problems was not conveyed effectively through the hierarchy.[20] The Space Shuttle Program had altered its structure by outsourcing to contractors, which added to communication problems. The Commission recommended many changes to remedy these problems, and NASA made many of them. However, the Board found that those post-*Challenger* changes were undone over time by management actions.[21] NASA administrators, reacting to government pressures, transferred more functions and responsibilities to the private sector. The change was cost-efficient, but personnel cuts reduced oversight of contractors at the same time that the agency's dependence upon contractor engineering judgment increased. When high-risk technology is the product and lives are at stake, safety, oversight, and communication flows are critical. The Board found that the Shuttle Program's normal chain of command and matrix system did not perform a check-and-balance function on either foam or O-rings.

The Flight Readiness Review process might have reversed the disastrous trend of normalizing O-ring erosion and foam debris hits, but it didn't. In fact, the Rogers Commission found that the Flight Readiness process only affirmed the pre-*Challenger* engineering risk assessments.[22] Equally troubling, the Board found that the Flight Readiness process, which is built on consensus verified by signatures of all responsible parties, in effect renders no one accountable. Although the process was altered after *Challenger*, these changes did not erase the basic problems that were built into the structure of the Flight Readiness Review.[23] Managers at the top were dependent on engineers at the bottom for their engineering analysis and risk assessments. Information was lost as engineering risk analyses moved through the process. At succeeding stages, management awareness of anomalies, and therefore risks, was reduced either because of the need to be increasingly brief and concise as all the parts of the system came together, or because of the need to produce consensus decisions at each level. The Flight Readiness process was designed to assess hardware and take corrective actions that would transform known problems into acceptable flight risks, and that is precisely what it did. The 1986 House Committee on Science and Technology concluded during its investigation into *Challenger* that Flight Readiness Reviews had performed exactly as they were designed, but that they could not be expected to replace engineering analysis, and therefore they "cannot be expected to prevent a flight because of a design flaw that Project management had already determined an acceptable risk."[24] Those words, true for the history of O-ring erosion, also hold true for the history of foam debris.

The last line of defense against errors is usually a safety system. But the previous policy decisions by leaders described in Chapter 5 also impacted the safety structure and contributed to both accidents. Neither in the O-ring erosion nor the foam debris problems did NASA's safety system attempt to reverse the course of events. In 1986, the Rogers Commission called it "The Silent Safety System."[25] Pre-*Challenger* budget shortages resulted in safety personnel cutbacks. Without clout or independence, the safety personnel who remained were ineffective. In the case of *Columbia*, the Board found the same problems were reproduced and for an identical reason: when pressed for cost reduction, NASA attacked its own safety system. The faulty assumption that supported this strategy prior to *Columbia* was that a reduction in safety staff would not result in a reduction of safety, because contractors would assume greater safety responsibility. The effectiveness of those remaining staff safety engineers was blocked by their dependence on the very Program they were charged to supervise. Also, the Board found many safety units with unclear roles and responsibilities that left crucial gaps. Post-*Challenger* NASA still had no systematic procedure for identifying and monitoring trends. The Board was surprised at how long it took NASA to put together trend data in response to Board requests for information. Problem reporting and tracking systems were still overloaded or underused, which undermined their very purpose. Multiple job titles disguised the true extent of safety personnel shortages. The Board found cases in which the same person was occupying more than one safety position – and in one instance at least three positions – which compromised any possibility of safety organization independence because the jobs were established with built-in conflicts of interest.

8.4 Organization, Culture, and Unintended Consequences

A number of changes to the Space Shuttle Program structure made in response to policy decisions had the unintended effect of perpetuating dangerous aspects of pre-*Challenger* culture and continued the pattern of normalizing things that were not supposed to happen. At the same time that NASA leaders were emphasizing the importance of safety, their personnel cutbacks sent other signals. Streamlining and downsizing, which scarcely go unnoticed by employees, convey a message that efficiency is an important goal. The Shuttle/Space Station partnership affected both programs. Working evenings and weekends just to meet the International Space Station Node 2 deadline sent a signal to employees that schedule is important. When paired with the "faster, better, cheaper" NASA motto of the 1990s and cuts that dramatically decreased safety personnel, efficiency becomes a strong signal and safety a weak one. This kind of doublespeak by top administrators affects people's decisions and actions without them even realizing it.[26]

Changes in Space Shuttle Program structure contributed to the accident in a second important way. Despite the constraints that the agency was under, prior to both accidents NASA appeared to be immersed in a culture of invincibility, in stark contradiction to post-accident reality. The Rogers Commission found a NASA blinded by its "Can-Do" attitude,[27] a cultural artifact of the Apollo era that was inappropriate in a Space Shuttle Program so strapped by schedule pressures and shortages that spare parts had to be cannibalized from one vehicle to launch another.[28] This can-do attitude bolstered administrators' belief in an achievable launch rate, the belief that they had an operational system, and an unwillingness to listen to outside experts. The Aerospace Safety and Advisory Panel in a 1985 report told NASA that the vehicle was not operational and NASA should stop treating it as if it were.[29] The Board found that even after the loss of *Challenger*, NASA was guilty of treating an experimental vehicle as if it were operational and of not listening to outside experts. In a repeat of the pre-*Challenger* warning, the 1999 Shuttle Independent Assessment Team report reiterated that "the Shuttle was not an 'operational' vehicle in the usual meaning of the term."[30] Engineers and program planners were also affected by "Can-Do," which, when taken too far, can create a reluctance to say that something cannot be done.

How could the lessons of *Challenger* have been forgotten so quickly? Again, history was a factor. First, if success is measured by launches and landings,[31] the machine appeared to be working successfully prior to both accidents. *Challenger* was the 25th launch. Seventeen years and 87 missions passed without major incident. Second, previous policy decisions again had an impact. NASA's Apollo-era research and development culture and its prized deference to the technical expertise of its working engineers was overridden in the Space Shuttle era by "bureaucratic accountability" – an allegiance to hierarchy, procedure, and following the chain of command.[32] Prior to *Challenger*, the can-do culture was a result not just of years of apparently successful launches, but of the cultural belief that the Shuttle Program's many structures, rigorous procedures, and detailed system of rules were responsible for those successes.[33] The Board noted that the pre-*Challenger* layers of processes, boards, and panels that had produced a false sense of confidence in the system and its level of safety returned in full force prior to *Columbia*. NASA made many changes to the Space Shuttle Program structure after *Challenger*. The fact that many changes had been made supported a belief in the safety of the system, the invincibility of organizational and technical systems, and ultimately, a sense that the foam problem was understood.

8.5 History as Cause: Two Accidents

Risk, uncertainty, and history came together when unprecedented circumstances arose prior to both accidents. For *Challenger*, the weather prediction for launch time the next day was for cold temperatures that were out of the engineering experience base. For *Columbia*, a large foam hit – also outside the experience base – was discovered after launch. For the first case, all the discussion was pre-launch; for the second, it was post-launch. This initial difference determined the shape these two decision sequences took, the number of people who had information about the problem, and the locations of the involved parties.

For *Challenger*, engineers at Morton-Thiokol,[34] the Solid Rocket Motor contractor in Utah, were concerned about the effect of the unprecedented cold temperatures on the rubber O-rings.[35] Because launch was scheduled for the next morning, the new condition required a reassessment of the engineering analysis presented at the Flight Readiness Review two weeks prior. A teleconference began at 8:45 p.m. Eastern Standard Time (EST) that included 34 people in three locations: Morton-Thiokol in Utah, Marshall, and Kennedy. Thiokol engineers were recommending a launch delay. A reconsideration of a Flight Readiness Review risk assessment the night before a launch was as unprecedented as the predicted cold temperatures. With no ground rules or procedures to guide their discussion, the participants automatically reverted to the centralized, hierarchical, tightly structured, and procedure-bound model used in Flight Readiness Reviews. The entire discussion and decision to launch began and ended with this group of 34 engineers. The phone conference linking them together concluded at 11:15 p.m. EST after a decision to accept the risk and fly.

For *Columbia*, information about the foam debris hit was widely distributed the day after launch. Time allowed for videos of the strike, initial assessments of the size and speed of the foam, and the approximate location of the impact to be dispersed throughout the agency. This was the first debris impact of this magnitude. Engineers at the Marshall, Johnson, Kennedy, and Langley centers showed initiative and jumped on the problem without direction from above. Working groups and e-mail groups formed spontaneously. The size of Johnson's Debris Assessment Team alone neared and in some instances exceeded the total number of participants in the 1986 *Challenger* teleconference. Rather than a tightly constructed exchange of information completed in a few hours, time allowed for the development of ideas and free-wheeling discussion among the engineering ranks. The early post-launch discussion among engineers and all later decision-making at management levels were decentralized, loosely organized, and with little form. While the spontaneous and decentralized exchanging of information was evidence that NASA's original technical culture was alive and well, the diffuse form and lack of structure in the rest of the proceedings would have several negative consequences.

In both situations, all new information was weighed and interpreted against past experience. Formal categories and cultural beliefs provide a consistent frame of reference in which people view and interpret information and experiences.[36] Pre-existing definitions of risk shaped the actions taken and not taken. Worried engineers in 1986 and again in 2003 found it impossible to reverse the Flight Readiness Review risk assessments that foam and O-rings did not pose safety-of-flight concerns. These engineers could not prove that foam strikes and cold temperatures were unsafe, even though the previous analyses that declared them safe had been incomplete and were based on insufficient data and testing. Engineers' failed attempts were not just a matter of psychological frames and interpretations. The obstacles these engineers faced were political and organizational. They were rooted in NASA history and the decisions of leaders that had altered NASA culture, structure, and the structure of the safety system and affected the social context of decision-making for both accidents. In the following comparison of these critical decision scenarios for *Columbia* and *Challenger*, the systemic problems in the NASA organization are in italics, with the system effects on decision-making following.

NASA had conflicting goals of cost, schedule, and safety. Safety lost out as the mandates of an "operational system" increased the schedule pressure. Scarce resources went to problems that were defined as more serious, rather than to foam strikes or O-ring erosion.

In both situations, upper-level managers and engineering teams working the O-ring and foam strike problems held opposing definitions of risk. This was demonstrated immediately, as engineers reacted with urgency to the immediate safety implications: Thiokol engineers scrambled to put together an engineering assessment for the teleconference, Langley Research Center engineers initiated simulations of landings that were run after hours at Ames Research Center, and Boeing analysts worked through the weekend on the debris impact analysis. But key managers were responding to additional demands of cost and schedule, which competed with their safety concerns. NASA's conflicting goals put engineers at a disadvantage before these new situations even arose. In neither case did they have good data as a basis for decision-making. Because both problems had been previously normalized, resources sufficient for testing or hardware were not dedicated. The Space Shuttle Program had not produced good data on the correlation between cold temperature and O-ring resilience or good data on the potential effect of bipod ramp foam debris hits.[37]

Cultural beliefs about the low risk O-rings and foam debris posed, backed by years of Flight Readiness Review decisions and successful missions, provided a frame of reference against which the engineering analyses were judged. When confronted with the engineering risk assessments, top Shuttle Program managers held to the previous Flight Readiness Review assessments. In the *Challenger* teleconference, where engineers were recommending that NASA delay the launch, the Marshall Solid Rocket Booster Project manager, Lawrence Mulloy, repeatedly challenged the contractor's risk assessment and restated Thiokol's engineering rationale for previous flights.[38] STS-107 Mission Management Team Chair Linda Ham made many statements in meetings reiterating her understanding that foam was a maintenance problem and a turnaround issue, not a safety-of-flight issue.

The effects of working as a manager in a culture with a cost/efficiency/safety conflict showed in managerial responses. In both cases, managers' techniques focused on the information that tended to support the expected or desired result at that time. In both cases, believing the safety of the mission was not at risk, managers drew conclusions that minimized the risk of delay.[39] At one point, Marshall's Mulloy, believing in the previous Flight Readiness Review assessments, unconvinced by the engineering analysis, and concerned about the schedule implications of the 53-degree temperature limit on launch the engineers proposed, said, "My God, Thiokol, when do you want me to launch, next April?"[40] Reflecting the overall goal of keeping to the Node 2 launch schedule, Ham's priority was to avoid the delay of STS–114, the next mission after STS-107. Ham was slated as Manager of Launch Integration for STS-114 – a dual role promoting a conflict of interest and a single-point failure, a situation that should be avoided in all organizational as well as technical systems.

NASA's culture of bureaucratic accountability emphasized chain of command, procedure, following the rules, and going by the book. While rules and procedures were essential for coordination, they had an unintended but negative effect. Allegiance to hierarchy and procedure had replaced deference to NASA engineers' technical expertise.

In both cases, engineers initially presented concerns as well as possible solutions – a request for images, a recommendation to place temperature constraints on launch. Management did not listen to what their engineers were telling them. Instead, rules and procedures took priority. For *Columbia*, program managers turned off the Kennedy engineers' initial request for Department of Defense imagery, with apologies to Defense Department representatives for not having followed "proper channels." In addition, NASA administrators asked for and promised corrective action to prevent such a violation of protocol from recurring. Debris Assessment Team analysts at Johnson were asked by managers to demonstrate a "mandatory need" for their imagery request, but were not told how to do that. Both *Challenger* and *Columbia* engineering teams were held to the usual quantitative standard of proof. But it was a reverse of the usual circumstance: instead of having to prove it was safe to fly, they were asked to prove that it was unsafe to fly.

In the *Challenger* teleconference, a key engineering chart presented a qualitative argument about the relationship between cold temperatures and O-ring erosion that engineers were asked to prove. Thiokol's Roger Boisjoly said, "I had no data to quantify it. But I did say I knew it was away from goodness in the current data base."[41] Similarly, the Debris Assessment Team was asked to prove that the foam hit was a threat to flight safety, a determination that only the imagery they were requesting could help them make. Ignored by management was the qualitative data that the engineering teams did have: both instances were outside the experience base. In stark contrast to the requirement that engineers adhere to protocol and hierarchy was management's failure to apply this criterion to their own activities. The Mission Management Team did not meet on a regular schedule during the mission, proceeded in a loose format that allowed informal influence and status differences to shape their decisions, and allowed unchallenged opinions and assumptions to prevail, all the while holding the engineers who were making risk assessments to higher standards. In highly uncertain circumstances, when lives were immediately at risk, management failed to defer to its engineers and failed to recognize that different data standards – qualitative, subjective, and intuitive – and different processes – democratic rather than protocol and chain of command – were more appropriate.

The organizational structure and hierarchy blocked effective communication of technical problems. Signals were overlooked, people were silenced, and useful information and dissenting views on technical issues did not surface at higher levels. What was *communicated to parts of the organization was that O-ring erosion and foam debris were not problems.*

Structure and hierarchy represent power and status. For both *Challenger* and *Columbia*, employees' positions in the organization determined the weight given to their information, by their own judgment and in the eyes of others. As a result, many signals of danger were missed. Relevant information that could have altered the course of events was available but was not presented.

Early in the *Challenger* teleconference, some engineers who had important information did not speak up. They did not define themselves as qualified because of their position: they were not in an appropriate specialization, had not recently worked the O-ring problem, or did not have access to the "good data" that they assumed others more involved in key discussions would have.[42] Geographic locations also resulted in missing signals. At one point, in light of Marshall's objections, Thiokol managers in Utah requested an "off-line caucus" to discuss their data. No consensus was reached, so a "management risk decision" was made. Managers voted and engineers did not. Thiokol managers came back on line, saying they had reversed their earlier NO-GO recommendation, decided to accept risk, and would send new engineering charts to back their reversal. When a Marshall administrator asked, "Does anyone have anything to add to this?," no one spoke. Engineers at Thiokol who still objected to the decision later testified that they were intimidated by management authority, were accustomed to turning their analysis over to managers and letting them decide, and did not have the quantitative data that would empower them to object further.[43]

In the more decentralized decision process prior to *Columbia*'s re-entry, structure and hierarchy again were responsible for an absence of signals. The initial request for imagery came from the "low status" Kennedy Space Center, bypassed the Mission Management Team, and went directly to the Department of Defense separate from the all-powerful Shuttle Program. By using the Engineering Directorate avenue to request imagery, the Debris Assessment Team was working at the margins of the hierarchy. But some signals were missing even when engineers traversed the appropriate channels. The Mission Management Team Chair's position in the hierarchy governed what information she would or would not receive. Information was lost as it traveled up the hierarchy. A demoralized Debris Assessment Team did not include a slide about the need for better imagery in their presentation to the Mission Evaluation Room. Their presentation included the Crater analysis, which they reported as incomplete and uncertain. However, the Mission Evaluation Room manager perceived the Boeing analysis as rigorous and quantitative. The choice of headings, arrangement of information, and size of bullets on the key chart served to highlight what management already believed. The uncertainties and assumptions that signaled danger dropped out of the information chain when the Mission Evaluation Room manager condensed the Debris Assessment Team's formal presentation to an informal verbal brief at the Mission Management Team meeting.

As what the Board calls an "informal chain of command" began to shape STS-107's outcome, location in the structure empowered some to speak and silenced others. For example, a Thermal Protection System tile expert, who was a member of the Debris Assessment Team but had an office in the more prestigious Shuttle Program, used his personal network to shape the Mission Management Team view and snuff out dissent. The informal hierarchy among and within Centers was also influential. Early identifications of problems by Marshall and Kennedy may have contributed to the Johnson-based Mission Management Team's indifference to concerns about the foam strike. The engineers and managers circulating e-mails at Langley were peripheral to the Shuttle Program, not structurally connected to the proceedings, and therefore of lower status. When asked in a post-accident press conference why they didn't voice their concerns to Shuttle Program management, the Langley engineers said that people "need to stick to their expertise."[44] Status mattered. In its absence, numbers were the great equalizer. One striking exception: the Debris Assessment Team tile expert was so influential that his word was taken as gospel, though he lacked the requisite expertise, data, or analysis to evaluate damage to RCC. For those with lesser standing, the requirement for data was stringent and inhibiting, which resulted in information that warned of danger not being passed up the chain. As in the teleconference, Debris Assessment Team engineers did not speak up when the Mission Management Team Chair asked if anyone else had anything to say. Not only did they not have the numbers, they also were intimidated by the Mission Management Team Chair's position in the hierarchy and the conclusions she had already made. Debris Assessment Team members signed off on the Crater analysis, even though they had trouble understanding it. They still wanted images of *Columbia*'s left wing.

In neither impending crisis did management recognize how structure and hierarchy can silence employees and follow through by polling participants, soliciting dissenting opinions, or bringing in outsiders who might have a different perspective or useful information. In perhaps the ultimate example of engineering concerns not making their way upstream, *Challenger* astronauts were told that the cold temperature was not a problem, and *Columbia* astronauts were told that the foam strike was not a problem.

NASA structure changed as roles and responsibilities were transferred to contractors, which increased the dependence on the private sector for safety functions and risk assessment while simultaneously reducing the in-house capability to spot safety issues.

A critical turning point in both decisions hung on the discussion of contractor risk assessments. Although both Thiokol and Boeing engineering assessments were replete with uncertainties, NASA ultimately accepted each. Thiokol's initial recommendation against the launch of *Challenger* was at first criticized by Marshall as flawed and unacceptable. Thiokol was recommending an unheard-of delay on the eve of a launch, with schedule ramifications and NASA-contractor relationship repercussions. In the Thiokol off-line caucus, a senior vice president who seldom participated in these engineering discussions championed the Marshall engineering rationale for flight. When he told the managers present to "Take off your engineering hat and put on your management hat," they reversed the position their own engineers had taken.[45] Marshall engineers then accepted this assessment, deferring to the expertise of the contractor. NASA was dependent on Thiokol for the risk assessment, but the decision process was affected by the contractor's dependence on NASA. Not willing to be responsible for a delay, and swayed by the strength of Marshall's argument, the contractor did not act in the best interests of safety. Boeing's Crater analysis was performed in the context of the Debris Assessment Team, which was a collaborative effort that included Johnson, United Space Alliance, and Boeing. In this case, the decision process was also affected by NASA's dependence on the contractor. Unfamiliar with Crater, NASA engineers and managers had to rely on Boeing for interpretation and analysis, and did not have the training necessary to evaluate the results. They accepted Boeing engineers' use of Crater to model a debris impact 400 times outside validated limits.

NASA's safety system lacked the resources, independence, personnel, and authority to successfully apply alternate perspectives to developing problems. Overlapping roles and responsibilities across multiple safety offices also undermined the possibility of a reliable system of checks and balances.

NASA's "Silent Safety System" did nothing to alter the decision-making that immediately preceded both accidents. No safety representatives were present during the *Challenger* teleconference – no one even thought to call them.[46] In the case of *Columbia*, safety representatives were present at Mission Evaluation Room, Mission Management Team, and Debris Assessment Team meetings. However, rather than critically question or actively participate in the analysis, the safety representatives simply listened and concurred.

8.6 Changing NASA's Organizational System

The echoes of *Challenger* in *Columbia* identified in this chapter have serious implications. These repeating patterns mean that flawed practices embedded in NASA's organizational system continued for 20 years and made substantial contributions to both accidents. The Columbia Accident Investigation Board noted the same problems as the Rogers Commission. An organization system failure calls for corrective measures that address all relevant levels of the organization, but the Board's investigation shows that for all its cutting-edge technologies, "diving-catch" rescues, and imaginative plans for the technology and the future of space exploration, NASA has shown very little understanding of the inner workings of its own organization.

NASA managers believed that the agency had a strong safety culture, but the Board found that the agency had the same conflicting goals that it did before *Challenger*, when schedule concerns, production pressure, cost-cutting and a drive for ever-greater efficiency – all the signs of an "operational" enterprise – had eroded NASA's ability to assure mission safety. The belief in a safety culture has even less credibility in light of repeated cuts of safety personnel and budgets – also conditions that existed before *Challenger*. NASA managers stated confidently that everyone was encouraged to speak up about safety issues and that the agency was responsive to those concerns, but the Board found evidence to the contrary in the responses to the Debris Assessment Team's request for imagery, to the initiation of the imagery request from Kennedy Space Center, and to the "we were just 'what-iffing'" e-mail concerns that did not reach the Mission Management Team. NASA's bureaucratic structure kept important information from reaching engineers and managers alike. The same NASA whose engineers showed initiative and a solid working knowledge of how to get things done fast had a managerial culture with an allegiance to bureaucracy and cost-efficiency that squelched the engineers' efforts. When it came to managers' own actions, however, a different set of rules prevailed. The Board found that Mission Management Team decision-making operated outside the rules even as it held its engineers to a stifling protocol. Management was not able to recognize that in unprecedented conditions, when lives are on the line, flexibility and democratic process should take priority over bureaucratic response.[47]

During the *Columbia* investigation, the Board consistently searched for causal principles that would explain both the technical and organizational system failures. These principles were needed to explain *Columbia* and its echoes of *Challenger*. They were also necessary to provide guidance for NASA. The Board's analysis of organizational causes in Chapters 5, 6, and 7 supports the following principles that should govern the changes in the agency's organizational system. The Board's specific recommendations, based on these principles, are presented in Part Three.

Leaders create culture. It is their responsibility to change it. Top administrators must take responsibility for risk, failure, and safety by remaining alert to the effects their decisions have on the system. Leaders are responsible for establishing the conditions that lead to their subordinates' successes or failures. The past decisions of national leaders – the White House, Congress, and NASA Headquarters – set the *Columbia* accident in motion by creating resource and schedule strains that compromised the principles of a high-risk technology organization. The measure of NASA's success became how much costs were reduced and how efficiently the schedule was met. But the Space Shuttle is not now, nor has it ever been, an operational vehicle. We cannot explore space on a fixed-cost basis. Nevertheless, due to International Space Station needs and scientific experiments that require particular timing and orbits, the Space Shuttle Program seems likely to continue to be schedule-driven. National leadership needs to recognize that NASA must fly only when it is ready. As the White House, Congress, and NASA Headquarters plan the future of human space flight, the goals and the resources required to achieve them safely must be aligned.

Changes in organizational structure should be made only with careful consideration of their effect on the system and their possible unintended consequences. Changes that make the organization more complex may create new ways that it can fail.[48] When changes are put in place, the risk of error initially increases, as old ways of doing things compete with new. Institutional memory is lost as personnel and records are moved and replaced. Changing the structure of organizations is complicated by external political and budgetary constraints, the inability of leaders to conceive of the full ramifications of their actions, the vested interests of insiders, and the failure to learn from the past.[49]

Nonetheless, changes must be made. The Shuttle Program's structure is a source of problems, not just because of the way it impedes the flow of information, but because it has had effects on the culture that contradict safety goals. NASA's blind spot is it believes it has a strong safety culture. Program history shows that the loss of a truly independent, robust capability to protect the system's fundamental requirements and specifications inevitably compromised those requirements, and therefore increased risk. The Shuttle Program's structure created power distributions that need new structuring, rules, and management training to restore deference to technical experts, empower engineers to get resources they need, and allow safety concerns to be freely aired.

Strategies must increase the clarity, strength, and presence of signals that challenge assumptions about risk. Twice in NASA history, the agency embarked on a slippery slope that resulted in catastrophe. Each decision, taken by itself, seemed correct, routine, and indeed, insignificant and unremarkable. Yet in retrospect, the cumulative effect was stunning. In both pre-accident periods, events unfolded over a long time and in small increments rather than in sudden and dramatic occurrences. NASA's challenge is to design systems that maximize the clarity of signals, amplify weak signals so they can be tracked, and account for missing signals. For both accidents there were moments when management definitions of risk might have been reversed were it not for the many missing signals – an absence of trend analysis, imagery data not obtained, concerns not voiced, information overlooked or dropped from briefings. A safety team must have equal and independent representation so that managers are not again lulled into complacency by shifting definitions of risk. It is obvious but worth acknowledging that people who are marginal and powerless in organizations may have useful information or opinions that they don't express. Even when these people are encouraged to speak, they find it intimidating to contradict a leader's strategy or a group consensus. Extra effort must be made to contribute all relevant information to discussions of risk. These strategies are important for all safety aspects, but especially necessary for ill-structured problems like O-rings and foam debris. Because ill-structured problems are less visible and therefore invite the normalization of deviance, they may be the most risky of all.

Challenger launches on the ill-fated STS-33/51-L mission on January 28, 1986. The Orbiter would be destroyed 73 seconds later.

ENDNOTES FOR CHAPTER 8

The citations that contain a reference to "CAIB document" with CAB or CTF followed by seven to eleven digits, such as CAB001-0010, refer to a document in the Columbia Accident Investigation Board database maintained by the Department of Justice and archived at the National Archives.

1 Turner studied 85 different accidents and disasters, noting a common pattern: each had a long incubation period in which hazards and warning signs prior to the accident were either ignored or misinterpreted. He called these "failures of foresight." Barry Turner, *Man-made Disasters*, (London: Wykeham, 1978); Barry Turner and Nick Pidgeon, *Man-made Disasters*, 2nd ed. (Oxford: Butterworth Heinneman,1997).

2 Changing personnel is a typical response after an organization has some kind of harmful outcome. It has great symbolic value. A change in personnel points to individuals as the cause and removing them gives the false impression that the problems have been solved, leaving unresolved organizational system problems. See Scott Sagan, *The Limits of Safety*. Princeton: Princeton University Press, 1993.

3 Diane Vaughan, *The Challenger Launch Decision: Risky Technology, Culture, and Deviance at NASA* (Chicago: University of Chicago Press. 1996).

4 William H. Starbuck and Frances J. Milliken, "Challenger: Fine-tuning the Odds until Something Breaks." Journal of Management Studies 23 (1988), pp. 319-40.

5 *Report of the Presidential Commission on the Space Shuttle Challenger Accident*, (Washington: Government Printing Office, 1986), Vol. II, Appendix H.

6 Alex Roland, "The Shuttle: Triumph or Turkey?" *Discover*, November 1985: pp. 29-49.

7 *Report of the Presidential Commission*, Vol. I, Ch. 6.

8 Turner, *Man-made Disasters*.

9 Vaughan, *The Challenger Launch Decision*, pp. 243-49, 253-57, 262-64, 350-52, 356-72.

10 Turner, *Man-made Disasters*.

11 U.S. Congress, House, *Investigation of the Challenger Accident*, (Washington: Government Printing Office, 1986), pp. 149.

12 *Report of the Presidential Commission*, Vol. I, p. 148; Vol. IV, p. 1446.

13 Vaughan, *The Challenger Launch Decision*, p. 235.

14 *Report of the Presidential Commission*, Vol. I, pp. 1-3.

15 Howard E. McCurdy, "The Decay of NASA's Technical Culture," *Space Policy* (November 1989), pp. 301-10.

16 *Report of the Presidential Commission*, Vol. I, pp. 164-177.

17 *Report of the Presidential Commission*, Vol. I, Ch. VII and VIII.

18 *Report of the Presidential Commission*, Vol. I, pp. 140.

19 For background on culture in general and engineering culture in particular, see Peter Whalley and Stephen R. Barley, "Technical Work in the Division of Labor: Stalking the Wily Anomaly," in Stephen R. Barley and Julian Orr (eds.) *Between Craft and Science*, (Ithaca: Cornell University Press, 1997) pp. 23-53; Gideon Kunda, *Engineering Culture: Control and Commitment in a High-Tech Corporation*, (Philadelphia: Temple University Press, 1992); Peter Meiksins and James M. Watson, "Professional Autonomy and Organizational Constraint: The Case of Engineers," *Sociological Quarterly* 30 (1989), pp. 561-85; Henry Petroski, *To Engineer is Human: The Role of Failure in Successful Design* (New York: St. Martin's, 1985); Edgar Schein. *Organization Culture and Leadership*, (San Francisco: Jossey-Bass, 1985); John Van Maanen and Stephen R. Barley, "Cultural Organization," in Peter J. Frost, Larry F. Moore, Meryl Ries Louise, Craig C. Lundberg, and Joanne Martin (eds.) *Organization Culture*, (Beverly Hills: Sage, 1985).

20 *Report of the Presidential Commission*, Vol. I, pp. 82-111.

21 Harry McDonald, Report of the Shuttle Independent Assessment Team.

22 *Report of the Presidential Commission*, Vol. I, pp. 145-148.

23 Vaughan, *The Challenger Launch Decision*, pp. 257-264.

24 U. S. Congress, House, *Investigation of the Challenger Accident*, (Washington: Government Printing Office, 1986), pp. 70-71.

25 *Report of the Presidential Commission*, Vol. I, Ch.VII.

26 Mary Douglas, *How Institutions Think* (London: Routledge and Kegan Paul, 1987); Michael Burawoy, *Manufacturing Consent* (Chicago: University of Chicago Press, 1979).

27 *Report of the Presidential Commission*, Vol. I, pp. 171-173.

28 *Report of the Presidential Commission*, Vol. I, pp. 173-174.

29 National Aeronautics and Space Administration, Aerospace Safety Advisory Panel, "National Aeronautics and Space Administration Annual Report: Covering Calendar Year 1984," (Washington: Government Printing Office, 1985).

30 Harry McDonald, Report of the Shuttle Independent Assessment Team.

31 Richard J. Feynman, "Personal Observations on Reliability of the Shuttle," *Report of the Presidential Commission*, Appendix F:1.

32 Howard E. McCurdy, "The Decay of NASA's Technical Culture," *Space Policy* (November 1989), pp. 301-10; See also Howard E. McCurdy, *Inside NASA* (Baltimore: Johns Hopkins University Press, 1993).

33 Diane Vaughan, "The Trickle-Down Effect: Policy Decisions, Risky Work, and the Challenger Tragedy," *California Management Review*, 39, 2, Winter 1997.

34 Morton subsequently sold its propulsion division of Alcoa, and the company is now known as ATK Thiokol Propulsion.

35 *Report of the Presidential Commission*, pp. 82-118.

36 For discussions of how frames and cultural beliefs shape perceptions, see, e.g., Lee Clarke, "The Disqualification Heuristic: When Do Organizations Misperceive Risk?" in *Social Problems and Public Policy*, vol. 5, ed. R. Ted Youn and William F. Freudenberg, (Greenwich, CT: JAI, 1993); William Starbuck and Frances Milliken, "Executive Perceptual Filters – What They Notice and How They Make Sense," in *The Executive Effect*, Donald C. Hambrick, ed. (Greenwich, CT: JAI Press, 1988); Daniel Kahneman, Paul Slovic, and Amos Tversky, eds. *Judgment Under Uncertainty: Heuristics and Biases* (Cambridge: Cambridge University Press, 1982); Carol A. Heimer, "Social Structure, Psychology, and the Estimation of Risk." *Annual Review of Sociology 14 (1988): 491-519; Stephen J. Pfohl, Predicting Dangerousness* (Lexington, MA: Lexington Books, 1978).

37 *Report of the Presidential Commission*, Vol. IV: 791; Vaughan, *The Challenger Launch Decision*, p. 178.

38 *Report of the Presidential Commission*, Vol. I, pp. 91-92; Vol. IV, p. 612.

39 *Report of the Presidential Commission*, Vol. I, pp. 164-177; Chapter 6, this Report.

40 *Report of the Presidential Commission*, Vol. I, p. 90.

41 *Report of the Presidential Commission*, Vol. IV, pp. 791. For details of teleconference and engineering analysis, see Roger M. Boisjoly, "Ethical Decisions: Morton Thiokol and the Space Shuttle Challenger Disaster," *American Society of Mechanical Engineers*, (Boston: 1987), pp. 1-13.

42 Vaughan, *The Challenger Launch Decision*, pp. 358-361.

43 *Report of the Presidential Commission*, Vol. I, pp. 88-89, 93.

44 Edward Wong, "E-Mail Writer Says He was Hypothesizing, Not Predicting Disaster," *New York Times*, 11 March 2003, Sec. A-20, Col. 1 (excerpts from press conference, Col. 3).

45 *Report of the Presidential Commission*, Vol. I, pp. 92-95.

46 *Report of the Presidential Commission*, Vol. I, p. 152.

47 Weick argues that in a risky situation, people need to learn how to "drop their tools:" learn to recognize when they are in unprecedented situations in which following the rules can be disastrous. See Karl E. Weick, "The Collapse of Sensemaking in Organizations: The Mann Gulch Disaster." *Administrative Science Quarterly* 38, 1993, pp. 628-652.

48 Lee Clarke, *Mission Improbable: Using Fantasy Documents to Tame Disaster*, (Chicago: University of Chicago Press, 1999); Charles Perrow, *Normal Accidents*, op. cit.; Scott Sagan, *The Limits of Safety*, op. cit.; Diane Vaughan, "The Dark Side of Organizations," *Annual Review of Sociology*, Vol. 25, 1999, pp. 271-305.

49 Typically, after a public failure, the responsible organization makes safety the priority. They sink resources into discovering what went wrong and lessons learned are on everyone's minds. A boost in resources goes to safety to build on those lessons in order to prevent another failure. But concentrating on rebuilding, repair, and safety takes energy and resources from other goals. As the crisis ebbs and normal functioning returns, institutional memory grows short. The tendency is then to backslide, as external pressures force a return to operating goals. William R. Freudenberg, "Nothing Recedes Like Success? Risk Analysis and the Organizational Amplification of Risks," *Risk: Issues in Health and Safety* 3, 1: 1992, pp. 1-35; Richard H. Hall, *Organizations: Structures, Processes, and Outcomes*, (Prentice-Hall. 1998), pp. 184-204; James G. March, Lee S. Sproull, and Michal Tamuz, "Learning from Samples of One or Fewer," *Organization Science*, 2, 1: February 1991, pp. 1-13.

Part Three

A Look Ahead

When it's dark, the stars come out ... The same is true with people. When the tragedies of life turn a bright day into a frightening night, God's stars come out and these stars are families who say although we grieve deeply as do the families of Apollo 1 *and* Challenger *before us, the bold exploration of space must go on. These stars are the leaders in Government and in NASA who will not let the vision die. These stars are the next generation of astronauts, who like the prophets of old said, "Here am I, send me."*

– Brig. Gen. Charles Baldwin, STS-107 Memorial Ceremony at the National Cathedral, February 6, 2003

As this report ends, the Board wants to recognize the outstanding people in NASA. We have been impressed with their diligence, commitment, and professionalism as the agency has been working tirelessly to help the Board complete this report. While mistakes did lead to the accident, and we found that organizational and cultural constraints have worked against safety margins, the NASA family should nonetheless continue to take great pride in their legacy and ongoing accomplishments. As we look ahead, the Board sincerely hopes this report will aid NASA in safely getting back to human space flight.

In Part Three the Board presents its views and recommendations for the steps needed to achieve that goal, of continuing our exploration of space, in a manner with improved safety.

Chapter 9 discusses the near-term, mid-term and long-term implications for the future of human space flight. For the near term, NASA should submit to the Return-to-Flight Task Force a plan for implementing the return-to-flight recommendations. For the mid-term, the agency should focus on: the remaining Part One recommendations, the Part Two recommendations for organizational and cultural changes, and the Part Three recommendation for recertifying the Shuttle for use to 2020 or beyond. In setting the stage for a debate on the long-term future of human space flight, the Board addresses the need for a national vision to direct the design of a new Space Transportation System.

Chapter 10 contains additional recommendations and the significant "look ahead" observations the Board made in the course of this investigation that were not directly related to the accident, but could be viewed as "weak signals" of future problems. The observations may be indications of serious future problems and must be addressed by NASA.

Chapter 11 contains the recommendations made in Parts One, Two and Three, all issued with the resolve to continue human space flight.

Columbia *in the Vehicle Assembly Building at the Kennedy Space Center being readied for STS-107 in late 2002.*

CHAPTER 9

Implications for the Future of Human Space Flight

> *And while many memorials will be built to honor* Columbia*'s crew, their greatest memorial will be a vibrant space program with new missions carried out by a new generation of brave explorers.*
>
> – Remarks by Vice President Richard B. Cheney, Memorial Ceremony at the National Cathedral, February 6, 2003

The report up to this point has been a look backward: a single accident with multiple causes, both physical and organizational. In this chapter, the Board looks to the future. We take the insights gained in investigating the loss of *Columbia* and her crew and seek to apply them to this nation's continuing journey into space. We divide our discussion into three timeframes: 1) short-term, NASA's return to flight after the *Columbia* accident; 2) mid-term, what is needed to continue flying the Shuttle fleet until a replacement means for human access to space and for other Shuttle capabilities is available; and 3) long-term, future directions for the U.S. in space. The objective in each case is for this country to maintain a human presence in space, but with enhanced safety of flight.

In this report we have documented numerous indications that NASA's safety performance has been lacking. But even correcting all those shortcomings, it should be understood, will not eliminate risk. All flight entails some measure of risk, and this has been the case since before the days of the Wright Brothers. Furthermore, the risk is not distributed evenly over the course of the flight. It is greater by far at the beginning and end than during the middle.

This concentration of risk at the endpoints of flight is particularly true for crew-carrying space missions. The Shuttle Program has now suffered two accidents, one just over a minute after takeoff and the other about 16 minutes before landing. The laws of physics make it extraordinarily difficult to reach Earth orbit and return safely. Using existing technology, orbital flight is accomplished only by harnessing a chemical reaction that converts vast amounts of stored energy into speed. There is great risk in placing human beings atop a machine that stores and then burns millions of pounds of dangerous propellants. Equally risky is having humans then ride the machine back to Earth while it dissipates the orbital speed by converting the energy into heat, much like a meteor entering Earth's atmosphere. No alternatives to this pathway to space are available or even on the horizon, so we must set our sights on managing this risky process using the most advanced and versatile techniques at our disposal.

Columbia launches as STS-107 on January 16, 2003.

Because of the dangers of ascent and re-entry, because of the hostility of the space environment, and because we are still relative newcomers to this realm, operation of the Shuttle and indeed all human spaceflight must be viewed as a developmental activity. It is still far from a routine, operational undertaking. Throughout the *Columbia* accident investigation, the Board has commented on the widespread but erroneous perception of the Space Shuttle as somehow comparable to civil or military air transport. They are not comparable; the inherent risks of spaceflight are vastly higher, and our experience level with spaceflight is vastly lower. If Shuttle operations came to be viewed as routine, it was, at least in part, thanks to the skill and dedication of those involved in the program. They have made it look easy, though in fact it never was. The Board urges NASA leadership, the architects of U.S. space policy, and the American people to adopt a realistic understanding of the risks and rewards of venturing into space.

9.1 Near-Term: Return to Flight

The Board supports return to flight for the Space Shuttle at the earliest date consistent with an overriding consideration: safety. The recognition of human spaceflight as a developmental activity requires a shift in focus from operations and meeting schedules to a concern for the risks involved. Necessary measures include:

- Identifying risks by looking relentlessly for the next eroding O-ring, the next falling foam; obtaining better data, analyzing and spotting trends.
- Mitigating risks by stopping the failure at its source; when a failure does occur, improving the ability to tolerate it; repairing the damage on a timely basis.
- Decoupling unforeseen events from the loss of crew and vehicle.
- Exploring all options for survival, such as provisions for crew escape systems and safe havens.
- Barring unwarranted departures from design standards, and adjusting standards only under the most rigorous, safety-driven process.

The Board has recommended improvements that are needed before the Shuttle Program returns to flight, as well as other measures to be adopted over the longer term – what might be considered "continuing to fly" recommendations. To ensure implementation of these longer-term recommendations, the Board makes the following recommendation, which should be included in the requirements for return-to-flight:

R9.1-1 Prepare a detailed plan for defining, establishing, transitioning, and implementing an independent Technical Engineering Authority, independent safety program, and a reorganized Space Shuttle Integration Office as described in R7.5-1, R7.5-2, and R7.5-3. In addition, NASA should submit annual reports to Congress, as part of the budget review process, on its implementation activities.

The complete list of the Board's recommendations can be found in Chapter 11.

9.2 Mid-Term: Continuing to Fly

It is the view of the Board that the present Shuttle is not inherently unsafe. However, the observations and recommendations in this report are needed to make the vehicle safe enough to operate in the coming years. In order to continue operating the Shuttle for another decade or even more, which the Human Space Flight Program may find necessary, these significant measures must be taken:

- Implement all the recommendations listed in Part One of this report that were not already accomplished as part of the return-to-flight reforms.
- Institute all the organizational and cultural changes called for in Part Two of this report.
- Undertake complete recertification of the Shuttle, as detailed in the discussion and recommendation below.

The urgency of these recommendations derives, at least in part, from the likely pattern of what is to come. In the near term, the recent memory of the *Columbia* accident will motivate the entire NASA organization to scrupulous attention to detail and vigorous efforts to resolve elusive technical problems. That energy will inevitably dissipate over time. This decline in vigilance is a characteristic of many large organizations, and it has been demonstrated in NASA's own history. As reported in Part Two of this report, the Human Space Flight Program has at times compromised safety because of its organizational problems and cultural traits. That is the reason, in order to prevent the return of bad habits over time, that the Board makes the recommendations in Part Two calling for changes in the organization and culture of the Human Space Flight Program. These changes will take more time and effort than would be reasonable to expect prior to return to flight.

Through its recommendations in Part Two, the Board has urged that NASA's Human Space Flight Program adopt the characteristics observed in high-reliability organizations. One is separating technical authority from the functions of managing schedules and cost. Another is an independent Safety and Mission Assurance organization. The third is the capability for effective systems integration. Perhaps even more challenging than these organizational changes are the cultural changes required. Within NASA, the cultural impediments to safe and effective Shuttle operations are real and substantial, as documented extensively in this report. The Board's view is that cultural problems are unlikely to be corrected without top-level leadership. Such leadership will have to rid the system of practices and patterns that have been validated simply because they have been around so long. Examples include: the tendency to keep knowledge of problems contained within a Center or program; making technical decisions without in-depth, peer-reviewed technical analysis; and an unofficial hierarchy or caste system created by placing excessive power in one office. Such factors interfere with open communication, impede the sharing of lessons learned, cause duplication and unnecessary expenditure of resources, prompt resistance to external advice, and create a burden for managers, among other undesirable outcomes. Collectively, these undesirable characteristics threaten safety.

Unlike return-to-flight recommendations, the Board's management and cultural recommendations will take longer to implement, and the responses must be fine-tuned and adjusted during implementation. The question of how to follow up on NASA's implementation of these more subtle, but equally important recommendations remains unanswered. The Board is aware that response to these recommendations will be difficult to initiate, and they will encounter some degree of institutional resistance. Nevertheless, in the Board's view, they are so critical to safer operation of the Shuttle fleet that they must be carried out completely. Since NASA is an independent agency answerable only to the White House and Congress, the ultimate responsibility for enforcement of the recommended corrective actions must reside with those governmental authorities.

Recertification

Recertification is a process to ensure flight safety when a vehicle's actual utilization exceeds its original design life; such a baseline examination is essential to certify that vehicle for continued use, in the case of the Shuttle to 2020 and possibly beyond. This report addresses recertification as a mid-term issue.

Measured by their 20 or more missions per Orbiter, the Shuttle fleet is young, but by chronological age – 10 to 20 years each – it is old. The Board's discovery of mass loss in RCC panels, the deferral of investigation into signs of metal corrosion, and the deferral of upgrades all strongly suggest that a policy is needed requiring a complete recertification of the Space Shuttle. This recertification must be rigorous and comprehensive at every level (i.e., material, component, subsystem, and system); the higher the level, the more critical the integration of lower-level components. A post-*Challenger*, 10-year review was conducted, but it lacked this kind of rigor, comprehensiveness and, most importantly, integration at the subsystem and system levels.

Aviation industry standards offer ample measurable criteria for gauging specific aging characteristics, such as stress and corrosion. The Shuttle Program, by contrast, lacks a closed-loop feedback system and consequently does not take full advantage of all available data to adjust its certification process and maintenance practices. Data sources can include experience with material and component failures, non-conformances (deviations from original specifications) discovered during Orbiter Maintenance Down Periods, Analytical Condition Inspections, and Aging Aircraft studies. Several of the recommendations in this report constitute the basis for a recertification program (such as the call for nondestructive evaluation of RCC components). Chapters 3 and 4 cite instances of waivers and certification of components for flight based on analysis rather than testing. The recertification program should correct all those deficiencies.

Finally, recertification is but one aspect of a Service Life Extension Program that is essential if the Shuttle is to continue operating for another 10 to 20 years. While NASA has such a program, it is in its infancy and needs to be pursued with vigor. The Service Life Extension Program goes beyond the Shuttle itself and addresses critical associated components in equipment, infrastructure, and other areas. Aspects of the program are addressed in Appendix D.15.

The Board makes the following recommendation regarding recertification:

R9.2-1 Prior to operating the Shuttle beyond 2010, develop and conduct a vehicle recertification at the material, component, subsystem, and system levels. Recertification requirements should be included in the Service Life Extension Program.

9.3 LONG-TERM: FUTURE DIRECTIONS FOR THE U.S. IN SPACE

The Board in its investigation has focused on the physical and organizational causes of the *Columbia* accident and the recommended actions required for future safe Shuttle operation. In the course of that investigation, however, two realities affecting those recommendations have become evident to the Board. One is the lack, over the past three decades, of any national mandate providing NASA a compelling mission requiring human presence in space. President John Kennedy's 1961 charge to send Americans to the moon and return them safely to Earth "before this decade is out" linked NASA's efforts to core Cold War national interests. Since the 1970s, NASA has not been charged with carrying out a similar high priority mission that would justify the expenditure of resources on a scale equivalent to those allocated for Project Apollo. The result is the agency has found it necessary to gain the support of diverse constituencies. NASA has had to participate in the give and take of the normal political process in order to obtain the resources needed to carry out its programs. NASA has usually failed to receive budgetary support consistent with its ambitions. The result, as noted throughout Part Two of the report, is an organization straining to do too much with too little.

A second reality, following from the lack of a clearly defined long-term space mission, is the lack of sustained government commitment over the past decade to improving U.S. access to space by developing a second-generation space transportation system. Without a compelling reason to do so, successive Administrations and Congresses have not been willing to commit the billions of dollars required to develop such a vehicle. In addition, the space community has proposed to the government the development of vehicles such as the National Aerospace Plane and X-33, which required "leapfrog" advances in technology; those advances have proven to be unachievable. As Apollo 11 Astronaut Buzz Aldrin, one of the members of the recent Commission on the Future of the United States Aerospace Industry, commented in the Commission's November 2002 report, "Attempts at developing breakthrough space transportation systems have proved illusory."[1] The Board believes that the country should plan for future space transportation capabilities without making them dependent on technological breakthroughs.

Lack of a National Vision for Space

In 1969 President Richard Nixon rejected NASA's sweeping vision for a post-Apollo effort that involved full development of low-Earth orbit, permanent outposts on the moon, and initial journeys to Mars. Since that rejection, these objectives have reappeared as central elements in many proposals setting forth a long-term vision for the U.S. Space program. In 1986 the National Commission on Space proposed "a pioneering mission for 21st-century America: To lead the exploration and development of the space frontier, advancing science, technology, and enterprise, and building institutions and systems that make accessible vast new resources and support human settlements beyond Earth orbit, from the highlands of the Moon to the plains of Mars."[2] In 1989, on the 20th anniversary of the first lunar landing, President George H.W. Bush proposed a Space Exploration Initiative, calling for "a sustained program of manned exploration of the solar system."[3] Space advocates have been consistent in their call for sending humans beyond low-Earth orbit as the appropriate objective of U.S. space activities. Review committees as diverse as the 1990 Advisory Committee on the Future of the U.S. Space Program, chaired by Norman Augustine, and the 2001 International Space Station Management and Cost Evaluation Task Force have suggested that the primary justification for a space station is to conduct the research required to plan missions to Mars and/or other distant destinations. However, human travel to destinations beyond Earth orbit has not been adopted as a national objective.

The report of the Augustine Committee commented, "It seems that most Americans do support a viable space program for the nation – but no two individuals seem able to agree upon *what* that space program should be."[4] The Board observes that none of the competing long-term visions for space have found support from the nation's leadership, or indeed among the general public. The U.S. civilian space effort has moved forward for more than 30 years without a guiding vision, and none seems imminent. In the past, this absence of a strategic vision in itself has reflected a policy decision, since there have been many opportunities for national leaders to agree on ambitious goals for space, and none have done so.

The Board does observe that there is one area of agreement among almost all parties interested in the future of U.S. activities in space: *The United States needs improved access for humans to low-Earth orbit as a foundation for whatever directions the nation's space program takes in the future.* In the Board's view, a full national debate on how best to achieve such improved access should take place in parallel with the steps the Board has recommended for returning the Space Shuttle to flight and for keeping it operating safely in coming years. Recommending the content of this debate goes well beyond the Board's mandate, but we believe that the White House, Congress, and NASA should honor the memory of *Columbia*'s crew by reflecting on the nation's future in space and the role of new space transportation capabilities in enabling whatever space goals the nation chooses to pursue.

All members of the Board agree that America's future space efforts must include human presence in Earth orbit, and eventually beyond, as outlined in the current NASA vision. Recognizing the absence of an agreed national mandate cited above, the current NASA strategic plan stresses an approach of investing in "transformational technologies" that will enable the development of capabilities to serve as "stepping stones" for whatever path the nation may decide it wants to pursue in space. While the Board has not reviewed this plan in depth, this approach seems prudent. Absent any long-term statement of what the country wants to accomplish in space, it is difficult to state with any specificity the requirements that should guide major public investments in new capabilities. The Board does believe that NASA and the nation should give more attention to developing a new "concept of operations" for future activities – defining the range of activities the country intends to carry out in space – that could provide more specificity than currently exists. Such a concept does not necessarily require full agreement on a future vision, but it should help identify the capabilities required and prevent the debate from focusing solely on the design of the next vehicle.

Developing a New Space Transportation System

When the Space Shuttle development was approved in 1972, there was a corresponding decision not to fund technologies for space transportation other than those related to the Shuttle. This decision guided policy for more than 20 years, until the National Space Transportation Policy of 1994 assigned NASA the role of developing a next-generation, advanced-technology, single-stage-to-orbit replacement for the Space Shuttle. That decision was flawed for several reasons. Because the United States had not funded a broad portfolio of space transportation technologies for the preceding three decades, there was a limited technology base on which to base the choice of this second-generation system. The technologies chosen for development in 1996, which were embodied in the X-33 demonstrator, proved not yet mature enough for use. Attracted by the notion of a growing private sector market for space transportation, the Clinton Administration hoped this new system could be developed with minimal public investment – the hope was that the private sector would help pay for the development of a Shuttle replacement.

In recent years there has been increasing investment in space transportation technologies, particularly through NASA's Space Launch Initiative effort, begun in 2000. This investment has not yet created a technology base for a second-generation reusable system for carrying people to orbit. Accordingly, in 2002 NASA decided to reorient the Space Launch Initiative to longer-term objectives, and to introduce the concept of an Orbital Space Plane as an interim complement to the Space Shuttle for space station crew-carrying responsibilities. The Integrated Space Transportation Plan also called for using the Space Shuttle for an extended period into the future. The Board has evaluated neither NASA's Integrated Space Transportation Plan nor the detailed requirements of an Orbital Space Plane.

Even so, based on its in-depth examination of the Space Shuttle Program, the Board has reached an inescapable conclusion: *Because of the risks inherent in the original design of the Space Shuttle, because that design was based in many aspects on now-obsolete technologies, and because the Shuttle is now an aging system but still developmental in character, it is in the nation's interest to replace the Shuttle as soon as possible as the primary means for transporting humans to and from Earth orbit.* At least in the mid-term, that replacement will be some form of what NASA now characterizes as an Orbital Space Plane. The design of the system should give overriding priority to crew safety, rather than trade safety against other performance criteria, such as low cost and reusability, or against advanced space operation capabilities other than crew transfer.

This conclusion implies that whatever design NASA chooses should become the primary means for taking people to and from the International Space Station, not just a complement to the Space Shuttle. And it follows from the same conclusion that there is urgency in choosing that design, after serious review of a "concept of operations" for human space flight, and bringing it into operation as soon as possible. This is likely to require a significant commitment of resources over the next several years. The nation must not shy from making that commitment. The International Space Station is likely to be the major destination for human space travel for the next decade or longer. The Space Shuttle would continue to be used when its unique capabilities are required, both with respect to space station missions such as experiment delivery and retrieval or other logistical missions, and with respect to the few planned missions not traveling to the space station. When cargo can be carried to the space station or other destinations by an expendable launch vehicle, it should be.

However, the Orbital Space Plane is seen by NASA as an interim system for transporting humans to orbit. NASA plans to make continuing investments in "next generation launch technology," with the hope that those investments will enable a decision by the end of this decade on what that next generation launch vehicle should be. This is a worthy goal, and should be pursued. *The Board notes that this approach can only be successful: if it is sustained over the decade; if by the time a decision to develop a new vehicle is made there is a clearer idea of how the new space transportation system fits into the nation's overall plans for space; and if the U.S. government is willing at the time a development decision is made to commit the substantial resources required to implement it.* One of the major problems with the way the Space Shuttle Program was carried out was an *a priori* fixed ceiling on development costs. That approach should not be repeated.

It is the view of the Board that *the previous attempts to develop a replacement vehicle for the aging Shuttle represent a failure of national leadership.* The cause of the failure was continuing to expect major technological advances in that vehicle. With the amount of risk inherent in the Space Shuttle, the first step should be to reach an agreement that the overriding mission of the replacement system is to move humans safely and reliably into and out of Earth orbit. To demand more would be to fall into the same trap as all previous, unsuccessful, efforts. That being said, it seems to the Board that past and future investments in space launch technologies should certainly provide by 2010 or thereabouts the basis for developing a system, significantly improved over one designed 40 years earlier, for carrying humans to orbit and enabling their work in space. Continued U.S. leadership in space is an important national objective. That leadership depends on a willingness to pay the costs of achieving it.

Final Conclusions

The Board's perspective assumes, of course, that the United States wants to retain a continuing capability to send people into space, whether to Earth orbit or beyond. The Board's work over the past seven months has been motivated by the desire to honor the STS-107 crew by understanding the cause of the accident in which they died, and to help the United States and indeed all spacefaring countries to minimize the risks of future loss of lives in the exploration of space. The United States should continue with a Human Space Flight Program consistent with the resolve voiced by President George W. Bush on February 1, 2003: *"Mankind is led into the darkness beyond our world by the inspiration of discovery and the longing to understand. Our journey into space will go on."*

Two proposals – a capsule (above) and a winged vehicle - for the Orbital Space Plane, courtesy of The Boeing Company.

ENDNOTES FOR CHAPTER 9

The citations that contain a reference to "CAIB document" with CAB or CTF followed by seven to eleven digits, such as CAB001-0010, refer to a document in the Columbia Accident Investigation Board database maintained by the Department of Justice and archived at the National Archives.

1. *Report on the Commission on the Future of the United States Aerospace Industry*, November 2002, p. 3-3.
2. National Commission on Space, *Pioneering the Space Frontier: An Exciting Vision of Our Next Fifty Years in Space, Report of the National Commission on Space* (Bantam Books, 1986), p. 2.
3. President George H. W. Bush, "Remarks on the 20th Anniversary of the Apollo 11 Moon Landing," Washington, D.C., July 20, 1989.
4. "Report of the Advisory Committee on the Future of the U.S. Space Program," December 1990, p. 2.

CHAPTER 10

Other Significant Observations

Although the Board now understands the combination of technical and organizational factors that contributed to the *Columbia* accident, the investigation did not immediately zero in on the causes identified in previous chapters. Instead, the Board explored a number of avenues and topics that, in the end, were not directly related to the cause of this accident. Nonetheless, these forays revealed technical, safety, and cultural issues that could impact the Space Shuttle Program, and, more broadly, the future of human space flight. The significant issues listed in this chapter are potentially serious matters that should be addresed by NASA because they fall into the category of "weak signals" that could be indications of future problems.

10.1 PUBLIC SAFETY

Shortly after the breakup of *Columbia* over Texas, dramatic images of the Orbiter's debris surfaced: an intact spherical tank in an empty parking lot, an obliterated office rooftop, mangled metal along roadsides, charred chunks of material in fields. These images, combined with the large number of debris fragments that were recovered, compelled many to proclaim it was a "miracle" that no one on the ground had been hurt.[1]

The *Columbia* accident raises some important questions about public safety. What were the chances that the general public could have been hurt by a breakup of an Orbiter? How safe are Shuttle flights compared with those of conventional aircraft? How much public risk from space flight is acceptable? Who is responsible for public safety during space flight operations?

Public Risk from *Columbia*'s Breakup

The Board commissioned a study to determine if the lack of reported injuries on the ground was a predictable outcome or simply exceptionally good fortune (see Appendix D.16). The study extrapolated from an array of data, including census figures for the debris impact area, the Orbiter's last reported position and velocity, the impact locations (latitude and longitude), and the total weight of all recovered debris, as well as the composition and dimensions of many debris pieces.[2]

Based on the best available evidence on *Columbia*'s disintegration and ground impact, the lack of serious injuries on the ground was the expected outcome for the location and time at which the breakup occurred.[3]

NASA and others have developed sophisticated computer tools to predict the trajectory and survivability of spacecraft debris during re-entry.[4] Such tools have been used to assess the risk of serious injuries to the public due to spacecraft re-entry, including debris impacts from launch vehicle malfunctions.[5] However, it is impossible to be certain about what fraction of *Columbia* survived to impact the ground. Some 38 percent of *Columbia*'s dry (empty) weight was recovered, but there is no way to determine how much still lies on the ground. Accounting for the inherent uncertainties associated with the amount of ground debris and the number of people outdoors,[6] there was about a 9- to 24-percent chance of at least one person being seriously injured by the disintegration of the Orbiter.[7]

Debris fell on a relatively sparsely populated area of the United States, with an average of about 85 inhabitants per square mile. Orbiter re-entry flight paths often pass over much more populated areas, including major cities that average more than 1,000 inhabitants per square mile. For example, the STS-107 re-entry profile passed over Sacramento, California, and Albuquerque, New Mexico. The Board-sponsored study concluded that, given the unlikely event of a similar Orbiter breakup over a densely populated area such as Houston, the most likely outcome would be one or two ground casualties.

Space Flight Risk Compared to Aircraft Operations

A recent study of U.S. civil aviation accidents found that between 1964 and 1999, falling aircraft debris killed an average of eight people per year.[8] In comparison, the National Center for Health Statistics reports that between 1992 and 1994, an average of 65 people in the United States were killed each year by lightning strikes. The aviation accident study revealed a decreasing trend in the annual number of "groundling" fatalities, so that an average of about four fatalities per year are predicted in the near future.[9] The probability of a U.S. resident being killed by aircraft debris is now less than one in a million over a 70-year lifetime.[10]

The history of U.S. space flight has a flawless public safety record. Since the 1950s, there have been hundreds of U.S. space launches without a single member of the public being injured. Comparisons between the risk to the public from space flight and aviation operations are limited by two factors: the absence of public injuries resulting from U.S. space flight operations, and the relatively small number of space flights (hundreds) compared to aircraft flights (billions).[11] Nonetheless, it is unlikely that U.S. space flights will produce many, if any, public injuries in the coming years based on (1) the low number of space flight operations per year, (2) the flawless public safety record of past U.S. space launches, (3) government-adopted space flight safety standards,[12] and (4) the risk assessment result that, even in the unlikely event of a similar Orbiter breakup over a major city, less than two ground casualties would be expected. In short, the risk posed to people on the ground by U.S. space flight operations is small compared to the risk from civil aircraft operations.

The government has sought to limit public risk from space flight to levels comparable to the risk produced by aircraft. U.S. space launch range commanders have agreed that the public should face no more than a one-in-a-million chance of fatality from launch vehicle and unmanned aircraft operations.[13] This aligns with Federal Aviation Administration (FAA) regulations that individuals be exposed to no more than a one-in-a-million chance of serious injury due to commercial space launch and re-entry operations.[14]

NASA has not actively followed public risk acceptability standards used by other government agencies during past Orbiter re-entry operations. However, in the aftermath of the *Columbia* accident, the agency has attempted to adopt similar rules to protect the public. It has also developed computer tools to predict the survivability of spacecraft debris during re-entry. Such tools have been used to assess the risk of public casualties attributable to spacecraft re-entry, including debris impacts from commercial launch vehicle malfunctions.[15]

Responsibility for Public Safety

The Director of the Kennedy Space Center is responsible for the ground and flight safety of Kennedy Space Center people and property for all launches.[16] The Air Force provides the Director with written notification of launch area risk estimates for Shuttle ascents. The Air Force routinely computes the risk that Shuttle ascents[17] pose to people on and off Kennedy grounds from potential debris impacts, toxic exposures, and explosions.[18]

However, no equivalent collaboration exists between NASA and the Air Force for re-entry risk. FAA rules on commercial space launch activities do not apply "where the Government is so substantially involved that it is effectively directing or controlling the launch." Based on the lack of a response, in tandem with NASA's public statements and informal replies to Board questions, the Board determined that NASA made no documented effort to assess public risk from Orbiter re-entry operations prior to the *Columbia* accident. The Board believes that NASA should be legally responsible for public safety during all phases of Shuttle operations, including re-entry.

Findings:

F10.1-1 The *Columbia* accident demonstrated that Orbiter breakup during re-entry has the potential to cause casualties among the general public.

F10.1-2 Given the best information available to date, a formal risk analysis sponsored by the Board found that the lack of general-public casualties from *Columbia*'s break-up was the expected outcome.

F10.1-3 The history of U.S. space flight has a flawless public safety record. Since the 1950s, hundreds of space flights have occurred without a single public injury.

F10.1-4 The FAA and U.S. space launch ranges have safety standards designed to ensure that the general public is exposed to less than a one-in-a-million chance of serious injury from the operation of space launch vehicles and unmanned aircraft.

F10.1-5 NASA did not demonstrably follow public risk acceptability standards during past Orbiter re-entries. NASA efforts are underway to define a national policy for the protection of public safety during all operations involving space launch vehicles.

Observations:

O10.1-1 NASA should develop and implement a public risk acceptability policy for launch and re-entry of space vehicles and unmanned aircraft.

O10.1-2 NASA should develop and implement a plan to mitigate the risk that Shuttle flights pose to the general public.

O10.1-3 NASA should study the debris recovered from *Columbia* to facilitate realistic estimates of the risk to the public during Orbiter re-entry.

10.2 CREW ESCAPE AND SURVIVAL

The Board has examined crew escape systems in historical context with a view to future improvements. It is important to note at the outset that *Columbia* broke up during a phase of flight that, given the current design of the Orbiter, offered no possibility of crew survival.

The goal of every Shuttle mission is the safe return of the crew. An escape system—a means for the crew to leave a vehicle in distress during some or all of its flight phases and return safely to Earth – has historically been viewed as one "technique" to accomplish that end. Other methods include various abort modes, rescue, and the creation of a safe haven (a location where crew members could remain unharmed if they are unable to return to Earth aboard a damaged Shuttle).

While crew escape systems have been discussed and studied continuously since the Shuttle's early design phases, only two systems have been incorporated: one for the developmental test flights, and the current system installed after the *Challenger* accident. Both designs have extremely limited capabilities, and neither has ever been used during a mission.

Developmental Test Flights

Early studies assumed that the Space Shuttle would be operational in every sense of the word. As a result, much like commercial airliners, a Shuttle crew escape system was considered unnecessary. NASA adopted requirements for rapid emergency egress of the crew in early Shuttle test flights. Modified SR-71 ejection seats for the two pilot positions were installed on the Orbiter test vehicle *Enterprise,* which was carried to an altitude of 25,000 feet by a Boeing 747 Shuttle Carrier Aircraft during the Approach and Landing Tests in 1977.[19]

Essentially the same system was installed on *Columbia* and used for the four Orbital Test Flights during 1981-82. While this system was designed for use during first-stage ascent and in gliding flight below 100,000 feet, considerable doubt emerged about the survivability of an ejection that would expose crew members to the Solid Rocket Booster exhaust plume. Regardless, NASA declared the developmental test flight phase complete after STS-4, *Columbia*'s fourth flight, and the ejection seat system was deactivated. Its associated hardware was removed during modification after STS-9. All Space Shuttle missions after STS-4 were conducted with crews of four or more, and no escape system was installed until after the loss of *Challenger* in 1986.

Before the *Challenger* accident, the question of crew survival was not considered independently from the possibility of catastrophic Shuttle damage. In short, NASA believed if the Orbiter could be saved, then the crew would be safe. Perceived limits of the use of escape systems, along with their cost, engineering complexity, and weight/payload tradeoffs, dissuaded NASA from implementing a crew escape plan. Instead, the agency focused on preventing the loss of a Shuttle as the sole means for assuring crew survival.

Post-*Challenger*: the Current System

NASA's rejection of a crew escape system was severely criticized after the loss of *Challenger*. The Rogers Commission addressed the topic in a recommendation that combined the issues of launch abort and crew escape:[20]

> *Launch Abort and Crew Escape. The Shuttle Program management considered first-stage abort options and crew escape options several times during the history of the program, but because of limited utility, technical infeasibility, or program cost and schedule, no systems were implemented. The Commission recommends that NASA:*
>
> - *Make all efforts to provide a crew escape system for use during controlled gliding flight.*
> - *Make every effort to increase the range of flight conditions under which an emergency runway landing can be successfully conducted in the event that two or three main engines fail early in ascent.*

In response to this recommendation, NASA developed the current "pole bailout" system for use during controlled, subsonic gliding flight (see Figure 10.2-1). The system requires crew members to "vent" the cabin at 40,000 feet (to equalize the cabin pressure with the pressure at that altitude), jettison the hatch at approximately 32,000 feet, and then jump out of the vehicle (the pole allows crew members to avoid striking the Orbiter's wings).

Figure 10.2-1. A demonstration of the pole bailout system. The pole is extending from the side of a C-141 simulating the Orbiter, with a crew member sliding down the pole so that he would fall clear of the Orbiter's wing during an actual bailout.

Current Human-Rating Requirements

In June 1998, Johnson Space Center issued new Human-Rating Requirements applicable to "all future human-rated spacecraft operated by NASA." In July 2003, shortly before this report was published, NASA issued further Human-Rating Requirements and Guidelines for Space Flight Systems, over the signature of the Associate Administrator for Safety and Mission Assurance. While these new requirements "... shall not supersede more stringent requirements imposed by individual NASA organizations ..." NASA has informed the Board that the earlier – and in some cases more prescriptive – Johnson Space Center requirements have been cancelled.

NASA's 2003 Human-Rating Requirements and Guidelines for Space Flight Systems laid out the following principles regarding crew escape and survival:

> 2.5.4 *Crew survival*
>
> 2.5.4.1 *As part of the design process, program management (with approval from the CHMO* [Chief Health and Medical Officer], *AA for OSF* [Associate Administrator for the Office of Spaceflight], *and AA for SMA* [Associate Administrator for Safety and Mission Assurance] *shall establish, assess, and document the program requirements for an acceptable life cycle cumulative probability of safe crew and passenger return. This probability requirement can be satisfied through the use of all available mechanisms including nominal mission completion, abort, safe haven, or crew escape.*
>
> 2.5.4.2 *The cumulative probability of safe crew and passenger return shall address all missions planned for the life of the program, not just a single space flight system for a single mission.*

The overall probability of crew and passenger survival must meet the minimum program requirements (as defined in section 2.5.4.1) for the stated life of a space flight systems program.[21] This approach is required to reflect the different technical challenges and levels of operational risk exposure on various types of missions. For example, low-Earth-orbit missions represent fundamentally different risks than does the first mission to Mars. Single-mission risk on the order of 0.99 for a beyond-Earth-orbit mission may be acceptable, but considerably better performance, on the order of 0.9999, is expected for a reusable low-Earth-orbit design that will make 100 or more flights.

> 2.6 *Abort and Crew Escape*
>
> 2.6.1 *The capability for rapid crew and occupant egress shall be provided during all pre-launch activities.*
>
> 2.6.2 *The capability for crew and occupant survival and recovery shall be provided on ascent using a combination of abort and escape.*
>
> 2.6.3 *The capability for crew and occupant survival and recovery shall be provided during all other phases of flight (including on-orbit, reentry, and landing) using a combination of abort and escape, unless comprehensive safety and reliability analyses indicate that abort and escape capability is not required to meet crew survival requirements.*
>
> 2.6.4 *Determinations regarding escape and abort shall be made based upon comprehensive safety and reliability analyses across all mission profiles.*

These new requirements focus on general crew survival rather than on particular crew escape systems. This provides a logical context for discussions of tradeoffs that will yield the best crew-survival outcome. Such tradeoffs include "mass-trades" – for example, an escape system could add weight to a vehicle, but in the process cause payload changes that require additional missions, thereby inherently increasing the overall exposure to risk.

Note that the new requirements for crew escape appear less prescriptive than Johnson Space Center Requirement 7, which deals with "safe crew extraction" from pre-launch to landing.[22]

In addition, the extent to which NASA's 2003 requirements will retroactively apply to the Space Shuttle is an open question:

> *The Governing Program Management Council (GPMC) will determine the applicability of this document to programs and projects in existence (e.g., heritage expendable and reusable launch vehicles and evolved expendable launch vehicles), at or beyond implementation, at the time of the issuance of this document.*

Recommendations of the NASA Aerospace Safety Advisory Panel

The issue of crew escape has long been a matter of concern to NASA's Aerospace Safety Advisory Panel. In its 2002 Annual Report, the panel noted that NASA Program Guidelines on Human Rating require escape systems for all flight vehicles, but the guidelines do not apply to the Space Shuttle. The Panel considered it appropriate, in view of the Shuttle's proposed life extension, to consider upgrading the vehicle to comply with the guidelines.[23]

> *Recommendation 02-9: Complete the ongoing studies of crew escape design options. Either document the reasons for not implementing the NASA Program Guidelines on Human Rating or expedite the deployment of such capabilities.*

The Board shares the concern of the NASA Aerospace Safety Advisory Panel and others over the lack of a crew escape system for the Space Shuttle that could cover the widest possible range of flight regimes and emergencies. At the same time, a crew escape system is just one element to be optimized for crew survival. Crucial tradeoffs in risk, complexity, weight, and operational utility must be made when considering a Shuttle escape system. Designs for future vehicles and possible retrofits should be evaluated in this context. The sole objective must be the highest probability of a crew's safe return regardless if that is due to successful mission completions, vehicle-intact aborts, safe haven/rescues, escape systems, or some combination of these scenarios.

Finally, a crew escape system cannot be considered separately from the issues of Shuttle retirement/replacement, separation of cargo from crew in future vehicles, and other considerations in the development – and the inherent risks of space flight.

Space flight is an inherently dangerous undertaking, and will remain so for the foreseeable future. While all efforts must be taken to minimize its risks, the White House, Congress, and the American public must acknowledge these dangers and be prepared to accept their consequences.

Observations:

O10.2-1 Future crewed-vehicle requirements should incorporate the knowledge gained from the *Challenger* and *Columbia* accidents in assessing the feasibility of vehicles that could ensure crew survival even if the vehicle is destroyed.

10.3 SHUTTLE ENGINEERING DRAWINGS AND CLOSEOUT PHOTOGRAPHS

In the years since the Shuttle was designed, NASA has not updated its engineering drawings or converted to computer-aided drafting systems. The Board's review of these engineering drawings revealed numerous inaccuracies. In particular, the drawings do not incorporate many engineering changes made in the last two decades. Equally troubling was the difficulty in obtaining these drawings: it took up to four weeks to receive them, and, though some photographs were available as a short-term substitute, closeout photos took up to six weeks to obtain. (Closeout photos are pictures taken of Shuttle areas before they are sealed off for flight.) The Aerospace Safety Advisory Panel noted similar difficulties in its 2001 and 2002 reports.

The Board believes that the Shuttle's current engineering drawing system is inadequate for another 20 years' use. Widespread inaccuracies, unincorporated engineering updates, and significant delays in this system represent a significant dilemma for NASA in the event of an on-orbit crisis that requires timely and accurate engineering information. The dangers of an inaccurate and inaccessible drawing system are exacerbated by the apparent lack of readily available closeout photographs as interim replacements (see Appendix D.15).

Findings:

F10.3-1 The engineering drawing system contains outdated information and is paper-based rather than computer-aided.

F10.3-2 The current drawing system cannot quickly portray Shuttle sub-systems for on-orbit troubleshooting.

F10.3-3 NASA normally uses closeout photographs but lacks a clear system to define which critical sub-systems should have such photographs. The current system does not allow the immediate retrieval of closeout photos.

Recommendations:

R10.3-1 Develop an interim program of closeout photographs for all critical sub-systems that differ from engineering drawings. Digitize the closeout photograph system so that images are immediately available for on-orbit troubleshooting.

R10.3-2 Provide adequate resources for a long-term program to upgrade the Shuttle engineering drawing system including:

- Reviewing drawings for accuracy
- Converting all drawings to a computer-aided drafting system
- Incorporating engineering changes

10.4 INDUSTRIAL SAFETY AND QUALITY ASSURANCE

The industrial safety programs in place at NASA and its contractors are robust and in good health. However, the scope and depth of NASA's maintenance and quality assurance programs are troublesome. Though unrelated to the *Columbia* accident, the major deficiencies in these programs uncovered by the Board could potentially contribute to a future accident.

Industrial Safety

Industrial safety programs at NASA and its contractors—covering safety measures "on the shop floor" and in the workplace – were examined by interviews, observations, and reviews. Vibrant industrial safety programs were found in every area examined, reflecting a common interview comment: "If anything, we go overboard on safety." Industrial safety programs are highly visible: they are nearly always a topic of work center meetings and are represented by numerous safety campaigns and posters (see Figure 10.4-1).

Figure 10.4-1. Safety posters at NASA and contractor facilities.

Initiatives like Michoud's "This is Stupid" program and the United Space Alliance's "Time Out" cards empower employees to halt any operation under way if they believe industrial safety is being compromised (see Figure 10.4-2). For example, the Time Out program encourages and even rewards workers who report suspected safety problems to management.

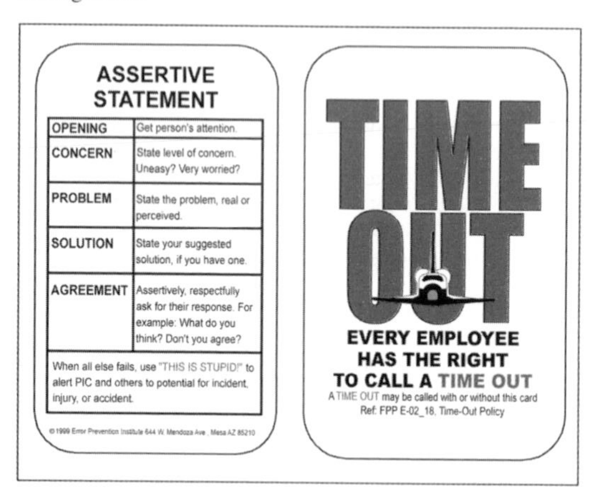

Figure 10.4-2. The "This is Stupid" card from the Michoud Assembly Facility and the "Time Out" card from United Space Alliance.

NASA similarly maintains the Safety Reporting System, which creates lines of communication through which anonymous inputs are forwarded directly to headquarters (see Figure 10.4-3). The NASA Shuttle Logistics Depot focus on safety has been recognized as an Occupational Safety and Health Administration Star Site for its participation in the Voluntary Protection Program. After the Shuttle Logistics Depot was recertified in 2002, employees worked more than 750 days without a lost-time mishap.

Quality Assurance

Quality Assurance programs – encompassing steps to encourage error-free work, as well as inspections and assessments of that work – have evolved considerably in scope over the past five years, transitioning from intensive, comprehensive inspection regimens to much smaller programs based on past risk analysis.

As described in Part Two, after the Space Flight Operations Contract was established, NASA's quality assurance role at Kennedy Space Center was significantly reduced. In the course of this transition, Kennedy reduced its inspections – called Government Mandatory Inspection Points – by more than 80 percent. Marshall Space Flight Center cut its inspection workload from 49,000 government inspection points and 821,000 contractor inspections in 1990 to 13,700 and 461,000, respectively, in 2002. Similar cutbacks were made at most NASA centers.

Inspection requirements are specified in the Quality Planning Requirements Document (also called the Mandatory Inspections Document). United Space Alliance technicians must document an estimated 730,000 tasks to complete a single Shuttle maintenance flow at Kennedy Space Center. Nearly every task assessed as Criticality Code 1, 1R (redundant), or 2 is always inspected, as are any systems not verifiable by operational checks or tests prior to final preparations for flight.

Nearly everyone interviewed at Kennedy indicated that the current inspection process is both inadequate and difficult to expand, even incrementally. One example was a long-standing request to add a main engine final review before transporting the engine to the Orbiter Processing Facility for installation. This request was first voiced two years before the launch of STS-107, and has been repeatedly denied due to inadequate staffing. In its place, NASA Mission Assurance conducts a final "informal" review. Adjusting government inspection tasks is constrained by institutional dogma that the status quo is based on strong engineering logic, and should need no adjustment. This mindset inhibits the ability of Quality Assurance to respond to an aging system, changing workforce dynamics, and improvement initiatives.

The Quality Planning Requirements Document, which defines inspection requirements, was well formulated but is not routinely reviewed. Indeed, NASA seems reluctant to add or subtract government inspections, particularly at Kennedy. Additions and subtractions are rare, and generally occur only as a response to obvious problems. For instance, NASA augmented wiring inspections after STS-93 in 1999, when a short circuit shut down two of *Columbia*'s Main Engine Controllers. Interviews confirmed that the current Requirements Document lacks numerous critical items, but conversely demands redundant and unnecessary inspections.

The NASA/United Space Alliance Quality Assurance processes at Kennedy are not fully integrated with each other, with Safety, Health, and Independent Assessment, or with Engineering Surveillance Programs. Individually, each plays a vital role in the control and assessment of the Shuttle as it comes together in the Orbiter Processing Facility and Vehicle Assembly Building. Were they to be carefully integrated, these programs could attain a nearly comprehensive quality control process. Marshall has a similar challenge. It

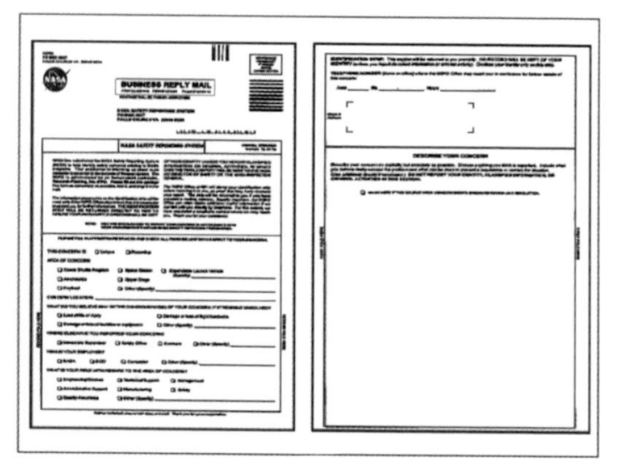

Figure 10.4-3. NASA Safety Reporting System Form.

is responsible for managing several different Shuttle systems through contractors who maintain mostly proprietary databases, and therefore, integration is limited. The main engine program overcomes this challenge by being centrally organized under a single Mission Assurance Division Chief who reports to the Marshall Center Director. In contrast, Kennedy has a separate Mission Assurance office working directly for each program, a separate Safety, Health, and Independent Assessment office under the Center Director, and separate quality engineers under each program. Observing the effectiveness of Marshall, and other successful Mission Assurance programs (such as at Johnson Space Center), a solution may be the consolidation of the Kennedy Space Center Quality Assurance program under one Mission Assurance office, which would report to the Center Director.

While reports by the 1986 Rogers Commission, 2000 Shuttle Independent Assessment Team, and 2003 internal Kennedy Tiger Team all affirmed the need for a strong and independent Quality Assurance Program, Kennedy's Program has taken the opposite tack. Kennedy's Quality Assurance program discrepancy-tracking system is inadequate to nonexistent.

Robust as recently as three years ago, Kennedy no longer has a "closed loop" system in which discrepancies and their remedies circle back to the person who first noted the problem. Previous methods included the NASA Corrective Action Report, two-way memos, and other tools that helped ensure that a discrepancy would be addressed and corrected. The Kennedy Quality Program Manager cancelled these programs in favor of a contractor-run database called the Quality Control Assessment Tool. However, it does not demand a closed-loop or reply deadline, and suffers from limitations on effective data entry and retrieval.

Kennedy Quality Assurance management has recently focused its efforts on implementing the International Organization for Standardization (ISO) 9000/9001, a process-driven program originally intended for manufacturing plants. Board observations and interviews underscore areas where Kennedy has diverged from its Apollo-era reputation of setting the standard for quality. With the implementation of International Standardization, it could devolve further. While ISO 9000/9001 expresses strong principles, they are more applicable to manufacturing and repetitive-procedure industries, such as running a major airline, than to a research-and-development, non-operational flight test environment like that of the Space Shuttle. NASA technicians may perform a specific procedure only three or four times a year, in contrast with their airline counterparts, who perform procedures dozens of times each week. In NASA's own words regarding standardization, "ISO 9001 is not a management panacea, and is never a replacement for management taking responsibility for sound decision making." Indeed, many perceive International Standardization as emphasizing process over product.

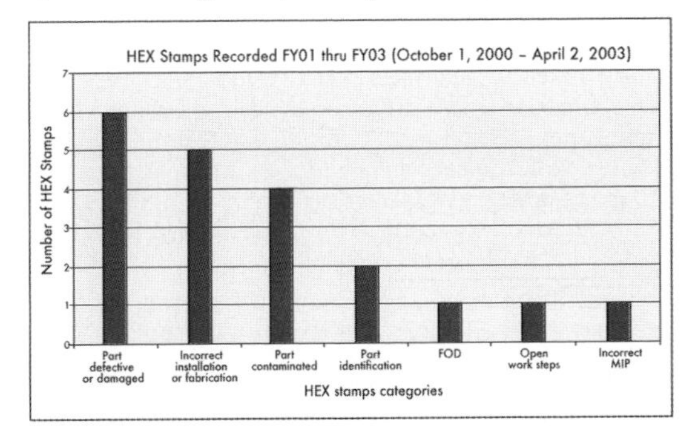

Figure 10.4-4. Rejection, or "Hex" stamps issued from October 2000 through April 2003.

Efforts by Kennedy Quality Assurance management to move its workforce towards a "hands-off, eyes-off" approach are unsettling. To use a term coined by the 2000 Shuttle Independent Assessment Team Report, "diving catches," or last-minute saves, continue to occur in maintenance and processing and pose serious hazards to Shuttle safety. More disturbingly, some proverbial balls are not caught until after flight. For example, documentation revealed instances where Shuttle components stamped "ground test only" were detected both before and after they had flown. Additionally, testimony and documentation submitted by witnesses revealed components that had flown "as is" without proper disposition by the Material Review Board prior to flight, which implies a growing acceptance of risk. Such incidents underscore the need to expand government inspections and surveillance, and highlight a lack of communication between NASA employees and contractors.

Another indication of continuing problems lies in an opinion voiced by many witnesses that is confirmed by Board tracking: Kennedy Quality Assurance management discourages inspectors from rejecting contractor work. Inspectors are told to cooperate with contractors to fix problems rather than rejecting the work and forcing contractors to resubmit it. With a rejection, discrepancies become a matter of record; in this new process, discrepancies are not recorded or tracked. As a result, discrepancies are currently not being tracked in any easily accessible database.

Of the 141,127 inspections subject to rejection from October 2000 through March 2003, only 20 rejections, or "hexes," were recorded, resulting in a statistically improbable discrepancy rate of .014 percent (see Figure 10.4-4). In interviews, technicians and inspectors alike confirmed the dubiousness of this rate. NASA's published rejection rate therefore indicates either inadequate documentation or an underused system. Testimony further revealed incidents of quality assurance inspectors being played against each other to accept work that had originally been refused.

Findings:

F10.4-1 Shuttle System industrial safety programs are in good health.

F10.4-2 The Quality Planning Requirements Document, which defines inspection conditions, was well formulated. However, there is no requirement that it be routinely reviewed.

F10.4-3 Kennedy Space Center's current government mandatory inspection process is both inadequate and difficult to expand, which inhibits the ability of Quality Assurance to process improvement initiatives.

F10.4-4 Kennedy's quality assurance system encourages inspectors to allow incorrect work to be corrected without being labeled "rejected." These opportunities hide "rejections," making it impossible to determine how often and on what items frequent rejections and errors occur.

Observations:

O10.4-1 Perform an independently led, bottom-up review of the Kennedy Space Center Quality Planning Requirements Document to address the entire quality assurance program and its administration. This review should include development of a responsive system to add or delete government mandatory inspections.

O10.4-2 Kennedy Space Center's Quality Assurance programs should be consolidated under one Mission Assurance office, which reports to the Center Director.

O10.4-3 Kennedy Space Center quality assurance management must work with NASA and perhaps the Department of Defense to develop training programs for its personnel.

O10.4-4 Kennedy Space Center should examine which areas of International Organization for Standardization 9000/9001 truly apply to a 20-year-old research and development system like the Space Shuttle.

10.5 MAINTENANCE DOCUMENTATION

The Board reviewed *Columbia's* maintenance records for any documentation problems, evidence of maintenance flaws, or significant omissions, and simultaneously investigated the organizations and management responsible for this documentation. The review revealed both inaccurate data entries and a widespread inability to find and correct these inaccuracies.

The Board asked Kennedy Space Center and United Space Alliance to review documentation for STS-107, STS-109, and *Columbia*'s most recent Orbiter Major Modification. A NASA Process Review Team, consisting of 445 NASA engineers, contractor engineers, and Quality Assurance personnel, reviewed some 16,500 Work Authorization Documents, and provided a list of Findings (potential relationships to the accident), Technical Observations (technical concerns or process issues), and Documentation Observations (minor errors). The list contained one Finding related to the External Tank bipod ramp. None of the Observations contributed to the accident.

The Process Review Team's sampling plan resulted in excellent observations.[24] The number of observations is relatively low compared to the total amount of Work Authorization Documents reviewed, ostensibly yielding a 99.75 percent accuracy rate. While this number is high, a closer review of the data reveals some of the system's weaknesses. Technical Observations are delineated into 17 categories. Five of these categories are of particular concern for mishap prevention and reinforce the need for process improvements. The category entitled "System configuration could damage hardware" is listed 112 times. Categories that deal with poor incorporation of technical guidance are of particular interest due to the Board's concern over the backlog of unincorporated engineering orders. Finally, a category entitled "paper has open work steps," indicates that the review system failed to catch a potentially significant oversight 310 times in this sample. (The complete results of this review may be found in Appendix D.14.)

The current process includes three or more layers of oversight before paperwork is scanned into the database. However, if review authorities are not aware of the most common problems to look for, corrections cannot be made. Routine sampling will help refine this process and cut errors significantly.

Observations:

O10.5-1 Quality and Engineering review of work documents for STS-114 should be accomplished using statistical sampling to ensure that a representative sample is evaluated and adequate feedback is communicated to resolve documentation problems.

O10.5-2 NASA should implement United Space Alliance's suggestions for process improvement, which recommend including a statistical sampling of all future paperwork to identify recurring problems and implement corrective actions.

O10.5-3 NASA needs an oversight process to statistically sample the work performed and documented by Alliance technicians to ensure process control, compliance, and consistency.

10.6 ORBITER MAINTENANCE DOWN PERIOD/ ORBITER MAJOR MODIFICATION

During the Orbiter Major Modification process, Orbiters are removed from service for inspections, maintenance, and modification. The process occurs every eight flights or three years.

Orbiter Major Modifications combine with Orbiter flows (preparation of the vehicle for its next mission) and include Orbiter Maintenance Down Periods (not every Orbiter Maintenance Down Period includes an Orbiter Major Modification). The primary differences between an Orbiter Major Modification and an Orbiter flow are the larger number of requirements and the greater degree of intrusiveness of a modification (a recent comparison showed 8,702 Orbiter Major Modification requirements versus 3,826 flow requirements).

Ten Orbiter Major Modifications have been performed to date, with an eleventh in progress. They have varied from 6 to 20 months. Because missions do not occur at the rate the Shuttle Program anticipated at its inception, it is endlessly challenged to meet numerous calendar-based requirements. These must be performed regardless of the lower flight rate, which contributes to extensive downtime. The Shuttle Program has explored the possibility of extending Orbiter Major Modification cycles to once every 12 flights or six years. This initiative runs counter to the industry norm of increasing the frequency of inspections as systems age, and should be carefully scrutinized, particularly in light of the high-performance Orbiters' demands.

Orbiter Major Modifications underwent a significant change when they were relocated from the Boeing facility in Palmdale, California, (where the Orbiters had been manufactured) to Kennedy Space Center in September 2002. The major impetus for this change was budget shortages in Fiscal Years 2002 and 2003. The move capitalizes on many advantages at Kennedy, including lower labor and utility costs and more efficient use of existing overhead, while eliminating expensive, underused, and redundant capabilities at Palmdale. However, the move also created new challenges: for instance, it complicates the integration of planning and scheduling, and forces the Space Shuttle Program to maintain a fluid workforce in which employees must repeatedly change tasks as they shift between Orbiter Major Modifications, flows, and downtime.

Throughout the history of Orbiter Major Modifications, a major area of concern has been their wide variability in content and duration. *Columbia*'s last Orbiter Major Modification is just the most recent example of overruns due to technical surprises and management difficulties. It exceeded the schedule by 186 days. While many factors contributed to this delay, the two most prominent were the introduction of a major wiring inspection one month after Orbiter Major Modification roll-in, and what an internal NASA assessment cited as "poor performance on the parts of NASA, USA [United Space Alliance], and Boeing."

While the Shuttle Program has made efforts to correct these problems, there is still much to be done. The transfer to Kennedy creates a steep learning curve both for technicians and managers. Planning and scheduling the integration of all three Orbiters, as well as ground support systems maintenance, is critical to limit competition for resources. Moreover, estimating the "right" amount of work required on each Orbiter continues to be a challenge. For example, 20 modifications were planned for *Discovery*'s modification; the number has since grown to 84. Such changes introduce turmoil and increase the potential for mistakes.

An Air Force "benchmarking" visit in June 2003 highlighted the need for better planning and more scheduling stability. It further recommended improvements to the requirements feedback process and incorporating service life extension actions into Orbiter Major Modifications.

Observations:

O10.6-1 The Space Shuttle Program Office must make every effort to achieve greater stability, consistency, and predictability in Orbiter Major Modification planning, scheduling, and work standards (particularly in the number of modifications). Endless changes create unnecessary turmoil and can adversely impact quality and safety.

O10.6-2 NASA and United Space Alliance managers must understand workforce and infrastructure requirements, match them against capabilities, and take actions to avoid exceeding thresholds.

O10.6-3 NASA should continue to work with the U.S. Air Force, particularly in areas of program management that deal with aging systems, service life extension, planning and scheduling, workforce management, training, and quality assurance.

O10.6-4 The Space Shuttle Program Office must determine how it will effectively meet the challenges of inspecting and maintaining an aging Orbiter fleet before lengthening Orbiter Major Maintenance intervals.

10.7 Orbiter Corrosion

Removing and replacing Thermal Protection System tiles sometimes results in damage to the anti-corrosion primer that covers the Orbiters' sheet metal skin. Tile replacement often occurs without first re-priming the primed aluminum substrate. The current repair practice allows Room Temperature Vulcanizing adhesive to be applied over a bare aluminum substrate (with no Koropon corrosion-inhibiting compound) when bonding tile to the Orbiter.

A video borescope of *Columbia* prior to STS-107 found corrosion on the lower forward fuselage skin panel and stringer areas. Corrosion on visible rivets and on the sides and feet of stringer sections was also uncovered during borescope inspections, but was not repaired.

Other corrosion concerns focus on the area between the crew module and outer hull, which is a difficult area to access for inspection and repair. At present, corrosion in this area is only monitored with borescope inspections. There is also concern that unchecked corrosion could progress from internal areas to external surfaces through fastener holes, joints, or directly through the skin. If this occurs beneath the tile, the tile system bond line could degrade.

Long-Term Corrosion Detection

Limited accessibility renders some corrosion damage difficult to detect. Approximately 90 percent of the Orbiter structure (excluding the tile-covered outer mold line) can be inspected for corrosion.[25] Corrosion in the remaining 10 percent may remain undetected for the life of the vehicle.

NASA has recently outlined a $70 million, 19-year program to assess and mitigate corrosion. The agency foresees inspection intervals based on trends in the Problem Resolution and Corrective Action database, exposure to the environment, and refurbishment programs. Development of a correlation between corrosion initiation, growth, and environmental exposure requires the judicious use of long-term test data. Moreover, some corrosion problems are uncovered during non-corrosion inspections. The risk of undetected corrosion may increase as other inspections are removed or intervals between inspections are extended.

Observations:

O10.7-1 Additional and recurring evaluation of corrosion damage should include non-destructive analysis of the potential impacts on structural integrity.

O10.7-2 Long-term corrosion detection should be a funding priority.

O10.7-3 Develop non-destructive evaluation inspections to find hidden corrosion.

O10.7-4 Inspection requirements for corrosion due to environmental exposure should first establish corrosion rates for Orbiter-specific environments, materials, and structural configurations. Consider applying Air Force corrosion prevention programs to the Orbiter.

10.8 Brittle Fracture of A-286 Bolts

Investigators sought to determine the cause of brittle fractures in the A-286 steel bolts that support the wing's lower carrier panels, which provide direct access to the interior of the Reinforced Carbon-Carbon (RCC) panels. Any misalignment of the carrier panels affects the continuity of airflow under the wing and can cause a "rough wing" (see Chapter 4). In the end, 57 of the 88 A-286 bolts on *Columbia*'s wings were recovered; 22 had brittle fractures. The fractures occurred equally in two groups of bolts in the same locations on each wing. Investigators determined that liquid metal embrittlement caused by aluminum vapor created by *Columbia*'s breakup could have contributed to these fractures, but the axial loads placed on the bolts when they separated from the carrier panel/box beam at temperatures approaching 2,000 degrees Fahrenheit likely caused the failures.

Findings:

F10.8-1 The present design and fabrication of the lower carrier panel attachments are inadequate. The bolts can readily pull through the relatively large holes in the box beams.

F10.8-2 The current design of the box beam in the lower carrier panel assembly exposes the attachment bolts to a rapid exchange of air along the wing, which enables the failure of numerous bolts.

F10.8-3 Primers and sealants such as Room Temperature Vulcanizing 560 and Koropon may accelerate corrosion, particularly in tight crevices.

F10.8-4 The negligible compressive stresses that normally occur in A-286 bolts help protect against failure.

Observations:

O10.8-1 Teflon (material) and Molybdenum Disulfide (lubricant) should not be used in the carrier panel bolt assembly.

O10.8-2 Galvanic coupling between aluminum and steel alloys must be mitigated.

O10.8-3 The use of Room Temperature Vulcanizing 560 and Koropon should be reviewed.

O10.8-4 Assuring the continued presence of compressive stresses in A-286 bolts should be part of their acceptance and qualification procedures.

10.9 Hold-Down Post Cable Anomaly

Each of the two Solid Rocket Boosters is attached to the Mobile Launch Platform by four "hold down" bolts. A five-inch diameter restraint nut that contains two pyrotechnic initiators secures each of these bolts. The initiators sever the nuts when the Solid Rocket Boosters ignite, allowing the Space Shuttle stack to lift off. During launch, STS-112 suffered a failure in the Hold-Down Post and External Tank Vent Arm Systems that control the firing of initiators in each Solid Rocket Booster restraint nut. NASA had been warned that a recurrence of this type of failure could cause catastrophic failure of the Shuttle stack (see Appendix D.15).

The signal to fire the initiators begins in the General Purpose Computers and goes to both of the Master Events Controllers on the Orbiter. Master Events Controller 1 communicates this signal to the A system cable, and Master Events Controller 2 feeds the B system. The cabling then goes through the T–0 umbilical (that connects fluid and electrical connections between the launch pad and the Orbiter) to the Pyrotechnics Initiator Controllers and then to the initiators. (There are 16 Pyrotechnics Initiator Controllers for Hold Down Post Systems A and B, and four for the External Tank Vent Arm Systems A and B.) The Hold Down Post System A is hard-wired to one of the initiators on each of the four restraint nuts (eight total) while System B is hard-wired to the other initiator on each nut. The A and B systems also send a duplicate signal to the External Tank Vent Arm System. Either Master Events Controller will operate if the other or the intervening cabling fails.

A post-launch review of STS-112 indicated that the System A Hold-Down Post and External Tank Vent Arm System Pyrotechnics Initiator Controllers did not discharge. Initial troubleshooting revealed no malfunction, leading to the conclusion that the failure was intermittent. A subsequent investigation recommended the following:

- All T–0 Ground Cables will be replaced after every flight.
- The T–0 interface to the Pyrotechnics Initiator Controllers rack cable (Kapton) is in redesign.
- All Orbiter T–0 Connector Savers have been replaced.
- Pyrotechnic connectors will be pre-screened with pin-retention tests, and the connector saver mate process will be verified using videoscopes.

However, prelaunch testing procedures have not changed and may not be able to identify intermittent failures.

Findings:

F10.9-1 The Hold-Down Post External Tank Vent Arm System is a Criticality 1R (redundant) system. Before the anomaly on STS-112, and despite the high-criticality factor, the original cabling for this system was used repeatedly until it was visibly damaged. Replacing these cables after every flight and removing the Kapton will prevent bending and manipulation damage.

F10.9-2 NASA is unclear about the potential for damage if the system malfunctions, or even if one nut fails to split. Several program managers were asked: What if the A system fails, and a B-system initiator fails simultaneously? The consensus was that the system would continue to burn on the pad or that the Solid Rocket Booster would rip free of the pad, causing potentially catastrophic damage to the Solid Rocket Booster skirt and nozzle maneuvering mechanism. However, they agree that the probability of this is extremely low.

F10.9-3 With the exception of STS-112's anomaly, numerous bolt hang-ups, and occasional Master Events Controller failures, these systems have a good record. In the early design stages, risk-mitigating options were considered, including strapping with either a wire that crosses over the nut from the A to B side, or with a toggle circuit that sends a signal to the opposite side when either initiator fires. Both options would eliminate the potential of a catastrophic dual failure. However, they could also create new failure potentials that may not reduce overall system risk. Today's test and troubleshooting technology may have improved the ability to test circuits and potentially prevent intermittent failures, but it is not clear if NASA has explored these options.

Observation:

O10.9-1 NASA should consider a redesign of the system, such as adding a cross-strapping cable, or conduct advanced testing for intermittent failure.

10.10 Solid Rocket Booster External Tank Attachment Ring

In Chapter 4, the Board noted how NASA's reliance on "analysis" to validate Shuttle components led to the use of flawed bolt catchers. NASA's use of this flawed "analysis" technique is endemic. The Board has found that such analysis was invoked, with potentially dire consequences, on the Solid Rocket Booster External Tank Attach Ring. Tests showed that the tensile strength of several of these rings was well below minimum safety requirements. This problem was brought to NASA's attention shortly before the launch of STS-107. To accommodate the launch schedule, the External Tanking Meeting chair, after a cursory briefing without a full technical review, reduced the Attach Rings' minimum required safety factor of 1.4 (that is, able to withstand 1.4 times the maximum load ever expected in operations) to 1.25. Though NASA has formulated short- and long-term corrections, its long-term plan has not yet been authorized.

Observation:

O10.10-1 NASA should reinstate a safety factor of 1.4 for the Attachment Rings—which invalidates the use of ring serial numbers 16 and 15 in their present state—and replace all deficient material in the Attachment Rings.

10.11 Test Equipment Upgrades

Visits to NASA facilities (both government and contractor operated, as well as contractor facilities) and interviews with technicians revealed the use of 1970s-era oscilloscopes and other analog equipment. Currently available equipment is digital, and in other venues has proved to be less costly, easier to maintain, and more reliable and accurate. With the Shuttle forecast to fly through 2020, an upgrade to digital equipment would avoid the high maintenance, lack of parts, and dubious accuracy of equipment currently used. New equipment would require certification for its uses, but the benefit in accuracy, maintainability, and longevity would likely outweigh the drawbacks of certification costs.

Observation:

O10.11-1 Assess NASA and contractor equipment to determine if an upgrade will provide the reliability and accuracy needed to maintain the Shuttle through 2020. Plan an aggressive certification program for replaced items so that new equipment can be put into operation as soon as possible.

10.12 Leadership/Managerial Training

Managers at many levels in NASA, from GS-14 to Associate Administrator, have taken their positions without following a recommended standard of training and education to prepare them for roles of increased responsibility. While NASA has a number of in-house academic training and career development opportunities, the timing and strategy for management and leadership development differs across organizations. Unlike other sectors of the Federal Government and the military, NASA does not have a standard agency-wide career planning process to prepare its junior and mid-level managers for advanced roles. These programs range from academic fellowships to civil service education programs to billets in military-sponsored programs, and will allow NASA to build a strong corps of potential leaders for future progression.

Observation:

O10.12-1 NASA should implement an agency-wide strategy for leadership and management training that provides a more consistent and integrated approach to career development. This strategy should identify the management and leadership skills, abilities, and experiences required for each level of advancement. NASA should continue to expand its leadership development partnerships with the Department of Defense and other external organizations.

ENDNOTES FOR CHAPTER 10

The citations that contain a reference to "CAIB document" with CAB or CTF followed by seven to eleven digits, such as CAB001-0010, refer to a document in the Columbia Accident Investigation Board database maintained by the Department of Justice and archived at the National Archives.

[1] "And stunningly, in as much as this was tragic and horrific through a loss of seven very important lives, it is amazing that there were no other collateral damage happened as a result of it. No one else was injured. All of the claims have been very, very minor in dealing with these issues." NASA Administrator Sean O'Keefe, testimony before the United States Senate Committee on Commerce, Science, and Transportation, May 14, 2003.

[2] An intensive search of over a million acres in Texas and Louisiana recovered 83,900 pieces of *Columbia* debris weighing a total of 84,900 pounds. (Over 700,000 acres were searched on foot, and 1.6 million acres were searched with aircraft.) The latitude and longitude was recorded for more than 75,000 of these pieces. The majority of the recovered items were no larger than 0.5 square feet. More than 40,000 items could not be positively identified but were classified as unknown tile, metal, composite, plastic, fabric, etc. Details about the debris reconstruction and recovery effort are provided in Appendix E.5, S. Altemis, J. Cowart, W. Woodworth, "STS-107 Columbia Reconstruction Report," NSTS-60501, June 30, 2003. CAIB document CTF076-20302182.

[3] The precise probability is uncertain due to many factors, such as the amount of debris that burned up during re-entry, and the fraction of the population that was outdoors when the *Columbia* accident occurred.

[4] "User's Guide for Object Reentry Survival Analysis Tool (ORSAT), Version 5.0, Volume I-Methodology, Input Description, and Results," JSC-28742, July 1999; W. Alior, "What Can We Learn From Recovered Debris," Aerospace Corp, briefing presented to CAIB, on March 13, 2003.

[5] "Reentry Survivability Analysis of Delta IV Launch Vehicle Upper Stage," JSC-29775, June 2002.

[6] Analysis of the recovered debris indicates that relatively few pieces posed a threat to people indoors. See Appendix D.16.

[7] Detailed information about individual fragments, including weight in most cases, was not available for the study. Therefore, some engineering discretion was needed to develop models of individual weights, dimensions, aerodynamic characteristics, and conditions of impact. This lack of information increases uncertainty in the accuracy of the final results. The study should be revisited after the fragment data has been fully characterized.

[8] K.M. Thompson, R.F. Rabouw, and R.M. Cooke, "The Risk of Groundling Fatalities from Unintentional Airplane Crashes," *Risk Analysis*, Vol. 21, No. 6, 2001.

[9] Ibid.

[10] The civil aviation study indicates that the risk to groundlings is significantly higher in the vicinity of an airport. The average annual risk of fatality within 0.2 miles of a busy (top 100) airport is about 1 in a million.

[11] Thompson, "The Risk of Groundling Fatalities;" Code of Federal Regulations (CFR) 14 CFR Part 415, 415, and 417, "Licensing and Safety Requirements for Launch: Proposed Rule," *Federal Register* Vol. 67, No. 146, July 30, 2002, p. 49495.

[12] Code of Federal Regulations (CFR) 14 CFR Part 415 Launch License, *Federal Register* Vol. 64, No. 76, April 21, 1999; Range Commanders Council Standard 321-02, "Common Risk Criteria for National Test Ranges," published by the Secretariat of the RCC U.S. Army White Sands Missile Range, NM 88002-5110, June 2002; "Mitigation of Orbital Debris," Notice of Proposed Rulemaking by the Federal Communications Commission, FCC 02-80, *Federal Register* Vol. 67, No. 86, Friday, May 3, 2002.

[13] Air Force launch safety standards define a Hazardous Launch Area, a controlled surface area and airspace, where individual risk of serious injury from a launch vehicle malfunction during the early phase of flight exceeds one in a million. Only personnel essential to the launch operation are permitted in this area. "Eastern and Western Range Requirements 127-1," March 1995, pp. 1-12 and Fig. 1-6.

[14] Code of Federal Regulations (CFR) 14 CFR Part 431, Launch and Reentry of a Reusable Launch Vehicle, Section 35 paragraphs (a) and (b), *Federal Register* Vol. 65, No. 182, September 19, 2000, p. 56660.

[15] "Reentry Survivability Analysis of Delta IV Launch Vehicle Upper Stage," JSC-29775, June 2002.

[16] M. Tobin, "Range Safety Risk Assessments For Kennedy Space Center," October 2002. CAIB document CTF059-22802288; "Space Shuttle Program Requirements Document," NSTS-07700, Vol. I, change no. 76, Section 5-1. CAIB document CAB024-04120475.

[17] Here, ascent refers to (1) the Orbiter from liftoff to Main Engine Cut Off (MECO), (2) the Solid Rocket Boosters from liftoff to splashdown, and (3) the External Tank from liftoff to splashdown.

[18] Pete Cadden, "Shuttle Launch Area Debris Risk," October 2002. CAIB document CTF059-22682279.

[19] See Dennis R. Jenkins, *Space Shuttle: The History of the National Space Transportation System – The First 100 Missions* (Cape Canaveral, FL, Specialty Press, 2001), pp. 205-212 for a complete description of the Approach and Landing Tests and other testing conducted with *Enterprise*.

[20] *Report of the Presidential Commission on the Space Shuttle Challenger Accident* (Washington: Government Printing Office, 1986).

[21] The pre-declared time period or number of missions over which the system is expected to operate without major redesign or redefinition.

[22] "A crew escape system shall be provided on Earth to Orbit vehicles for safe crew extraction and recovery from in-flight failures across the flight envelope from pre-launch to landing. The escape system shall have a probability of successful crew return of 0.99."

[23] *Report of the Aerospace Safety Advisory Panel Annual Report* for 2002, (Washington: Government Printing Office, March 2002). CAIB document CTF014-25882645.

[24] Charlie Abner, "KSC Processing Review Team Final Summary," June 16, 2003. CAIB document CTF063-11801276.

[25] Julie Kramer, et al., "Minutes from CAIB / Engineering Meeting to Discuss CAIB Action / Request for Information B1-000193," April 24, 2003. CAIB document CTF042-00930095.

CHAPTER 11

Recommendations

It is the Board's opinion that good leadership can direct a culture to adapt to new realities. NASA's culture must change, and the Board intends the following recommendations to be steps toward effecting this change.

Recommendations have been put forth in many of the chapters. In this chapter, the recommendations are grouped by subject area with the Return-to-Flight [RTF] tasks listed first within the subject area. Each Recommendation retains its number so the reader can refer to the related section for additional details. These recommendations are not listed in priority order.

PART ONE – THE ACCIDENT

Thermal Protection System

R3.2-1 Initiate an aggressive program to eliminate all External Tank Thermal Protection System debris-shedding at the source with particular emphasis on the region where the bipod struts attach to the External Tank. [RTF]

R3.3-2 Initiate a program designed to increase the Orbiter's ability to sustain minor debris damage by measures such as improved impact-resistant Reinforced Carbon-Carbon and acreage tiles. This program should determine the actual impact resistance of current materials and the effect of likely debris strikes. [RTF]

R3.3-1 Develop and implement a comprehensive inspection plan to determine the structural integrity of all Reinforced Carbon-Carbon system components. This inspection plan should take advantage of advanced non-destructive inspection technology. [RTF]

R6.4-1 For missions to the International Space Station, develop a practicable capability to inspect and effect emergency repairs to the widest possible range of damage to the Thermal Protection System, including both tile and Reinforced Carbon-Carbon, taking advantage of the additional capabilities available when near to or docked at the International Space Station.

For non-Station missions, develop a comprehensive autonomous (independent of Station) inspection and repair capability to cover the widest possible range of damage scenarios.

Accomplish an on-orbit Thermal Protection System inspection, using appropriate assets and capabilities, early in all missions.

The ultimate objective should be a fully autonomous capability for all missions to address the possibility that an International Space Station mission fails to achieve the correct orbit, fails to dock successfully, or is damaged during or after undocking. [RTF]

R3.3-3 To the extent possible, increase the Orbiter's ability to successfully re-enter Earth's atmosphere with minor leading edge structural sub-system damage.

R3.3-4 In order to understand the true material characteristics of Reinforced Carbon-Carbon components, develop a comprehensive database of flown Reinforced Carbon-Carbon material characteristics by destructive testing and evaluation.

R3.3-5 Improve the maintenance of launch pad structures to minimize the leaching of zinc primer onto Reinforced Carbon-Carbon components.

R3.8-1 Obtain sufficient spare Reinforced Carbon-Carbon panel assemblies and associated support components to ensure that decisions on Reinforced Carbon-Carbon maintenance are made on the basis of component specifications, free of external pressures relating to schedules, costs, or other considerations.

R3.8-2 Develop, validate, and maintain physics-based computer models to evaluate Thermal Protection System damage from debris impacts. These tools should provide realistic and timely estimates of any impact damage from possible debris from any source that may ultimately impact the Orbiter. Establish impact damage thresholds that trigger responsive corrective action, such as on-orbit inspection and repair, when indicated.

Imaging

R3.4-1 Upgrade the imaging system to be capable of providing a minimum of three useful views of the Space Shuttle from liftoff to at least Solid Rocket Booster separation, along any expected ascent azimuth. The operational status of these assets should be included in the Launch Commit Criteria for future launches. Consider using ships or aircraft to provide additional views of the Shuttle during ascent. [RTF]

R3.4-2 Provide a capability to obtain and downlink high-resolution images of the External Tank after it separates. [RTF]

R3.4-3 Provide a capability to obtain and downlink high-resolution images of the underside of the Orbiter wing leading edge and forward section of both wings' Thermal Protection System. [RTF]

R6.3-2 Modify the Memorandum of Agreement with the National Imagery and Mapping Agency to make the imaging of each Shuttle flight while on orbit a standard requirement. [RTF]

Orbiter Sensor Data

R3.6-1 The Modular Auxiliary Data System instrumentation and sensor suite on each Orbiter should be maintained and updated to include current sensor and data acquisition technologies.

R3.6-2 The Modular Auxiliary Data System should be redesigned to include engineering performance and vehicle health information, and have the ability to be reconfigured during flight in order to allow certain data to be recorded, telemetered, or both as needs change.

Wiring

R4.2-2 As part of the Shuttle Service Life Extension Program and potential 40-year service life, develop a state-of-the-art means to inspect all Orbiter wiring, including that which is inaccessible.

Bolt Catchers

R4.2-1 Test and qualify the flight hardware bolt catchers. [RTF]

Closeouts

R4.2-3 Require that at least two employees attend all final closeouts and intertank area hand-spraying procedures. [RTF]

Micrometeoroid and Orbital Debris

R4.2-4 Require the Space Shuttle to be operated with the same degree of safety for micrometeoroid and orbital debris as the degree of safety calculated for the International Space Station. Change the micrometeoroid and orbital debris safety criteria from guidelines to requirements.

Foreign Object Debris

R4.2-5 Kennedy Space Center Quality Assurance and United Space Alliance must return to the straightforward, industry-standard definition of "Foreign Object Debris" and eliminate any alternate or statistically deceptive definitions like "processing debris." [RTF]

PART TWO – WHY THE ACCIDENT OCCURRED

Scheduling

R6.2-1 Adopt and maintain a Shuttle flight schedule that is consistent with available resources. Although schedule deadlines are an important management tool, those deadlines must be regularly evaluated to ensure that any additional risk incurred to meet the schedule is recognized, understood, and acceptable. [RTF]

Training

R6.3-1 Implement an expanded training program in which the Mission Management Team faces potential crew and vehicle safety contingencies beyond launch and ascent. These contingencies should involve potential loss of Shuttle or crew, contain numerous uncertainties and unknowns, and require the Mission Management Team to assemble and interact with support organizations across NASA/Contractor lines and in various locations. [RTF]

Organization

R7.5-1 Establish an independent Technical Engineering Authority that is responsible for technical requirements and all waivers to them, and will build a disciplined, systematic approach to identifying, analyzing, and controlling hazards throughout the life cycle of the Shuttle System. The independent technical authority does the following as a minimum:

- Develop and maintain technical standards for all Space Shuttle Program projects and elements
- Be the sole waiver-granting authority for all technical standards
- Conduct trend and risk analysis at the sub-system, system, and enterprise levels
- Own the failure mode, effects analysis and hazard reporting systems
- Conduct integrated hazard analysis
- Decide what is and is not an anomalous event
- Independently verify launch readiness
- Approve the provisions of the recertification program called for in Recommendation R9.1-1.

The Technical Engineering Authority should be funded directly from NASA Headquarters, and should have no connection to or responsibility for schedule or program cost.

R7.5-2 NASA Headquarters Office of Safety and Mission Assurance should have direct line authority over the entire Space Shuttle Program safety organization and should be independently resourced.

R7.5-3 Reorganize the Space Shuttle Integration Office to make it capable of integrating all elements of the Space Shuttle Program, including the Orbiter.

PART THREE – A LOOK AHEAD

Organization

R9.1-1 Prepare a detailed plan for defining, establishing, transitioning, and implementing an independent Technical Engineering Authority, independent safety program, and a reorganized Space Shuttle Integration Office as described in R7.5-1, R7.5-2, and R7.5-3. In addition, NASA should submit annual reports to Congress, as part of the budget review process, on its implementation activities. [RTF]

Recertification

R9.2-1 Prior to operating the Shuttle beyond 2010, develop and conduct a vehicle recertification at the material, component, subsystem, and system levels. Recertification requirements should be included in the Service Life Extension Program.

Closeout Photos/Drawing System

R10.3-1 Develop an interim program of closeout photographs for all critical sub-systems that differ from engineering drawings. Digitize the close-out photograph system so that images are immediately available for on-orbit troubleshooting. [RTF]

R10.3-2 Provide adequate resources for a long-term program to upgrade the Shuttle engineering drawing system including:

- Reviewing drawings for accuracy
- Converting all drawings to a computer-aided drafting system
- Incorporating engineering changes

Columbia
United States

Part Four

Appendices

Sunrise from STS-107 on Flight Day 3

Columbia being transported to Launch Complex 39-A at the Kennedy Space Center, Florida, in preparation for STS-107.

APPENDIX A

The Investigation

A.1 ACTIVATION OF THE COLUMBIA ACCIDENT INVESTIGATION BOARD

At 8:59:32 a.m. Eastern Standard Time on Saturday, February 1, 2003, communication with the Shuttle *Columbia* was lost. Shortly after the planned landing time of 9:16 a.m., NASA declared a Shuttle Contingency and executed the Agency Contingency Action Plan for Space Flight Operations that had been established after the Space Shuttle *Challenger* accident in January 1986. As part of that plan, NASA Administrator Sean O'Keefe deployed NASA's Mishap Investigation Team, activated the Headquarters Contingency Action Team, and, at 10:30 a.m., activated the International Space Station and Space Shuttle Mishap Interagency Investigation Board.

The International Space Station and Space Shuttle Mishap Interagency Investigation Board is designated in Appendix D of the Agency Contingency Action Plan as an external investigating board that works to uncover the "facts, as well as the actual or probable causes of the Shuttle mishap" and to "recommend preventative and other appropriate actions to preclude the recurrence of a similar mishap."[1] The Board is composed of seven members and is chartered with provisions for naming a Chairman and additional members. The seven members take their position on the Board because they occupy specific government posts. At the time of the accident, these individuals included:

- Chief of Safety, U.S. Air Force: Major General Kenneth W. Hess
- Director, Office of Accident Investigation, Federal Aviation Administration: Steven B. Wallace
- Representative, U.S. Air Force Space Command: Brigadier General Duane W. Deal
- Commander, Naval Safety Center: Rear Admiral Stephen A. Turcotte
- Director, Aviation Safety Division, Volpe National Transportation Systems Center, Department of Transportation: Dr. James N. Hallock
- Representative, U.S. Air Force Materiel Command: Major General John L. Barry
- Director, NASA Field Center or NASA Program Associate Administrator (not related to mission): Vacant

Upon activating the Board, Administrator O'Keefe named Admiral Harold W. Gehman Jr., United States Navy (retired), as its Chair, and G. Scott Hubbard, Director of NASA Ames Research Center, as the NASA Field Center Director representative. In addition to these eight voting members, contingency procedures provided for adding two non-voting NASA representatives, who helped establish the Board during the first weeks of activity but then returned to their regular duties. They were Bryan D. O'Connor, NASA Associate Administrator for Safety and Mission Assurance, who served as an ex-officio Member of the Board, and Theron M. Bradley Jr., NASA Chief Engineer, who served as the Board's Executive Secretary. Upon the Board's activation, two NASA officials, David Lengyel and Steven Schmidt, were dispatched to provide for the Board's administrative needs. J. William Sikora, Chief Counsel of the Glenn Research Center in Cleveland, Ohio, was assigned as the counsel to the Board.

By noon on February 1, NASA officials notified most Board members of the mishap and issued tentative orders for the Board to convene the next day at Barksdale Air Force Base in Shreveport, Louisiana, where the NASA Mishap Investigation Team was coordinating the search for debris. At 5:00 p.m., available Board members participated in a teleconference with NASA's Headquarters Contingency Action Team. During that teleconference, Gehman proposed that the International Space Station and Space Shuttle Mishap Interagency Investigation Board be renamed the Columbia Accident Investigation Board. O'Keefe accepted this change and formally chartered the Board on Sunday, February 2, 2003.

On Sunday, Board members flew on government and commercial aircraft to Barksdale Air Force Base, where at 6:50 p.m. Central Standard Time the Board held its first official meeting. The Board initiated its investigation on Monday, February 3, at 8:00 a.m. Central Standard Time. On Tuesday morning, February 4, the Board toured the debris field in and around Nacogdoches, Texas, and observed a moment of silence. On Thursday, February 6, the Board relocated to the Johnson Space Center, eventually settling into its own offices off Center grounds. That evening, the Board formally relieved the NASA Headquarters Contingency Action Team of its interim responsibilities for initial accident investigation activities. The Board assumed operational control of the debris search and recovery efforts from NASA's Mishap Investigation Team, which functioned under the Board's direction until the completion of the search in early May.

A.2 Board Charter and Organization

During meetings that first week, Chairman Gehman and the Board proposed that its charter be rewritten. The original charter, derived from Appendix D of NASA's Contingency Action Plan, had a number of internal inconsistencies and provisions that the Board believed would impede the execution of its duties. Additionally, the Board was not satisfied that its initial charter adequately ensured independence from NASA. The Board resolved to:

- Have its own administrative and technical staff so that it could independently conduct testing and analysis and establish facts and conclusions
- Secure an adequate and independent budget to be overseen by the Board Chairman
- Establish and maintain records independent from NASA records
- Empower the Board Chairman to appoint new Board Members
- Provide the public with detailed updates on the progress of its investigation through frequent public hearings, press briefings, and by immediately releasing all significant information, with the exception of details relating to the death of the crew members and privileged witness statements taken under the condition of confidentiality
- Simultaneously release its report to Congress, the White House, NASA, the public, and the astronauts' families
- Allow Board members to voice any disagreements with Board conclusions in minority reports

With the full cooperation of Administrator O'Keefe, the Board's charter was rewritten to incorporate these principles. The new charter, which underwent three drafts, was signed and ratified by O'Keefe on February 18, 2003. In re-chartering the Board, O'Keefe waived the requirements specified in the Contingency Action Plan that the Board use standard NASA mishap investigation procedures and instead authorized the Board to pursue "whatever avenue you deem appropriate" to conduct the investigation.[2]

Additional Board Members

To manage its burgeoning investigative responsibilities, the Board added additional members, each of whom brought to the Board a needed area of expertise. On February 6, the Board appointed Roger E. Tetrault, retired Chairman and Chief Executive Officer of McDermott International. On February 15, the Board appointed Sheila E. Widnall, Ph.D., Institute Professor and Professor of Aeronautics and Astronautics at the Massachusetts Institute of Technology and former Secretary of the Air Force. On March 5, the Board appointed Douglas D. Osheroff, Ph.D., Nobel Laureate in Physics and Chair of the Stanford Physics Department; Sally K. Ride, Ph.D., Professor of Space Science at the University of California at San Diego and the nation's first woman in space; and John M. Logsdon, Ph.D., Director of the Space Policy Institute at George Washington University. This brought the total number of Board members to 13, coincidentally the same number as the Presidential Commission on the Space Shuttle *Challenger* Accident.

Board Organization

In the first week, the Board divided into four groups, each of which addressed separate areas of the investigation. Group I, consisting of General Barry, General Deal, and Admiral Turcotte, examined NASA management and treatment of materials, including Shuttle maintenance safety and mission assurance. Group II, consisting of General Hess, Mr. Wallace, and later Dr. Ride, scrutinized NASA training, operations, and the in-flight performance of ground crews and the Shuttle crew. Group III, consisting of Dr. Hallock, Mr. Hubbard, and later Mr. Tetrault, Dr. Widnall, and Dr. Osheroff, focused on engineering and technical analysis of the accident and resulting debris. Group IV, consisting of Dr. Logsdon, Dr. Ride, and Mr. Hubbard, examined how NASA history, budget, and institutional culture affected the operation of the Space Shuttle Program. Each group, with the approval of the Chairman, hired investigators and support staff and collaborated extensively with one another.

The Board also organized an internal staff of technical experts called the Independent Assessment Team. Under the leadership of James Mosquera, a senior nuclear engineer with the U. S. Navy, the Independent Assessment Team advised the Board when and where NASA analysis should be independently verified and, when needed, conducted fully independent tests on the Board's behalf.

A.3 Investigation Process and Scope

Decision to Pursue a Safety Investigation

During the first week of its investigation, the Board reviewed the structure and methodology of the Presidential Commission on the Space Shuttle *Challenger* Accident, the International Civil Aviation Organization standards used by the National Transportation Safety Board and the Federal Aviation Administration, and the accident investigation models under which the U.S. Air Force and Navy Safety Centers operate. Rather than assign formal blame or determine legal liability for the cause of the accident, the Board affirmed its charge to pursue both an accident investigation and a safety investigation, the primary aim of which would be to identify and correct threats to the safe operation of the Space Shuttle.

The Use of Privileged Witness Statements

With a principal focus on identifying and correcting threats to safe operations, safety investigations place a premium on obtaining full and complete disclosure about every aspect of an accident, even if that information may prove damaging or embarrassing to particular individuals or organizations. However, individuals who have made mistakes, know of negligence by others, or suspect potential flaws in their organizations are often afraid of being fired or even prosecuted if they speak out. To allay these fears, which can prevent the emergence of information that could save lives in the future, many safety investigations, including those by NASA and by the Air Force and Navy Safety Centers, grant witnesses complete confidentiality, as do internal affairs investigations by agency Inspector Generals. This confidentiality, which courts recognize as "privileged communication," allows witnesses to volunteer information that they would not otherwise provide and to speculate more openly about their organizations' flaws than they would in a public forum.

Given the stakes of the *Columbia* accident investigation, the most important being the lives of future astronauts, the Board decided to extend witnesses confidentiality, even though this confidentiality would necessitate that investigators redact some witness information before releasing it to the public.

Consistent with NASA Safety Program policy NPD 8621.1H Para 1.j, statements made to Board investigators under privilege were not made under legal oath. Investigators recorded and then transcribed interviews, with those interviewed affirming by their signatures the accuracy of the transcripts. The Board took extraordinary measures to ensure that privileged witness statements would remain confidential by restricting access to these statements to its 13 members and a small number of authorized support staff. Witness statements and information derived from them are exempt from disclosure under the Freedom of Information Act.

The existence of a safety investigation in which privileged statements are taken does not prevent an accounting of personal responsibility associated with an accident. It merely means that such an accounting must result from a separate investigation. In this instance, that responsibility has been left to the NASA administration and the Congressional committees that oversee the agency. To facilitate this separate investigation, the Board pledged to notify NASA and Congress if evidence of criminal activity or willful negligence is found in privileged statements or elsewhere. Additionally, the Board opened all its files to Congressional representatives, with the exception of privileged witness statements. Limited Congressional access to these statements is governed by a special written agreement between the oversight committees and the Board that preserves the Board's obligation to witnesses who have entrusted them with information on the condition of confidentiality.

Expanded Bounds of Board Investigation

Throughout the investigation, Chairman Gehman consulted regularly with members of Congress and the Administration to ensure that the Board met its responsibilities to provide the public with a full and open accounting of the *Columbia* accident. At the request of Congressional Oversight Committees, the Board significantly expanded the scope of its investigation to include a broad review of the Space Shuttle Program since its inception. In addition to establishing the accident's probable and contributing causes, the Board's report is intended to serve as the basis for an extended public policy debate over the future course of the Space Shuttle Program and the role it will play in the nation's manned space flight program.

A.4 Board Policies and Procedures

Authorizing Investigators

To maintain control over the investigation process, the Chairman established a system of written authorizations specifying individuals who were sanctioned to interview witnesses or perform other functions on behalf of the Board.

Consideration of Federal Advisory Committee Act Statutes

Not long after its activation, and well before adding additional members, the Board considered the applicability of the Federal Advisory Committee Act.[3] This statute requires advisory committees established by the President or a federal agency to provide formal public notice of their meetings as well as public access to their deliberations. In contrast to most committees governed by the Federal Advisory Committee Act, which meet a few times per year, the Board intended from the outset to conduct a full-time, fast-paced investigation, in which Board members themselves were active investigators who would shape the investigation's direction as it developed. The Board concluded that the formalities required by the Federal Advisory Committee Act are not compatible with the kind of investigation it was charged to complete. Nor did the Board find the Federal Advisory Committee Act statutes compatible with exercising operational responsibility for more than a hundred staff and thousands of debris searchers.

Though the Federal Advisory Committee Act did not apply to the Board's activities, the Board resolved to be faithful to the standards of openness the Act embodies. The Board held frequent press briefings and public hearings, released all significant findings immediately, and maintained a telephone hotline and a Web site, where users accessed Web pages more than 40,000,000 times. The Board also processed Freedom of Information Act requests according to procedures established in 14 C.F.R. Section 1206.

Board Members as Federal Employees

The possibility of litigation against Board members for their actions while on the Board, especially because the Space Flight Operations Contract would be a subject of investigation, made it necessary to bring Board Members within the protections that the Federal Tort Claims Act affords to federal employees. This and other considerations led the Board Chairman to determine that the Board should consist of full-time federal employees. As the Chairman named new Board members, the NASA Administrator honored the Board's determination and deemed them full-time federal employees.

Oversight of Board Activities

To ensure that the Board acted in an independent and unbiased manner in its investigation, the NASA Inspector General was admitted on request to any Board proceeding, except those involving privileged witness statements. The Board also allowed Congressional access to the Board's databases and offices in Houston and Washington, D.C., with special restrictions that preserved the integrity and confidentiality of privileged witness statements.

Financial Independence

To ensure the Board's financial independence, NASA established a separate operating budget for the Board's activities. This fund provided for Board operating expenses, including extensive testing and analysis and the acquisition of services by support staff and technical experts. With the exception of Chairman Gehman, whose salary was paid by the Office of Personnel Management, and those Board members who were already federal government or military employees, Board members were compensated by Congressionally appropriated funds administered by NASA.

Board Staffing and Administrative Support

Through a Government Services Administration-supervised bidding process, Valador, Inc., a service-disabled-veteran-owned professional services contractor, was selected to provide the Board's administrative and technical support. Under a Mission Operation and Business Improvement Systems contract, Valador arranged for the Board's support staff, technical experts, and information technology needs, including the Board Web site, http://*www.caib.us*. Valador also supported the Board's public hearings, press conferences, the public-input database, and the publication of the final report.

The Board was aided by public affairs officers; a budget manager; representatives from the National Transportation Safety Board, Federal Emergency Management Agency, Department of Defense, and the Department of Justice Civil Division, Office of Litigation Support; and Dr. James B. Bagian, an astronaut flight surgeon assigned from the Department of Veterans Affairs who worked with the NASA medical staff, Armed Forces Institute of Pathology, and the local medical examiner. A complete list of staff and consultants appears in Appendix B.2 and B.3.

Public Inputs

The Board established a system for inputs from the public that included a 24-hour hotline, mailing address, and online comment form linked to the Board's Web site. This enabled the submission of photographs, comments, technical papers, and other materials by the public, some of whom made submissions anonymously. Board staff logged every input into a database. To establish the relevance of every phone call, letter, e-mail, or online comment, investigators evaluated their significance and, if appropriate, followed up with the submitters. Of the 3,000 submissions the Board received, more than 750 resulted in actions by one of the Board's four investigative sub-groups, the Independent Assessment Team, or other Board staff.

Office of Governmental Affairs

As inquiries from Congress grew and the need to keep the Executive and Legislative branches updated on the investigation's progress became clear, the Board opened an Office of Governmental Affairs. Based in Washington, D.C., it served as the Board's liaison to the White House, departments within the Executive Branch, Congressional Oversight Committees, and members of Congress and their staffs. The office conducted numerous briefings, responded to Congressional inquiries, and ensured that the investigation met the needs of the Congressional Oversight Committees that plan to use the Board's report as the basis for a public policy debate on the future of the Space Shuttle Program.

A.5 INVESTIGATION INTERFACE WITH NASA

NASA mobilized hundreds of personnel to directly support the Board's investigation on a full-time basis. Initially, as part of the Contingency Action Plan activated on February 1, the Mishap Investigation Team went to Barksdale Air Force Base to coordinate the search for debris. NASA then deployed a Mishap Response Team to begin an engineering analysis of the accident. These groups consisted of Space Shuttle Program personnel and outside experts from NASA and contractor facilities.

As prescribed by its charter, the Board coordinated its investigation with NASA through a NASA Task Force Team, later designated the Columbia Task Force. This group was the liaison between the Board and the Mishap Response Team. As the investigation progressed, NASA modified the organizational structure of the Mishap Response Team to more closely align with Board structure and investigative paths, and NASA renamed it the NASA Accident Investigation Team. This team supported the Board's investigation, along with thousands of other NASA and contract personnel who worked in the fault tree teams described in Chapter 4 and on the debris search efforts described in Chapter 2.

Documents and Actions Requested From NASA

The close coordination of the NASA Investigation Team with the Board's sub-groups required a system for tracking documents and actions requested by the investigation. The Board and the Columbia Task Force each appointed representatives to track documents and manage their configuration.

Board investigators submitted more than 600 requests for action or information from NASA. Requests were submitted in writing, on a standardized form,[4] and signed by a Board member. Only Board members were authorized to sign such requests. Each request was given a priority and tracked in a database. Once answered by Columbia Task Force personnel, the Board member who submitted the request either noted by signature that the response was satisfactory or re-submitted the request for further action.

Reassignment of Certain NASA Personnel Involved in STS-107

On February 25, 2003, Chairman Gehman wrote to NASA Administrator O'Keefe, asking that he "reassign the top level Space Shuttle Program management personnel who were involved in the preparation and operation of the flight of STS-107 back to their duties and remove them from directly managing or supporting the investigation."[5] This letter expressed the Board's desire to prevent actual or perceived conflicts of interest between NASA personnel and the investigation. In response, O'Keefe reassigned several members of NASA's Columbia Task Force and Mishap Investigation Team and reorganized it along the same lines as the Board's groups. Additionally, Bryan O'Connor, an Ex-Officio Member to the Board, and Theron Bradley Jr., the Board's Executive Secretary, returned to their respective duties as Associate Administrator for Safety and Mission Assurance and Chief Engineer, and were not replaced. After O'Connor's departure, Colonel (Selectee) Michael J. Bloomfield, an active Shuttle Commander and the lead training astronaut, joined the Board as a representative from the Astronaut Office.

Handling of Debris and Impounded Materials

To ensure that all material associated with *Columbia*'s mission was preserved as evidence in the investigation, NASA officials impounded data, software, hardware, and facilities at NASA and contractor sites. At the Johnson Space Center in Houston, Texas, the door to the Mission Control Center was locked while flight control personnel created and archived backup copies of all original mission data and took statements from Mission Control personnel. At the Kennedy Space Center in Florida, mission facilities and related hardware, including Launch Pad Complex 39-A, were put under guard or stored in secure warehouses. Similar steps were taken at other key Shuttle facilities, including the Marshall Space Flight Center in Huntsville, Alabama, and the Michoud Assembly Facility near New Orleans, Louisiana. Impounded items and data were released only when the Board Chairman approved a formal request from the NASA Columbia Task Force.

Similarly, any testing performed on Shuttle debris was approved by the Board Chairman only after the Columbia Task Force provided a written request outlining the potential benefits of the testing and addressing any possible degradation of the debris that could affect the investigation. When testing of Shuttle debris or hardware occurred outside the secure debris hanger at the Kennedy Space Center, investigation personnel escorted the debris for the duration of the testing process or otherwise ensured the items' integrity and security.

A.6 Board Documentation System

The Columbia Accident Investigation Board Database Server

The sheer volume of documentation and research generated in the investigation required an electronic repository capable of storing hundreds of thousands of pages of technical information, briefing charts, hearing transcripts, government documents, witness statements, public inputs, and correspondence related to the *Columbia* accident.

For the first few months of its investigation, the Board used the Process-Based Mission Assurance (PBMA) system for many of its documentation needs. This Web-based action tracking and document management system, which is hosted on a server at the NASA Glenn Research Center, was developed and maintained by NASA Ames Research Center. The PBMA system was established as a repository for all data provided by NASA in response to the Board's Action/Request for Information process. It contained all information produced by the Columbia Task Force, as well as reports from NASA and other external groups, presentations to the Board, signed hardware release and test release forms, images, and schedule information.

However, the PBMA system had several critical limitations that eventually compelled the Board to establish its own server and databases. First, NASA owned the Mission Assurance system and was responsible for the documents it produced. The Board, seeking to maintain independence from NASA and the Columbia Task Force, found it unacceptable to keep its documentation on what was ultimately a NASA database. Second, the PBMA system is not full-text searchable, and did not allow investigators to efficiently cross-reference documents.

The Board wanted access to all the documents produced by the Columbia Task Force, while simultaneously maintaining its own secure and independent databases. To accomplish this, the Board secured the assistance of the Department of Justice Civil Division, Office of Litigation Support, which established the Columbia Accident Investigation Board Database Server. This server provided access to four document databases:

- Columbia Task Force Database: all the data in NASA's Process-Based Mission Assurance system, though independent from it.
- Columbia Accident Investigation Board Document Database: all documents gathered or generated by Board members, investigators, and support staff.
- Interview Database: all transcriptions of privileged witness interviews.
- Investigation Meeting Minutes Database: text of approved Board meeting minutes.

Although the Board had access to the Process-Based Mission Assurance system and therefore every document created by the Columbia Task Force, the Task Force did not have access to any of the Board's documents that were independently produced in the Board's four other databases. A security system allowed Board members to access these databases through the Board's Database Server using confidential IDs and passwords. In total, the Columbia Accident Investigation Board Database Server housed more than 450,000 pages that comprised more than 75,000 documents. The bulk of these are from NASA's Columbia Task Force Document Database, which holds over 45,000 documents totaling 270,000 pages.

To ensure that all documents received and generated by individual investigators became part of the permanent Columbia Accident Investigation Board archive, Department of Justice contractors had coordinators in each investigative group who gathered electronic or hard copies of all relevant investigation documents for inclusion in the Columbia Accident Investigation Board Document Database. Every page of hard copy received a unique tracking number, was imaged, converted to a digital format, and loaded onto the server. Documents submitted electronically were saved in Adobe PDF format and endorsed with a tracking number on each page. Where relevant, these document numbers are referenced in citations found in this report. The Columbia Accident Investigation Board Document database contains more than 30,000 documents comprising 180,000 pages.

Other significant holdings on the Columbia Accident Investigation Board Document Database Server include the Interview Database, which holds 287 documents comprising 6,300 pages, and the Investigative Meeting Minutes Database, which holds 72 documents totaling 598 pages.

Concordance

Acting on the recommendation of the Department of Justice, the Board selected Concordance as the software to manage all the electronic documents on the Columbia Accident Investigation Board Database Server. Concordance is a full-text, image-enabled document and transcript database accessible to authorized Board members on their office computers. Concordance allowed the Board to quickly search the data provided by the Columbia Task Force, as well as any documents created and stored in the four other databases. The Concordance application was on a server in a secure location in the Board office. Though connected to the Johnson Space Center backbone, it was exclusively managed and administered by the Department of Justice and contract staff from Aspen Systems Corporation. Department of Justice and contract staff trained users to search the database, and performed searches at the request of Board members and investigators. The Department of Justice and contract staff also assisted Congressional representatives in accessing the Columbia Accident Investigation Board Database Server.

Investigation Database Tools

In addition to these databases, several information management tools aided the Board's investigation, deliberation, and report writing.

Group Systems

Group Systems is a collaborative software tool that organizes ideas and information by narrowing in on key issues and possible solutions. It supports academic, government, and commercial organizations worldwide. The Board used Group Systems primarily to brainstorm topics for inclusion in the report outline and to classify information related to the accident.

Investigation Organizer

Investigation Organizer is a Web-based pre-decisional management and modeling tool designed by NASA to support mishap investigation teams. Investigation Organizer provides a central information repository that can be used by investigation teams to store digital products. The Board used Investigation Organizer to connect data from various sources to the outline that guided its investigation. Investigation Organizer was developed, maintained, and hosted by NASA Ames Research Center. Access to Board files on Investigation Organizer was restricted to Board members and authorized staff.

TechDoc

The Board drafted its final report with the assistance of TechDoc, a secure Web-based file management program that allowed the 13 Board members and the editorial staff to comment on report drafts. TechDoc requires two-factor authentication and is certified to store sensitive Shuttle engineering data that is governed by the International Traffic in Arms Reduction Treaty.

Official Photographer

The Board employed an official photographer, who took more than 5,000 digital images. These photographs, many of which have been electronically edited, document Board members and support staff at work in their offices and in the field in Texas, Florida, Alabama, Louisiana, and Washington, D.C.; at Shuttle debris collection, analysis, and testing; and at public hearings, press briefings, and Congressional hearings. Images captured by NASA photographers relevant to the investigation are available through NASA's Public Affairs Office.

National Archives and Records Administration

All appropriate Board documentation and products will be stored for submission to the National Archives and Records Administration, with the exception of documents originating in the Process-Based Mission Assurance system, which will be archived by NASA under standard agency procedures. Representatives of the Board will review all documentation prior to its transfer to the National Archives to safeguard privacy and national security. This preparation will include a review of all documents to ensure compliance with the Freedom of Information Act, the Trade Secrets Act, the Privacy Act, the International Traffic in Arms Reduction Treaty, and Export Administration Regulations. To gain access to the Board's documents, requests can be made to:

National Archives and Records Administration
Customer Services Division (NWCC)
Room 2400
8601 Adelphi Road
College Park, MD 20740-6011

The National Archives and Records Administration can be contacted at 301.837.3130. More information is available at http://www.nara.gov.

A.7 List of Public Hearings

The Board held public hearings to listen to and question expert witnesses. A list of these hearings, and the participating witnesses, follows; transcripts of the hearings are available in Appendix G.

March 6, 2003 Houston, Texas

Review of NASA's Organizational Structure and Recent Space Shuttle History

Lt. Gen. Jefferson D. Howell, Jr., Director, NASA Johnson Space Center
Mr. Ronald D. Dittemore, Manager, Space Shuttle Program
Mr. Keith Y. Chong, Engineer, Boeing Corporation
Dr. Harry McDonald, Professor, University of Tennessee

March 17, 2003, Houston, Texas

***Columbia* Re-entry Telemetry Data, and Debris Dispersion Timeline**

Mr. Paul S. Hill, Space Shuttle and International Space Station Flight Director, NASA Johnson Space Center
Mr. R. Douglas White, Director for Operations Requirements, Orbiter Element Department, United Space Alliance

Prior Orbital Debris Re-entry Data

Dr. William H. Ailor, Director, Center for Orbital and Re-entry Debris Studies, The Aerospace Corporation

March 18, 2003, Houston, Texas

Aero and Thermal Analysis of Columbia Re-entry Data

Mr. Jose M. Caram, Aerospace Engineer, Aeroscience and Flight Mechanics Division, NASA Johnson Space Center
Mr. Steven G. Labbe, Chief, Applied Aeroscience and Computational Fluid Dynamics Branch, NASA Johnson Space Center
Dr. John J. Bertin, Professor of Aerodynamics, United States Air Force Academy
Mr. Christopher B. Madden, Deputy Chief, Thermal Design Branch, NASA Johnson Space Center

March 25, 2003, Cape Canaveral, Florida

Launch Safety Considerations

Mr. Roy D. Bridges, Jr., Director, Kennedy Space Center

Role of the Kennedy Space Center in the Shuttle Program

Mr. William S. Higgins, Chief of Shuttle Processing Safety and Mission Assurance Division, Kennedy Space Center
Lt. Gen. Aloysius G. Casey, U.S. Air Force (Retired)

March 26, 2003, Cape Canaveral, Florida

Debris Collection, Layout, and Analysis, including Forensic Metallurgy

Mr. Michael U. Rudolphi, Deputy Director, Stennis Space Center
Mr. Steven J. Altemus, Shuttle Test Director, Kennedy Space Center
Dr. Gregory T. A. Kovacs, Associate Professor of Electronics, Stanford University
Mr. G. Mark Tanner, Vice President and Senior Consulting Engineer, Mechanical & Materials Engineering

April 7, 2003, Houston, Texas

Post-Flight Analysis, Flight Rules, and the Dynamics of Shedding Foam from the External Tank

Col. James D. Halsell, Jr., U.S. Air Force, NASA Astronaut, NASA Johnson Space Center
Mr. Robert E. Castle, Jr., Chief Engineer, Mission Operations Directorate, NASA Johnson Space Center
Mr. J. Scott Sparks, Department Lead, External Tank Issues, NASA Marshall Space Flight Center
Mr. Lee D. Foster, Technical Staff, Vehicle and Systems Development Department, NASA Marshall Space Flight Center

April 8, 2003, Houston, Texas

Shuttle Safety Concerns, Upgrade Issues, and Debris Strikes on the Orbiter

Mr. Richard D. Blomberg, Former Chairman, NASA Aerospace Safety Advisory Panel
Mr. Daniel R. Bell, Thermal Protection System Sub-System Manager for the Boeing Company at Kennedy Space Center
Mr. Gary W. Grant, Systems Engineer in the Thermal Management Group for the Boeing Company at Kennedy Space Center

April 23, 2003, Houston, Texas

Tradeoffs Made During the Shuttle's Initial Design and Development Period

Dr. Milton A. Silveira, Technical Advisor to the Program Director, Missile Defense Agency, Office of the Secretary of Defense
Mr. George W. Jeffs, Retired President of Aerospace and Energy Operations, Rockwell International Corporation
Prof. Aaron Cohen, Professor Emeritus of Mechanical Engineering, Texas A&M University
Mr. Owen G. Morris, Founder, CEO, and Chairman of Eagle Aerospace, Inc.
Mr. Robert F. Thompson, former Vice President of the Space Station Program for McDonnell Douglas

Managing Aging Aircraft

Dr. Jean R. Gebman, Senior Engineer, RAND Corporation
Mr. Robert P. Ernst, Head of the Aging Aircraft Program, Naval Air Systems Command

Risk Assessment and Management in Complex Organizations

Dr. Diane Vaughan, Professor, Department of Sociology at Boston College

May 6, 2003, Houston, Texas

MADS Timeline Update, Ascent Video

Dr. Gregory J. Byrne, Assistant Manager, Human Exploration Science, Astromaterials Research and Exploration Science Office at the Johnson Space Center
Mr. Steven Rickman, Chief of the Thermal Design Branch, Johnson Space Center, NASA
Dr. Brian M. Kent, Air Force Research Laboratory Research Fellow
David W. Whittle, Chairman of the Systems Safety Review Panel and Chairman of the Mishap Investigation Team in the Shuttle Program Office

June 12, 2003, Washington, DC

NASA Budgetary History and Shuttle Program Management

Mr. Allen Li, Director, Acquisition and Sourcing Management, General Accounting Office
Ms. Marcia S. Smith, Specialist in Aerospace and Telecommunications Policy, Congressional Research Service
Mr. Russell D. Turner, Former President and CEO, United Space Alliance
Mr. A. Thomas Young, Retired Aerospace Executive

ENDNOTES FOR APPENDIX A

[1] NASA Agency Contingency Action Plan for Space Flight Operations, January 2003, p. D-2.

[2] Guidelines per NASA Policy Guideline 8621.

[3] 5 U.S.C. App § §1 *et seq.* (1972).

[4] JSC Form 564 (March 24, 2003).

[5] Harold W. Gehman to Sean O'Keefe, February 25, 2003.

APPENDIX B

Board Member Biographies

ADMIRAL HAROLD W. GEHMAN JR. (RETIRED)

Chairman, Columbia Accident Investigation Board. Formerly Co-Chairman of the Department of Defense review of the attack on the U.S.S. *Cole*. Before retiring, Gehman served as the NATO Supreme Allied Commander, Atlantic, Commander in Chief of the U.S. Joint Forces Command, and Vice Chief of Naval Operations for the U.S. Navy. Gehman earned a B.S. in Industrial Engineering from Penn State University and is a retired four star Admiral.

MAJOR GENERAL JOHN L. BARRY

Executive Director for the Columbia Accident Investigation. Director, Plans and Programs, Headquarters Air Force Materiel Command, Wright-Patterson Air Force Base, Ohio. An honors graduate of the Air Force Academy with an MPA from Oklahoma University, Barry has an extensive background as a fighter pilot and Air Force commander: Squadron, Group and two Wings. A trained accident investigator, Barry has presided or served on numerous aircraft mishap boards. He was a White House Fellow at NASA during the Challenger mishap and was the White House liaison for NASA, served as the Military Assistant to the Secretary of Defense during Desert Storm and was the director of Strategic Planning for the U.S. Air Force.

BRIGADIER GENERAL DUANE W. DEAL

Commander, 21st Space Wing, Peterson Air Force Base, Colorado. Currently in his eighth commander position in the U.S. Air Force, Deal has served on or presided over 12 investigations of space launch and aircraft incidents. Formerly a Research Fellow with the RAND Corporation and Fellow of the Harvard Center for International Affairs, he has flown seven aircraft types as an Air Force pilot, including the SR-71 Blackbird, and served as a crew commander in two space systems. Deal holds a B.S. in Physics and a M.S. in Counseling Psychology from Mississippi State University, as well as a M.S. in Systems Management from the University of Southern California.

JAMES N. HALLOCK, PH.D.

Manager, Aviation Safety Division, Volpe National Transportation Systems Center, Massachusetts. He has worked in the Apollo Optics Group of the MIT Instrumentation Lab and was a physicist at the NASA Electronics Research Center, where he developed a spacecraft attitude determining system. He joined the DOT Transportation Systems Center (now the Volpe Center) in 1970. Hallock received B.S., M.S. and Ph.D. degrees in Physics from the Massachusetts Institute of Technology (MIT). He is an expert in aircraft wake vortex behavior and has conducted safety analyses on air traffic control procedures, aircraft certification, and separation standards, as well as developed aviation-information and decision-support systems.

MAJOR GENERAL KENNETH W. HESS

Commander, Air Force Safety Center, Kirtland Air Force Base, New Mexico, and Chief of Safety, United States Air Force, Headquarters U.S. Air Force, Washington, D.C. Hess entered the Air Force in 1969 and has flown operationally in seven aircraft types. He has commanded three Air Force wings – the 47th Flying Training Wing, 374th Airlift Wing, and 319th Air Refueling Wing – and commanded the U.S. 3rd Air Force, RAF Mildenhall, England. Hess also has extensive staff experience at the Joint Staff and U.S. Pacific Command. He holds a B.B.A. from Texas A&M University and a M.S. in Human Relations and Management from Webster College.

G. SCOTT HUBBARD

Director of the NASA Ames Research Center, California. Hubbard was the first Mars Program Director at NASA Headquarters, successfully restructuring the program after mission failures. Other NASA positions include Deputy Director for Research, Director of NASA's Astrobiology Institute, and Manager of the Lunar Prospector mission. Before joining NASA, he was Vice President of Canberra Semiconductor and Staff Scientist at the Lawrence Berkeley National Laboratory. Hubbard holds a B.A. in Physics-Astronomy from Vanderbilt University, and conducted graduate studies at the University of California, Berkeley. Hubbard is a Fellow of the American Institute of Aeronautics and Astronautics.

JOHN M. LOGSDON , PH.D.

Director, Space Policy Institute, Elliott School of International Affairs, The George Washington University, Washington, D.C., where he has been a faculty member since 1970. A former member of the NASA Advisory Council, and current member of the Commercial Space Transportation Advisory Committee and the International Academy of Astronautics, Logsdon is a Fellow of the American Institute of Aeronautics and Astronautics and the American Association for the Advancement of Science, and was the first Chair in Space History at the National Air and Space Museum. He received a B.S. in Physics from Xavier University and a Ph.D. in Political Science from New York University.

DOUGLAS D. OSHEROFF, PH.D.

J. G. Jackson and C. J. Wood Professor of Physics and Applied Physics, Stanford University, California. A 1996 Nobel Laureate in Physics for his joint discovery of superfluidity in helium-3, Osheroff is also a member of the National Academy of Sciences and a MacArthur Fellow. Osheroff has been awarded the Simon Memorial Prize and the Oliver Buckley Prize. He received a B.S. from the California Institute of Technology and a Ph.D. from Cornell University.

Sally T. Ride, Ph.D.

Professor of Physics, University of California, San Diego, and President and CEO of Imaginary Lines, Inc. The first American female astronaut in space, Ride served on the Presidential Commission investigating the Space Shuttle Challenger Accident. A Fellow of the American Physical Society and Board Member of the California Institute of Technology, she was formerly Director of NASA's Strategic Planning and served on the Space Studies Board and the President's Committee of Advisors on Science and Technology. Ride has received the Jefferson Award for Public Service and twice been awarded the National Spaceflight Medal. She received a B.S. in Physics, a B.A. in English, and a M.S. and Ph.D. in Physics from Stanford University.

Roger E. Tetrault

Retired Chairman and Chief Executive Officer, McDermott International. Tetrault has also served as Corporate Vice President and President of the Electric Boat Division and the Land Systems Division at General Dynamics, as well as Vice President and Group Executive of the Government Group at Babcock and Wilcox Company. He is a 1963 graduate of the U.S. Naval Academy and holds a MBA from Lynchburg College.

Rear Admiral Stephen A. Turcotte

Commander, Naval Safety Center, Virginia. Formerly Commanding Officer of the Jacksonville Naval Air Station and Deputy Commander of the Joint Task Force Southwest Asia, Turcotte has also commanded an aviation squadron and served on the Joint Staff (Operations Division). A decorated aviator, he has flown more than 5,500 hours in 15 different aircraft and has extensive experience in aircraft maintenance and operations. Turcotte holds a B.S. in Political Science from Marquette University NROTC and masters degrees in National Security and Strategic Studies from the Naval War College, and in Management from Salve Regina University.

Steven B. Wallace

Director, Office of Accident Investigation, Federal Aviation Administration, Washington, D.C. Wallace's previous FAA positions include Senior Representative at the U.S. Embassy in Rome, Italy, Manager of the Transport Airplane Directorate Standards Staff in Seattle, and Attorney/ Advisor in the New York and Seattle offices. He holds a B.S. in Psychology from Springfield College and a J.D. from St. John's University School of Law. Wallace is admitted to legal practice before New York State and Federal courts, and is a licensed commercial pilot with multiengine, instrument, and seaplane ratings.

Sheila E. Widnall, Ph.D.

Institute Professor and Professor of Aeronautics and Astronautics and Engineering Systems, Massachusetts Institute of Technology (MIT), Massachusetts. Widnall has served as Associate Provost, MIT, and as Secretary of the Air Force. She is currently Co-Chairman of the Lean Aerospace Initiative. A leading expert in fluid dynamics, Widnall received her B.S., M.S., and Ph.D. in Aeronautics and Astronautics from MIT.

Board Member photographs by Rick W. Stiles

The launch of STS-107 on January 16, 2003.

APPENDIX C

Board Staff

ADVISORS TO THE CHAIR

James P. Bagian, M.D.	Medical Consultant and Chief Flight Surgeon	Astronaut (ret.), Department of Veterans Affairs
Guion S. Bluford Jr.	Executive Director for Investigative Activities	Astronaut (ret.)
Dennis R. Jenkins	Investigator and Liaison to the Board	Consulting Engineer, Valador, Inc.

GROUP I: MANAGEMENT AND TREATMENT OF MATERIALS

Charles A. Babish	Investigator	Air Force Materiel Command
Col. Timothy D. Bair	Investigator	Air Force Materiel Command
Lt. Col. Lawrence M. Butkus, P.E., Ph.D.	Investigator	Air Force Academy
CDR Michael J. Francis	Investigator	Naval Safety Center
CAPT James R. Fraser, M.D.	Investigator	Naval Safety Center
John F. Lehman	Investigator	Defense Contract Management Agency
Lt. Col. Christopher S. Mardis	Investigator	Air Force Materiel Command
Col. David T. Nakayama	Investigator	Air Force Materiel Command
Clare A. Paul	Investigator	Air Force Research Laboratory
Maj. Lisa Sayegh, Ph.D.	Investigator	Air Force Materiel Command
CAPT John K. Schmidt, Ph.D.	Investigator	Naval Safety Center
John R. Vallaster	Investigator	Naval Safety Center
Capt. Steven J. Clark	Researcher	Air Force Materiel Command
1st Lt. Michael A. Daniels	Support Staff	Air Force Materiel Command
1st Lt. David L. Drummond	Support Staff	Air Force Space Command
Joshua W. Lane	Support Staff	Analytical Graphics, Inc.
Ed Mackey	Support Staff	Analytical Graphics, Inc.
Jana M. Price, Ph.D.	Support Staff	National Transportation Safety Board
Dana L. Schulze	Support Staff	National Transportation Safety Board
Stacy L. Walpole	Administrative Support	Valador, Inc.

Group II: Training, Operations, and In-Flight Performance

Lt. Col. Richard J. Burgess	Investigator	Air Force Safety Center
Daniel P. Diggins	Investigator	Federal Aviation Administration
Gregory J. Phillips	Investigator	National Transportation Safety Board
Lisa M. Reed	Investigator	Booz Allen Hamilton
Lt. Col. Donald J. White	Investigator	Air Force Safety Center
Diane Vaughan, Ph.D.	Researcher	Boston College
Maj. Tracy G. Dillinger, Ph.D.	Support Staff	Air Force Safety Center
Lt. Matthew E. Granger	Support Staff	Air Force Safety Center
Maj. David L. Kral	Support Staff	Air Force Safety Center
Helen E. Cunningham	Administrative Support	Valador, Inc.
Col. Donald W. Pitts	Consultant	Air Force Safety Center

Group III: Engineering and Technical Analysis

James O. Arnold. Ph.D.	Investigator	University of California, Santa Cruz
R. Bruce Darling, Ph.D.	Investigator	University of Washington
Lt. Col. Patrick A. Goodman	Investigator	Air Force Space Command
G. Mark Tanner, P.E.	Investigator	Valador, Inc. Consultant
Gregory T. Kovacs, Ph.D.	Investigator	Stanford University
Paul D. Wilde, Ph.D.	Investigator	Federal Aviation Administration
Douglas R. Cooke	Advisor	NASA Johnson Space Center
Capt. David J. Bawcom	Support Staff	Air Force Space Command
Robert E. Carvalho	Support Staff	NASA Ames Research Center
Lisa Chu-Thielbar	Support Staff	NASA Ames Research Center
Capt. Anne-Marie Contreras	Support Staff	Air Force Space Command
Jay H. Grinstead	Support Staff	NASA Ames Research Center
Richard M. Keller	Support Staff	NASA Ames Research Center
Lt. Col. Robert J. Primbs, Jr.	Support Staff	Air Force Space Command
Ian B. Sturken	Support Staff	NASA Ames Research Center
Y'Dhanna Daniels	Administrative Support	Honeywell Technology Solutions, Inc.

Group IV: Organization and Policy

Dwayne A. Day, Ph.D	Investigator	Valador, Inc. Consultant
David H. Onkst	Researcher	American University
Richard H. Buenneke	Consultant	The Aerospace Corporation
W. Henry Lambright, Ph.D.	Consultant	Syracuse University
Roger D. Launius, Ph.D.	Consultant	National Air and Space Museum
Howard E. McCurdy, Ph.D.	Consultant	American University
Jill B. Dyszynski	Research Assistant	George Washington University
Jonathan M. Krezel	Research Assistant	George Washington University
Chirag B. Vyas	Research Assistant	George Washington University

Independent Assessment Team

James P. Mosquera	Lead Investigator	U.S. Navy
Ronald K. Gress	Investigator	Valador, Inc. Consultant
James W. Smiley, Ph.D.	Investigator	Valador, Inc. Consultant
David B. Pye	Investigator	Valador, Inc. Consultant
CDR (Selectee) Johnny P. Wolfe	Investigator	Strategic Systems Program
John Bertin, Ph.D.	Consultant	Valador, Inc. Consultant
Tim Foster	Consultant	Valador, Inc. Consultant
Robert M. Hammond	Consultant	Valador, Inc.
Daniel J. Heimerdinger, Ph.D.	Consultant	Valador, Inc.
Arthur Heuer, Ph.D.	Consultant	Valador, Inc. Consultant
Michael W. Miller	Consultant	Valador, Inc. Consultant
Gary C. Olson	Consultant	Valador, Inc. Consultant
Jacqueline A. Stemen	Administrative Support	Valador, Inc.

NASA Representatives

Col. (Selectee) Michael J. Bloomfield	Astronaut Representative	USAF/NASA Astronaut Office
Theron M. Bradley, Jr.	Executive Secretary	NASA Headquarters
Robert W. Cobb	Observer	NASA Office of the Inspector General
Bryan D. O'Connor	Ex-Officio Board Member	NASA Headquarters
David M. Lengyel	Executive Secretary for Administration	NASA Headquarters
Steven G. Schmidt	Executive Secretary for Management	NASA Headquarters
J. William Sikora, Esq.	Board General Counsel	NASA Glenn Research Center

Editorial Team and Production Staff

Lester A. Reingold	Lead Editor	Valador, Inc. Consultant
Christopher M. Kirchhoff	Editor	Valador, Inc. Consultant
Patricia D. Trenner	Copy Editor	Air & Space/Smithsonian Magazine
Ariel H. Simon	Assistant Editor	Valador, Inc. Consultant
Joshua M. Limbaugh	Layout Artist	Valador, Inc.
Joseph A. Reid	Graphic Designer	Valador, Inc.
James M. Thoburn	Website and Public Database Lead	Valador, Inc.

Public Affairs

Laura J. Brown	Lead Public Affairs Officer	Federal Aviation Administration
Patricia L. Brach	Public Affairs Officer	Federal Emergency Management Agency
Paul I. Schlamm	Public Affairs Officer	National Transportation Safety Board
Terry Williams	Public Affairs Officer	National Transportation Safety Board
Lt. Col. Tyrone M. Woodyard	Public Affairs Officer	Air Force Office of the Chief of Staff
Rick W. Stiles	Photographer	Rick Stiles Photography
Marie T. Jones	Public Affairs Administrative Support	Valador, Inc.

Keith Carney	Press Conference/Hearing Support	Federal Networks, Inc.
Clifford Feldman	Press Conference/Hearing Support	Federal Networks, Inc.

Government Affairs

Thomas L. Carter	Lead Government Affairs	Government Relations Consultant
Matthew J. Martin	Government Affairs	Government Relations Consultant
Capt. David R. Young	Government Affairs	Kansas Air National Guard
Lt. Col. Wade J. Thompson	Government Affairs	Air Combat Command
Col. Jack F. Anthony	Department of Defense Liaison	Air Force Space Command
Paul E. Cormier, Esq.	Counsel to the Chairman	Government Relations Consultant
Frances C. Fisher	ANSER Liaison	ANSER, Inc.
Frank E. Hutchison	Special Assistant	ANSER, Inc.
Charles R. McKee	Special Assistant	ANSER, Inc.

Administrative Staff

Lt. Charles W. Ensinger	Administrative Support to the Chairman	Naval Safety Center
YNC(SS) Barry M. Fitzgibbons	Administrative Support to the Chairman	Naval Safety Center
Christine F. Cole	Administrative Support	NASA Johnson Space Center
Jana T. Schultz	Administrative Support	NASA Johnson Space Center
Sharon J. Martin	Budget Manager	Al-Razaq Computing Services
Anna K. "Kitty" Rogers	Lead Travel Coordinator	Valador, Inc. Consultant
Trudy Davis	Travel Coordinator	Valador, Inc. Consultant
Lillian M. Hudson	Travel Coordinator	Valador, Inc. Consultant
Anita I. Abrego	Investigation Software Support	NASA Ames Research Center
Kevin P. Bass	Investigation Software Support	NASA Ames Research Center
Robert J. Navarro	Investigation Software Support	NASA Ames Research Center
James F. Williams	Investigation Software Support	NASA Ames Research Center
James McMahon	Information Technology	NASA Marshall Space Flight Center
Michele O'Connell	Information Technology	Science Applications International Corporation
Robert L. Binkley	Information Technology	NASA Dryden Flight Research Center
Roberta B. Sherrard	Information Technology	NASA Dryden Flight Research Center
Paula B. Frankel	Recorder	Westover and Associates, Inc.
Mitchell L. Bage, Jr.	Scheduler	Blackhawk
Rudy G. Gazarek	Software Support	GroupSystems.com, Inc.
Patrick Garrett	Software Support	GroupSystems.com, Inc.
Douglas S. Griffen	Software Support	GroupSystems.com, Inc.

Documentation Support

Clarisse Abramidis	Director	U.S. Department of Justice, Office of Litigation Support
Norman L. Bailey	Information Technology Lead	Aspen Systems Corporation
Michael R. Broschat	Database Administrator	Aspen Systems Corporation

Jennifer L. Bukvics	Lead Project Manager	Aspen Systems Corporation
Bethany C. Frye	Paralegal	Aspen Systems Corporation
Donna J. Fudge	Senior Paralegal, Group II Coordinator	Aspen Systems Corporation
Elizabeth G. Henderson	Case Manager	U.S. Department of Justice, Office of Litigation Support
Ronald K. Hourihane	Network Administrator	Aspen Systems Corporation
Kenneth B. Hulsey	Senior Paralegal, IAT Coordinator	U.S. Department of Justice/ ASPEN Systems
Leo Kaplus	Network Administrator	Aspen Systems Corporation
Carl Kikuchi	Contracting Officer's Technical Representative	U.S. Department of Justice, Office of Litigation Support
Douglas P. McManus	Branch Chief and IT Manager	U.S. Department of Justice, Office of Litigation Support
Susan M. Plott	Project Supevisor, Group III Coordinator	Aspen Systems Corporation
Maxwell Prempeh	Database Administrator	Aspen Systems Corporation
Donald Smith	Scanner Operator	Aspen Systems Corporation
Ellen M. Tanner	Project Supervisor	U.S. Department of Justice/ ASPEN Systems
Vera M. Thorpe	Contract Director	U.S. Department of Justice/ ASPEN Systems
David L. Vetal	Lead Project Manager	Aspen Systems Corporation
Shannon S. Wiggins	Senior Paralegal, Group I Coordinator	Aspen Systems Corporation
Susan Corbin	TechDoc Support	NASA Kennedy Space Center
Carolyn Paquette	TechDoc Support	NASA Kennedy Space Center
Joseph Prevo	TechDoc Support	Prevo Tech

ADVISORS AND CONSULTANTS

John C. Clark	Advisor	National Transportation Safety Board
Vernon S. Ellingstad, Ph.D.	Advisor	National Transportation Safety Board
Jeff Guzzetti	Advisor	National Transportation Safety Board
Thomas E. Haueter	Advisor	National Transportation Safety Board
Tina L. Panontin Ph.D. P.E.	Advisor	NASA Ames Research Center
Max D. Alexander	Consultant	Air Force Research Laboratory
Anthony M. Calomino, Ph.D.	Consultant	Glenn Research Center
RADM Walter H. Cantrell, USN (Ret)	Consultant	Aerospace Safety Advisory Panel
Elisabeth Paté-Cornell, Ph.D	Consultant	Stanford University
Robert L. Crane, Ph.D.	Consultant	Air Force Research Laboratory
Peter J. Erbland, Ph.D.	Consultant	Air Force Research Laboratory
Jean R. Gebman, Ph.D.	Consultant	RAND Corporation
Leon R. Glicksman, Ph.D.	Consultant	Massachusetts Institute of Technology
Howard E. Goldstein	Consultant	Valador, Inc. Consultant
Deborah L. Grubbe	Consultant	DuPont Corporation
Mark F. Horstemeyer, Ph.D.	Consultant	Mississippi State University
Francis I. Hurwitz, Ph.D.	Consultant	Glenn Research Center
Sylvia M. Johnson, Ph.D.	Consultant	NASA Ames Research Center
Ralph L. Keeney, Ph.D.	Consultant	Duke University
Brian M. Kent, Ph.D.	Consultant	Air Force Research Laboratory
Daniel B. Leiser, Ph.D.	Consultant	NASA Ames Research Center

Nancy G. Leveson, Ph.D.	Consultant	Massachusetts Institute of Technology
M. Sam Mannan, Ph.D.	Consultant	Texas A&M University
Robert A. Mantz, Ph.D.	Consultant	Air Force Research Laboratory
Alan C. McMillan	Consultant	National Safety Council
Story Musgrave	Consultant	Astronaut (Ret.)
Theodore Nicholas, Ph.D.	Consultant	University of Dayton Research Institute
Larry P. Perkins	Consultant	Air Force Research Laboratory
Donald J. Rigali, Ph.D., P.E.	Consultant	Valador, Inc. Consultant
Karlene H. Roberts, Ph.D.	Consultant	University of California, Berkeley
John R. Scully, Ph.D.	Consultant	University of Virginia
George A. Slenski	Consultant	Air Force Research Laboratory
Roger W. Staehle, Ph.D.	Consultant	Roger W. Staehle Consulting
Ethiraj Venkatapathy, Ph.D.	Consultant	NASA Ames Research Center
Karl E. Weick, Ph.D.	Consultant	University of Michigan
Douglas A. Weigmann, Ph.D.	Consultant	University of Illinois, Urbana-Champaign
James Wick	Consultant	Intel Corporation
David D. Woods, Ph.D.	Consultant	Ohio State University

Valador, Inc., Board Support Contractor

Kevin T. Mabie	President and CEO	Valador, Inc.
Richard A. Kaplan	Senior Vice President of Engineering and CIO	Valador, Inc.
Neda Akbarzadeh	Contract Support	Valador, Inc.
Christy D. Blasingame	Contract Support	Valador, Inc.
Kim Hunt	Contract Support	Valador, Inc.
Charles M. Mitchell	Contract Support	Valador, Inc.
Karen Bircher	Transcriber	Valador, Inc.
Vicci Biondo	Transcriber	Valador, Inc.
Jeanette Hutcherson	Transcriber	Valador, Inc.
Caye Liles	Transcriber	Valador, Inc.
Jennifer North	Transcriber	Valador, Inc.
Barbara Rowe	Transcriber	Valador, Inc.
Terry Rogers	Transcriber	Valador, Inc.
Susan Vick	Transcriber	Valador, Inc.
Elizabeth Malek	Support Staff	Valador, Inc.
Sally McGrath	Support Staff	Valador, Inc.
Bridget D. Penk	Public Database Support	Valador, Inc.
Ray Weal	Public Database Support	Valador, Inc.
Robert Floodeen	Website Support	Valador, Inc.
James Helmlinger	Website Support	Valador, Inc.
Philippe E. Simard	Website Support	Valador, Inc.
Mario A. Loundermon	Graphic Designer	Valador, Inc.

NASA Press Conference on the Space Shuttle Columbia Sean O'Keefe, Administrator

NASA Facts
National Aeronautics and Space Administration
Washington, DC 20546 (202) 358-1600
For Release
Wednesday, August 27, 2003, 11:02 a.m.

MR. MAHONE: Good morning, and thank you for joining us here in Washington and from the centers across the country at our various NASA field centers. Before I introduce the NASA Administrator, I want to go over a few guidelines for this morning's press conference. We'll begin with questions here in Washington, and then go to the various NASA centers. Please wait for the microphone before asking your question, and don't forget to tell us your name and affiliation. Because of the large number of reporters who want to participate in today's briefing, please limit your inquiries to one question and one follow-up, and, please, please, no multi-part questions. Again, thank you for taking the time to join us today, and allow me to introduce the NASA Administrator, Sean O'Keefe.

MR. O'KEEFE: Thank you, Glenn, and good morning. Thank you all for spending time with us here this morning. Yesterday, we received the report of the Columbia Accident Investigation Board run by Admiral Hal Gehman, and shortly thereafter I had the opportunity to speak to several of our colleagues here throughout this agency to describe those initial findings and recommendations as well as to offer some views of what the direction will be from this point forward. And so if you'll permit me, let me draw a little bit from some of those comments here in the context of today's discussion with you, as well, as we start this and, of course, respond to your questions.

This is, I think, a very seminal moment in our agency's history. Over the 45 years of this extraordinary agency, it has been marked and defined in many respects by its extraordinary successes and the tragic failures in both contexts. And in each of those, in a tracing of the history of that 45 years, there is always an extended debate and discussion of the national policy as well as the focus of the charter and objective of exploration of what this agency was chartered and founded to do in 1958. And I expect that in this circumstance it will be no different. This is one of those moments in which there will certainly be a very profound debate, discussion, and I think a very inward look here within the agency of how we approach this important charter that we've been asked to follow on behalf of the American people to explore and discover on their behalf. In each of these defining moments as well, our strength and resolve as professionals has been tested, and certainly that will be the case in this circumstance, and it has been for these past seven months, to be sure. On February 1st, on the morning of that horrific tragedy that befell the NASA families and the families of the crew of Columbia, we pledged to the Columbia families that we would find the problem, fix it, and return to the exploration objectives that their loved ones had dedicated their lives to. The Board's effort and the report we received yesterday completes the first of those commitments and does it in an exemplary manner. They have succeeded in a very, very thorough coverage of all the factors which caused this accident and that led to this seminal moment, which is marked by a tragic failure. And their exceptional public service and their incredible diligence in working through this very difficult task I think will stand us in good stead for a long time to come as we evaluate those findings and recommendations as carefully as we know how.

As we begin to fulfill the second commitment that we made to the families to fix the problems, the very first important step in that direction is to accept those findings and to comply with the recommendations, and that is our commitment. We intend to do that without reservation. This report is a very, very valuable blueprint. It's a road map to achieving that second objective, to fix the problem. They've given us a head start in the course of their discussions over the last several months and in the course of their investigation, in the public testimony, in their press conferences, in all of their commentary, which has been very, very open in an extremely inclusive process as they have wrestled with the challenges of finding the problems that caused this particular horrible accident. And that candor, that openness, that release of their findings and recommendations during the course of the investigation has given us a very strong head start in the direction of fulfilling that second commitment. At this point, we have already developed a preliminary implementation plan, and we will update that, and we're about that process right now of updating to include all the findings and recommendations included in the report, in addition to those that were released and described very specifically during the course of their investigative procedures. But, again, much as the Chairman, Admiral Hal Gehman, observed throughout the course of those proceedings, what we will read and what we did read as of yesterday was precisely the same commentary that we had heard during the course of their investigative activities and in all of their public testimony that they've offered, which has been considerable and, again, very extensive, exhaustive. So as we implement those particular findings and recommendations, our challenge at this point will be to choose wisely as we select the options that are necessary to fully comply with each of those recommendations. We'll continually improve and upgrade that implementation plan in order to incorporate every aspect of knowing what's in the report, but also so much of what we have determined and seen as factors that need improvement and consistent upgrading throughout our own process within the NASA family. It's going to be a long road in order to do that, but it is necessary in order to fulfill that second commitment we've made to the families.

Now, the report covers hardware failures, to be sure, but it also covers human failures and how our culture needs to change to mitigate succumbing to these failings again. We get it. Clearly got the point. There is just no question that is one of their primary observations, that what we need to do, we need to be focused on, is to examine those cultural procedures, those systems, the way we do business, the principles and the values that we adhere to as a means to improve and constantly upgrade to focus on safety objectives as well as the larger task before us of exploring and discovering on behalf of the American people. But they've been very clear in their statements throughout the report in several instances, repetitively, and in the public commentary that the Chairman and members of the Board have offered following their efforts yesterday after the release of the report, that these must be institutional changes. And that's what we're committed to doing, and that will assure that over time those changes will be sustained, as those process, procedures, and systems are altered in order to reinvigorate the very strong ethos and culture of safety and exploration, those dual objectives that we have always pursued. That is what's going to withstand the test of time if we are successful in this effort, and we fully intend to be. So we will go forward now and with great resolve to follow this blueprint and do our best to make this a much stronger organization. In the process of doing so, it will involve the capacity and capability of all of us within this agency.

This is not about an individual program. It's not about an individual aspect or enterprise of what we pursue. It is about everything we do throughout this agency. There is so much of what has been observed in this report that really has tremendous bearing and tremendous purpose in defining everything we do throughout the agency. And so, therefore, we will approach it and have considered this to be an agency-wide issue that must be confronted in that regard. Now, this is a very different NASA today than it was on the 1st of February. Our lives are forever changed by this tragic event, but certainly not nearly as much as the lives of the Columbia families. This is forever for them. And so that resolve to find the problem which we have successfully done, thanks to the extraordinary efforts on the part of this Board, to fix those problems which we are now in pursuit of as the second commitment, and to return to the exploration objectives that their loved ones dedicated their lives to is something we take as an absolute solemn promise. We have to resolve and be as resolute and courageous in our efforts as they have been in working through this horrible tragedy. The time that we have spent, I think, over the course of since the accident, and certainly well before, in trying to work through those particular questions, again, are focused on institutional change. Since I arrived a little less than a year and a half ago, we have almost completely rebuilt the management team, and so it is a new, fresh perspective in looking at a range of challenges that we currently confront, and those changes have been ongoing of a management team as well as the institutional changes we have implemented and will continue to do in full compliance with this report.

The new management team began I think by evaluating initially on the first day that I arrived here the contingency planning effort that was necessary in the event of such a tragedy. It was the first thing I did on the first morning I arrived at this agency. And in reviewing that contingency plan of how we would respond to a disaster, to a tragic event, which I had hoped and was in the expectation and fond hope that I would never, ever have to utilize, we nonetheless improved that contingency planning effort by doing two things: First of all, reaching back to the Rogers Commission, the Challenger incident and accident, to incorporate in that contingency plan all the changes necessary in order to respond definitively. The second step we went through was to specifically benchmark it against best practices of any comparable organization, of which there are very, very few. And the only one that in my personal experience that I was aware or felt had any direct comparability to the risks and the stakes involved was the Navy nuclear program. And so from that first day, we upgraded that particular contingency plan based on the benchmarking procedures that we followed through with them. We then began a very vigorous effort by late spring, early summer of last year to begin a comprehensive benchmarking procedure against the submarine service as well as the naval reactors community, to, again, pick up best practices as well as to institutionally change the way we do business. And that process is ongoing as it had been a year ago as we continue to make those changes. That was a lesson I learned very specifically in my tenure as Navy Secretary better than ten years ago, was to look at those particular procedures and assure that we have incorporated as much of that, and that was a work in progress that will continue. But, again, the observation by Admiral Gehman and the members of the Board yesterday and replete throughout the report, it is not about changing boxes or individual faces in each of those positions. It is about the longer-term institutional changes that must be made. And, again, to that point we get it. It is about the culture of this agency, and we all throughout the agency view that as something that's applicable to the entire agency, not any individual element thereof. With that, I thank you again for the opportunity to get together this morning and, again, look forward to your questions and comments.

MR. MAHONE: Yes, sir?

QUESTION: Mr. Administrator, Matt Wald, New York Times. There are other organizations that have gone through this kind of change. Most have called for some outside help. I'm tempted to ask if you're read Diane Vaughn's book or called her up or if there are other specialists in safety culture who you would be bringing in at this time to help transform yourself, your agency.

MR. O'KEEFE: I appreciate that. Yes, indeed, we have read Dr. Vaughn's book, and there have been several folks here in headquarters as well as Johnson who have been in touch with her. Dr. Michael Greenfield spoke to her I think initially about four months ago, three months ago, shortly after her testimony before the Columbia Accident Investigation Board's hearings. The primary source of safety experts that we have been trying to encourage and have requested come in to assist with us, again, are from the naval reactors community. This is a very specific set of procedures they follow. It's a very exhaustive effort that they have gone through over a comparable period of time as the span of this agency, in order to upgrade their procedures as a consequence of incidents in the early phases of that program that gave them great pause. And so there's a report that I think was released about a month and a half ago which was the second step in that benchmarking procedure with the submarine service, which is the operational community, and the naval reactors community, which is the disciplinaires, if you will, over the technical requirements side, that we continue to solicit. Beyond that, there are certainly a number of folks that we have invited in and will continue to do so. I spent the better part of four hours last night with Admiral Gehman and most of the members of the Board asking them specifically for the folks that they had brought in as advisers to the Board on this particular question so we may be in contact with them in order to ask for their advice and assistance and contributions in this regard as we implement these recommendations on that front as well. So, yes, we're about that as well.

MR. MAHONE: Keith?

QUESTION: Keith Cowing, Nasawatch.com. Yesterday you read Gene Kranz's inspiring words that were issued to his troops after another accident. And, you know, that was then and this is now. You've got a workforce that has been downsized, bought out, they're jaded by innumerable management fads, and clearly it hasn't worked. I got an e-mail from somebody yesterday saying, "What's he going to do, actually make us—write us on the white board?" I mean, the cynicism is that high. What are you going to do this time that is demonstrably different than all these attempts before it, getting the agency motivated and beyond the cynicism and malaise that seems to have beset it?

MR. O'KEEFE: Well, it's going to require leadership at every level. This is not something that you direct or dictate. Again, in my experience, in my prior life as the Navy Secretary confronted with an incident, an event that really rocked that institution at that time, when I came in, in the post-Tailhook incident, it's not about just walking around telling everybody shape up or ship out. It really takes persistent, regular, constant leadership focus, and I think the folks that we have recruited and are in place now as the senior management team that, again, have been over the course of certainly this last seven months, to be sure, but over the previous year, have been recruited to those capacities specifically for that, are the kinds of people, I think, who not only get it but also are going to be the first start at that leadership objective. Throughout the agency we're going to have to persistently move through that, but I think it is staying with a very set of clear principles and values that we will continue to work through, and it's going to take time, but the time begins right now. And it has been in process, I think, for some period before this, but we will continue to redouble our efforts of that. But it's something that there is no one trick pony at this. It is not something that happens simply because I send out a memo. I'm not a Pollyanna on that point at all. It is something that really requires, I think, constant, unrelenting diligence, and that is another theme that I think comes out very resolutely in the Accident Investigation Board report, which is consistency as well as persistence and vigilance in the leadership direction in that regard. And that's what we are committed to doing.

MR. MAHONE: Yes, sir?

QUESTION: Thank you. I'm Larry Wheeler with Gannett News Service. I want to get back to the leadership question a little bit. I was wondering if you could share with us your thinking about how you motivate your leaders to follow through on this point that you said they get it. Two weeks ago, one of your senior managers had a press conference at Kennedy Space Center in which he, if I understand—if I recollect correctly, he denied that there was a culture in NASA or that he was aware that there was a culture in NASA. And this is the same senior manager who ran the Safety and Mission Assurance Program throughout the '90s, which has been highly criticized by the CAIB. Can you give us your thinking? How do you turn around that kind of thinking?

MR. O'KEEFE: Well, first of all, I think it's a—it's always a challenge to define with common specificity to which all accept of what the term "culture" means. And in my experience, again, as Navy Secretary, there were multiple cultures. There's the culture—there's a Navy culture, to be sure, and a naval service ethos. But there's also a surface sailor culture, an aviator's culture, a submariner's culture. And then, just to really get some extraordinary oomph into it, let's get the Marine Corps involved. They're part of the Navy Department as well. And the common distinctions between those are born of years of history as well as deep tradition. It is also true here. There is every single aspect of how this agency has formed over its 45 years and well before when at the beginning of the last century the NACA was formed to respond to aeronautics challenges at that time that were to be advanced. Every one of the centers, every one of the elements of what you see throughout this agency, can reach back and trace historical roots to each of those individual moments. And so in that regard, there are lots of different ways in which folks respond, but the overall, overarching, overriding NASA culture for this agency overall is a set of principles and discipline in order to pursue safety of program consideration, which has always been the case, in pursuit of those exploration objectives. Those are the kinds of things we need to redouble, and, again, as you define it very specifically in that regard, there is importance that I think we get great clarity of exactly what the definition is, and that's the part we get. There is an overriding culture which must dominate, and certainly we celebrate the history and traditions of every aspect of this agency, much as any other storied agency or institution does.

MR. MAHONE: Yes, sir?

QUESTION: Earl Lane with Newsday. A lot of what the report spoke about on culture, though, I think dealt with attitudes as much as institutions and talked about how lower-level engineers were reluctant to come forward with the concerns. And I'm wondering how you deal with that to get that message out, and is it perhaps time for a stand-down like the Navy sometimes does?

MR. O'KEEFE: Well, it is—to be sure, that's one of their findings and views, is that there is—there was evidence that they saw, even in the course of their investigation, in which reluctance dominated. And I think part of that is—or the two things we've really got to focus on in that direction is, first of all, reinforce that principle, which, again, we articulate regularly and I think we see evidence of all the time. There was a stand-down in June through October of last year in which an individual observed an anomaly on the fuel line for Atlantis. There was a crack on the fuel line that in turn stood down the fleet for that period of four months as we ran that to parade rest and determined exactly what the conclusions and solutions needed to be. So we've got to, again, continually identify that as the kind of behavior we want to encourage, and to the extent we do not see it evidenced or there is evidence in the opposite direction, to assure that we motivate and encourage folks to feel that sense of responsibility. And that's the second part as well, is that there is, I guess, a renewal of the view that I heard expressed best by Leroy Cain, the Flight Director on STS-107, who observed this is all of our responsibility. And so for those who are part of this agency, we have to renew that view, and for those we recruit to that have to have it understand as the first principle that we all must adhere to.

MR. MAHONE: Yes, Tracy?

QUESTION: Tracy Watson with USA Today. Administrator, did you have any hints before the accident that you had this kind of serious attitude and value problem at the agency?

MR. O'KEEFE: Well, to be sure, there's always cases in which there are folks who feel like there are certain aspects of what has occurred in the course of our history or in the course of events that are not as advantageous as others. And so I've had a very open policy of let's communicate whatever those concerns are, let's have an open dialogue throughout the agency on every matter. I've tried to be as open about that to include encouraging e-mails, of which I get lots of from lots of folks. So I've seen, I think, lots of evidence of folks who are feeling, you know, very empowered to offer their view and their concerns. And at the same time, I think it's also evidence of the fact that the process or the systems to permit that discussion isn't happening at every level. So there's two things you can draw from that that I have taken away, which is those who feel that it's necessary to respond in that regard really require other means because the systems may have broken down. So there is certainly some indicator of that, but certainly this was a wake-up call in yesterday's report to see how extensive that communications link that contributed during the course of this mission and operation needed to be improved to deal with precisely that set of problems. It wasn't for lack of people talking. It was for lack of people, I think, coordinating those observations effectively to serve up appropriate decision making about the challenges we were confronting at that time. And I think that's—you know, the upside of that is that there's ample evidence to suggest that folks are feeling like there is an opportunity to communicate and speak. It is also another question, though, of exactly at what level can they do so, and I think that's the point and the communications breakdown that is part of the culture and is part of the observation that was made by the Board, and the findings and recommendations speak to that very effectively.

MR. MAHONE: Yes, ma'am?

QUESTION: Marsha Dutton, Associated Press. The Board made—put quite a bit of emphasis on deadline pressure affecting decision making and even usurping safety, and that this pressure came from on high. And you're up there in the highness here, and I'm just wondering-[Laughter.]

QUESTION: —do you feel some accountability also for this accident since you've frequently made mention of the February 2004 date?

MR. O'KEEFE: Absolutely. I feel accountable for everything that goes on in this agency. That's a part of the responsibility and accountability I think you must accept in these capacities. No question about it. The Board, I think, was very specific in observing that schedules and milestone objectives and so forth are important management goals in order to achieve outcomes, and these are—this is an appropriate and necessary way to go about doing business. But their observation was that in this instance, this may have influenced managers, may have begun to influence managers to think in terms of different approaches in order to comply. And in that regard, I think we have-we've got to take great heart in the point that—and stock in the point that in order to pursue such appropriate management techniques and approaches in order to establish goals, objectives, and milestones, you must also assure that the checks and balances are in place to guarantee that paramount, number one objective, which is safety. In the course of my tenure here, there was not a single flight of a Shuttle that occurred when it was scheduled. Not one. And so as a consequence of that, I think the system has demonstrated the capacity to not only establish what those objectives would be, but also a capacity and a flexibility to adjust to those based on the realities and the pressures that may exist at the time. Now, the fact that that, again, observed by the Board as may have begun to influence a decision on the part of managers was a very important observation and one that we need to assure that, as we make these institutional changes, that we adhere to the same management principles of setting goals and objectives, but at the same time assuring that the checks and balances are in place they not override.

MR. MAHONE: Yes, sir?

QUESTION: Steven Young with spaceflightnow.com. You said a few months ago that you warned NASA employees this report was going to be ugly. I'm wondering: Was it ugly? And what effect do you think it's going to have on agency morale?

MR. O'KEEFE: Well, I think Admiral Gehman's observation, when asked the same question yesterday, was that, no, it's clinical and very straightforward. And there is no question about that. It is a very direct review. It is—again, the whole contingency planning effort that we went through on the prospect that something like this could happen ended up working exactly—better than we could have ever anticipated in that sense. That Board was activated that day. They met for the first time at 5:00 p.m. on that afternoon. So they were immediately about the business of investigating, and in concert with that, there was-there was nary a hint or suggestion that there was ever any point throughout the course of this seven months in which we sought to influence the outcome of that result. What we wanted was an unvarnished, straightforward assessment from them, and we got that. Now, I think the approach that we have talked about among our colleagues here in the agency is that it would be that straightforward approach, that that would be that direct commentary, and then in the process of reading through this, that we'd be deliberate about following—accepting those findings and complying with those recommendations in order to strengthen this organization in the future. I think we've got a very competent, very professional, extremely well considered work that didn't, you know, spare anything in risking, you know, the sensibilities or the emotions or sentiments of anybody in this agency. And that's exactly the way we expected it to be. That's what we wanted it to be. And that's what we asked for them to do. And they did it.

MR. MAHONE: we're going to take one more question here, and then we're going to go to our centers, and then we will come back here in just a few moments. Kathy?

QUESTION: Kathy Sawyer, the Washington Post. Mr. O'Keefe, the report pointed out that the schedule leading up to next February was going to be as challenging and fast-paced as the one that immediately preceded the Challenger launch in 1986. Were you aware of that? Did anybody come to you and say, hey, we're pressing too hard? And what do you feel about that now in light of events?

MR. O'KEEFE: Well, again, the scheduling and the manifest, as it were, the milestones and so forth that were set, was established by the Shuttle Program Office and the International Space Station program management at the request to specifically identify the optimum systems engineering approach for deployment of all of the components of the International Space Station. So they laid out the schedule. They established what those dates would be and milestone objectives would be. And, again, in the course of my tenure, there was not a single launch that occurred when it was actually scheduled. So I think the approach that we adhered to at the time, as well as continue to, I think, is to always set what our milestone objectives and goals, and clearly the establishment of the core configuration of the International Space Station was an objective that our international partners looked to, Members of Congress, all kinds of folks examined and viewed as one of the seminal aspects that needed to be achieved in order to permit then a wider debate of what that broader composition or configuration of the International Space Station could be. But you had to reach that point first. And so in dealing with that, the approach that the International Space Station and the Shuttle Program office devised was that schedule for the optimum engineering configuration necessary to do so, and the operational considerations were factored into it. And, again, at every single interval, at any point in which there appeared to be any anomaly, the flight schedule was adjusted, as it was for every single flight since I've been here. There has not been one that flew on the day on which the launch schedule dictated it should. And that's, again, appropriate, necessary. The stand-down that occurred from June to October of last year was a direct consequence of that. So all those factors, I think the paramount objective that we continue to look to is the safety objective. And, again, that's what the Board report points to, is that the checks and balances really needed to be reinforced, and we need to be mindful in the future that those be in place as we use that appropriate management tool, as they have identified it, of establishing goals, objectives, and milestones.

MR. MAHONE: Sir, we're going to go to Stennis first, so, Stennis Space Flight Center?

QUESTION: Hi, Administrator. This is Keith Darcy with the Times-Picayune out of New Orleans. Can you say how the return to flight process will affect the long-term flight schedule of the Shuttle, and specifically the production level at the external fuel tank plant in New Orleans.

MR. O'KEEFE: I wouldn't speculate at this moment. We've really—we've received the report yesterday, and what we have put together, again, is an implementation plan in its preliminary form based on everything that the Board identified in its public statements and commentary and in the written material they sent to us as preliminary findings over the course of the last several months. Now we have the benefit of the entire report. We're going to update and upgrade that implementation plan. We hope to release that here in the next ten days to two weeks so we can identify what those objectives are, informed by the report. We also have a number of factors and issues that we have identified within the agency that need to be adjusted prior to return to flight. And so as that unfolds in the weeks and months ahead, we'll be able to establish exactly what it will take in order to achieve that. But, again, the paramount, overriding factor in this case is going to be that we comply with those recommendations, and when we are fit to fly, that's when that milestone will be achieved on return to flight.

MR. MAHONE: We'll go to Langley. Langley?

QUESTION: This is Dave Schlect with the Daily Press. I have a question about the Safety Center being developed here at Langley. One of the Board's recommendations is to establish an independent technical engineering authority that would be the sole waiver-granting authority for all technical standards. It would decide what is and what is not an anomalous event and would independently verify launch readiness. How might the new NASA Engineering and Safety Center fulfill this recommendation?

MR. O'KEEFE: Well, we're sorting through that right now. The initial charter of the Safety Center has been formulated. As a matter of fact, Brian O'Connor is there at Langley today, working with General Roy Bridges, the Center Director at Langley, and others in order to begin working through the findings and recommendations of this report and how it will affect how we should adjust the charter of the NASA Engineering and Safety Center. The approach that is identified—and, again, we spent a lot of time last night talking to Admiral Gehman and his colleagues on the board—of exactly how we may consider various approaches here, and they were more in the listening mode of what that could be, because, again, they have not been dispositive about which options we should select other than to, again, reiterate that the recommendations are to, again, establish that independent technical authority for the control of requirements of the Space Shuttle Program. And that's a factor of whether or not that's part of the Engineering and Safety Center, which, frankly, could serve as more of a research and development, testing, trend analysis kind of center and an organization that can come in to regularly examine what our processes and procedures are with a fresh set of eyes all the time, and to have the influence during the course of operational activities to identify cases where they see anomalies that have some historical or trend assessment to it, that's the issue that we've really got to sort through, is whether or not you have both of those capacities inherent in the same organization or whether it should be two separate functions. In the time ahead, very short time ahead, that's, you know, the set of options we really need to sort through in order to comply with those recommendations, which I think are solid.

MR. MAHONE: Next would be the Glenn Research Center.

QUESTION: Mr. Administrator, Paul Winovsky (ph) from WOIO Television. I'm working on an assumption here that there's a backlog of science waiting to fly once safety concerns are handled. How will you go about prioritizing what flies in the payloads. For example, the combustion experiment developed here was destroyed on the last mission. Is the pipeline full? And how will you prioritize what goes into space next?

MR. O'KEEFE: That's a very good question. There are two approaches we're going to use to this. The first one is that if you go to the Kennedy Space Center today, the payload processing facility and all the International Space Station program elements that have arrived are stacked up in sequence and are being tested and checked out for deployment at the—as soon as the resumption of flight occurs. So there will be not a lot of confusion about exactly what that sequence will be. It's going to follow the pattern that, again, fits that optimum systems integration, engineering strategy that is best for the production—construction of the International Space Station to reach the core configuration. The science component will be drawn from an effort that we conducted through last summer and early fall, not quite a year ago, which was an effort to prioritize what the science performance will be aboard the International Space Station. We had a blue-ribbon panel of external scientists representing every single scientific discipline who came in to specifically organize what that priority sequence is. Until that time, it was a collection of priorities from every discipline, all of which ranked number one. And so when everything is number one, that means nothing is number one. So what the Board—what was referred to the re-map effort did last summer and fall that organized that prioritization set actually had a rank order that began with the number one and moved through by sequence, two, three, four, and five, and so that is the sequence in which we will organize the Space Station scientific objectives from this point forward, because that is the primary source of all the scientific micro-gravity experimentation that will be carried out in the future, is aboard the International Space Station. So we'll adhere to that blueprint very carefully.

MR. MAHONE: Sir, we have a question at the Kennedy Space Center.

QUESTION: Mr. O'Keefe, this is Jay Barbee with NBC News. In talking with the workers here and in Houston, I'm finding they are very encouraged with you at the helm. They believe at this time in NASA's history that you are the right man for the job. Now, they're encouraged by your honesty and your willingness to admit NASA's mistakes. But their concern is still communications. It has been stifled, and many with safety concerns have been intimidated into silence, in fear of losing their jobs. Can you today reassure any NASA or contractor employee if they speak up with safety concerns, even to members of the press, that they won't be fired, that they won't suffer setbacks in their careers?

MR. O'KEEFE: Absolutely. We get it, and that's what message has been transmitted and understood by every single leader and senior official in this agency, is that we need to promote precisely that attitude. So the answer is absolutely, unequivocally yes.

MR. MAHONE: Johnson Space Center?

QUESTION: Gina Treadgold with ABC News. Sir, you've said you take responsibility. Do you plan to step down as a result of this? Or do you feel any pressure to resign?

MR. O'KEEFE: Well, certainly I serve at the pleasure of the President of the United States, and I will adhere to his judgment always on any matter, including that one. And so, no, there is nothing that in my mind transcends that requirement, and I intend to be guided by his judgment in that regard.

MR. MAHONE: Marshall Space Flight Center?

QUESTION: Shelby Spires with the Huntsville Times. Given that the Board suggests that the external tank be blown with no foam loss, and engineers say this isn't possible, is NASA prepared to redesign the tank without foam and go to Congress to ask for the money to do this?

MR. O'KEEFE: We'll see. I mean, there may be an option down the road in which will be selecting to do something along those lines. Don't know. But the approach that I think very clearly articulated yesterday by the Accident Investigation Board membership was that there—just based on the current configuration and the safety considerations, the issue of foam loss per se is not something they find as being totally disqualifying. What they do find to be a problem and what was a contributor, to be sure, a causal effect based on what is the likely condition of what occurred in that first 81 seconds, was the departure of the bipod ramp from the—insulation from the external tank which struck the leading edge of the orbiter. That's the part that already we have eliminated as a factor that's going to be heating segments around that area to act as, instead of the insulation, so you will not find an insulated bipod ramp at that point on the external tank in the future. Exactly how much further that's going to need to go, that's one of the things that I think in the report they said very specifically we ought to aggressively develop a program to eliminate departure of any debris of insulation coming off the external tank. And that's the part that has already been tasked and that Bill Readdy, as part of the return to flight effort, has already charged our external tank management team over to look at. So we'll be looking to the results of that view, and all the options are on the table. We'll see what comes.

MR. MAHONE: We'll take two more questions from the centers, and then we'll come back here to headquarters, and we'll go to Dryden.

QUESTION: Mr. Administrator, this is Jim Steen with the L.A. Daily News. I was wondering if the folks at NASA are looking at the possibility of bringing Shuttle landings back to Edwards Air Force Base as a safety precaution. And I also wanted to know what role, if any, that Dryden Palmdale facility will have in your return to space operations.

MR. O'KEEFE: Well, in terms of the option of landing at Edwards, to be sure, that is an option we've always exerted and used anytime the weather conditions don't permit a return to the Kennedy Space Center in Florida. So we'll continue to do that, and anytime there is a condition which would dictate that we land on the west coast, that's exactly what we'll do. The challenge thereafter, once landing at Edwards, is to transport the orbiters across country, and that's something that, again, one of the quality assurance and risk management challenges of dealing with the Shuttle orbiters, is the more you touch it and the more you fiddle with it, the more likely is the prospect that you can damage it. And every time we do that, it gets more and more difficult to sort with, because, again, it's always launched from Cape Canaveral at the Kennedy Space Center. So, yes, Edwards will always be an option, and it's one that we are not deterred by that challenge if there are factors that dictate the consideration of landing there. In terms of the Dryden Center, there is no question that the flight operations activities that are continuing to go on there that cover a wide range of different supporting efforts that we go through for unmanned aerial vehicles for the Defense Department, for a wide range of different programs, no question we will continue to see that activity unabated there. And as circumstances dictate, there may be further flight test requirements that we would conduct there in support of return to flight activities for the Shuttle.

MR. MAHONE: We're having some technical difficulties at JPL, so we'll come back to headquarters. And, Mr. Administrator, if I could start off with Bill Harwood, we'll start with Bill.

QUESTION: Thanks, Glenn. Bill Harwood, CBS News. Well, just looking ahead to 114, I think the previous questioner was probably asking you about overflight to land at Edwards versus Kennedy, just for the record. My

QUESTION: Looking at 114, are you committed to not flying that flight until you have both a tile repair capability and an on-orbit RCC repair capability, realizing that it's the RCC that's obviously the long pole in the tent right now.

MR. O'KEEFE: Well, there's no question. The report very specifically divides the findings and recommendations into those areas which must be complied with prior to return to flight. We intend to take that with absolute conviction, no doubt about it, and we're committed to doing that. Among them is the point of an on-orbit repair capacity, and that's the range of options, because it could cover a wide set of circumstances. We've got to look at what is a responsible set of options in order to provide that repair capacity, and those are the things we're looking at right now as weighing all those options to figure out what's the most appropriate course on that. But it's one of the requirements within the—or what we view as a requirement within the report as a recommendation that must be complied with prior to return to flight, and we intend to adhere to that.

MR. MAHONE: Mike?

QUESTION: Mike Cabbage with the Orlando Sentinel. One of the things the Board made pretty clear in their report was that they have a concern that after you implement cultural changes, that NASA will sort of backslide the way that it did after the Rogers Commission. What can you do to make sure that cultural changes you put in place now will still be in effect 5, 10, 15 years from now?

MR. O'KEEFE: That's a point that we really have spent a lot of time. Again, last night the Board was generous with their time for several hours in sorting through, and that dominated the discussion in many ways, and they were consistent and repetitive in their responses to this, which is it can't be personality dependent. It's got to be a set of institutional changes that will withstand any change in leadership and management and so forth, and it's got to be a set of principles and values that are reiterated regularly that then become institutionalized. So, I mean, the measure of that is going to be, I think, over time if we see a real change in the mindset. But, importantly, I'm very mindful of the observations that several have made in the public, which is, yes, we've heard this before, and, yes, they've pledged to do these things. No question, that's a very clear criticism. All I can offer is I wasn't here at that time, and a lot of folks who were in senior management and leadership positions were not in those capacities at that time either. So we've got to move forward with the objective of adhering to what the Board has said, which is to be sure that it not turn on just the individual personalities involved, but instead become an institutional set of values and disciplines that will withstand that test of time. And that's going to be the real measure. It's something that, again, the jury's out. We'll see how far that goes, and I'm certain, I'm absolutely certain that you will be the judge of that.

MR. MAHONE: Frank?

QUESTION: Frank Sietzen (ph) with Aerospace America. Among the Board's report—recommendations yesterday was that the Space Shuttle be replaced as soon as possible. Admiral Gehman expressed his concern that there wasn't at least a design candidate on the drawing boards, he said. Given that, are you looking afresh at when and under what circumstances to retire the Shuttle? And what kind of mix of systems do you propose to do so with?

MR. O'KEEFE: Well, it's exactly one of the charges that is now slightly over 24 hours old that we do, so maybe I could—if I could ask you for another hour or two to get through that analysis, it would be helpful. But we are trying, I think, to sort through exactly what the implications would be there of a range of alternatives. The Board—what I read and what I saw in the report was very specific in saying that if there is an extension of the Shuttle operations beyond the beginning of the next decade, it must be recertified. And so establishing what those recertification requirements would be is part of what I read also to be one of their recommendations and findings, that we establish exactly how we would go about doing that, so that you make those judgments today so that later, when those decisions are made by all of our successors, that there not be just matters of convenience taken at the time to determine what the recertification requirements would be. So that's an aspect we've got to think about now in anticipation of tomorrow. And, finally, the approach we have pursued as a consequence of the President's amendment to last year's program submitted in November of last year to, as part of the integrated space transportation plan, is to begin an effort for a crew transfer vehicle that is focused on crew transfer capacity as a supplement to that capability that we have used for both crew transfer as well as heavy-lift cargo assets on the Shuttle. And so we're pursuing that. There is a very aggressive effort right now to be very specific and very deliberate about a very limited number of requirements, and I think we have followed through on what the Board observation on that point is, which is to make sure that those requirements are very straightforward and not so extensive that it requires either an invention, a suspension of the laws of physics, or the use of what Admiral Gehman referred to last night as a material referred to as "unobtanium" in the effort of trying to put together the alternative. So make sure it's realistic, is something that is technically doable now, and that is the set of very limited requirements that we have put together for a crew transfer vehicle that is the orbital space plane configuration. So we'll see what the results from the creative juices and innovation of the industry will be here in the weeks and very short months to follow.

MR. MAHONE: Debra?

QUESTION: I'm Debra Zabarenko. I work for Reuters. You've got a lot of big challenges contained in this report but for safety concerns, you can go to safety experts and systems analysts. For organizational problems, you can go to the folks who are expert there. But one thing the report said that NASA needs and does not now have is the kind of urgent mission that it had during the Cold War years. Are you going to be looking to the White House, to Congress? Where are you going to go for guidance on dealing with what seems to be one of the biggest underlying problems that the report remarked on?

MR. O'KEEFE: Absolutely. Again, as I mentioned at the very opening of my comments here this morning, in each of these events of great success and great tragedy it has been always attendant thereafter with a very extensive national policy debate. And sometimes that national policy debate has resulted in a set of objectives that are identified, and in other cases it has been unsatisfying. Our anticipation is this next national debate coming is one that we hope and we certainly plan for it to be a satisfying result. And how that sorts its way out between our colleagues within the administration as well as in Congress, and certainly the general public, is going to be a question that in the time ahead—and Congress has—the committees of jurisdiction have planned a set of very aggressive, very extensive public hearings in the weeks ahead that I expect will spark that debate. And the answer to your question I think will be resolved from that set of policy debates that will be shortly coming.

QUESTION: Do you agree with the report's estimation that that is something that NASA doesn't have right now, an urgent sense of mission?

MR. O'KEEFE: Nothing comparable to what drove us as a nation with the threat of the prospect of thermonuclear war by a bipolar, you know, opponent on the other side of this globe that existed in the early 1960s. No, we don't have anything nearly as earth-shattering in that. Thank God.

MR. MAHONE: Frank?

QUESTION: Frank Moring with Aviation Week. Another thing that the space program needs is money, and there's been some bad news lately from the—most recently from the Congressional Budget Office. What is your assessment of the budget prospects for the space program as this national debate gets underway? And, also, what do you see as the cost of meeting—in rough terms, of meeting the Gehman recommendations?

MR. O'KEEFE: Again, I would not even speculate on what

the national debate that will occur over the federal budget proposals would yield. That's going to be in the time ahead as well. That's happening currently. I think you pointed that very succinctly. As a member of this administration, we certainly are going to be valuing and evaluating those particular consequences in the context of what is necessary to proceed forward with compliance with these recommendations and what resource requirements we'll have. And certainly that debate will continue and will go on inside the administration as well as within the Congress. And so the results of that will be known in due time. In terms of what it's going to cost for us to implement, again, if you give me another hour on top of the one that I asked from Frank to figure out what the cost is beyond just evaluating a report 24 hours old, we might be able to get back to you. But at this juncture, I wouldn't even put an estimate or a price tag on that at this juncture.

MR. MAHONE: Okay. Brian?

QUESTION: Brian Berger with Space News. One of the points that the report made is that NASA has exhibited a tendency to bite off more than it can chew, have more ambition than budget. Can you fix Shuttle, can you complete Station, and undertake an ambitious effort like Project Prometheus on the same schedule that you've laid out so far?

MR. O'KEEFE: Well, again, this is not a new observation, as your point. It's one that was very clearly driven home to me in the course of my confirmation hearings, as a matter of fact, a year and a half ago by several Members of Congress, that we have had a history of trying to do too much with too little or not prioritizing sufficiently. And there are several different ways to go about looking at technology management. One is what is commonly referred to within, I think, the technology sector kind of approach, which is put a lot out there and let a thousand flowers bloom. And the ones that do come up and the ones that are considered to be of greatest value, those are the ones you pursue. Well, maybe that's the closest comparable management approach of technology that was pursued within this agency in the past. Upon my arrival here, in fairly short order we established that there were three mission objectives: understanding and protecting the home planet, exploring the universe and searching for life, and inspiring that next generation of explorers. And if it doesn't fall in those three mission categories, it doesn't belong here—not because it isn't a neat thing to do or would be interesting or whatever else. So in the course of the past year-plus, we've been really going through the process of winnowing down what are the programs that really participate and contribute to those three mission objectives very succinctly, and those that are neat ideas and good things to do, well, we try to find some other home for them somewhere else, but not here, because we're trying to be very disciplined and very selective about what we do. We've got to continue that effort and be more deliberate about it in the future, I think, in finding those efforts that fall within those categories. In terms of the very specific example that you cited of Project Prometheus and developing power generation and propulsion capabilities, that is something that comes right into our wheelhouse of the kinds of things we need to be doing, and it marks the technology kind of prowess of this agency that it's been known for four decades, which is to overcome those technical obstacles in order to achieve the next set of exploration objectives. And so that is there in the program. It's fully financed. You know, the money that's required and the resources necessary in order to do so have been approved within our administration, have been offered to Congress for their consideration. And we're underway with that effort because that's one of the serious long poles in the tent to pursuing future exploration objectives. And so that one fits very, very precisely within those three mission categories, without reservation or equivocation.

MR. MAHONE: Mark?

QUESTION: Thank you. Mark Carreau (ph) from the Houston Chronicle. I think I have a question and a follow-up, if that's okay. What do you contemplate—

MR. O'KEEFE: How can you have a follow-up when you haven't heard the answer yet? [Laughter.]

MR. O'KEEFE: Sorry. Go ahead.

QUESTION: Okay.

MR. O'KEEFE: Pardon me. I didn't mean to be flip.

QUESTION: That's okay, sir. Thank you. What do you contemplate doing or saying to your managers and workforce to explicitly uncouple schedule pressure to build the Space Station from the Shuttle recovery?

MR. O'KEEFE: Well, let me take the first part of that because I'm not sure—Shuttle recovery, do you mean return to flight?

QUESTION: Yes, sir.

MR. O'KEEFE: Okay. I'm sorry. Again, the point that I think was very clearly enunciated in the report that resonated with me is that this may have begun to influence the program manager's view of how you proceed to meet milestone objectives. Again, that's a useful, very valuable management tool that has to establish goals. It's a leadership principle. You have to have folks—again, it's part of the point that was raised in several other questions earlier, too, that you have to have goals, you have to have objectives, you have to enunciate what they are and when you intend to achieve them. That's part of any other aspect of what we do. The really profound point, I think, that the Board raised was that there was some mixed signal, miscommunication of that point, of which was more dominant. And so the checks and balances must be established, and they were very clear on that point repetitively in their—in every part of the report, that what we need to do is establish institutionally an ethos, a set of values, a discipline that really encourages folks to have an open communications loop, to express when they believe something to be not safe at that time to proceed with. Now, that may not rule the day. It may not be, well, in that case, since you've simply asserted it, it must be so. There really is a case in which we've got to demonstrate that it is safe, and that's a very different approach that now the burden of proof, I think, has to be reiterated in that direction as well. So as we move through this, establishing what those institutional checks and balances will be, and part—I think the answer to that one in this particular instance is assuring that that communication loop is very open and that there is resolution to each of the objectives or objections heard so that everybody is heard and that crisp decisions are made thereafter in terms of how to serve it up and follow through from there. Once you've heard it, your follow-up?

QUESTION: Yes, my follow-up is: Do you need the flexibility to deal with the Russians, contract with the Russians, or whatever, to give you this time so that you have the supplies aboard the Station? And how do you deal with your international partners' expectations of having their equipment aboard, that there's commitments made even above your level to try to do that that you have to respond to? And I'm wondering how you deal with the workforce, but also deal with that issue.

MR. O'KEEFE: That's a very important question and one that we've taken extremely seriously. But I'm very, very impressed with the response of our international partners and their capacity to really act like partners in an International Space Station effort. This is an endeavor pursued by 16 nations, and they have responded very, very definitively. So in working through all those issues, as recently as a month ago I met with all the heads of agencies of the International Space Station partnership, and we worked through all of the challenges that, as we sort through the months ahead and anticipate return to flight, that there be a lot of obligations and commitments. We're going to continue to look to them and to us to honor as we work through this. And we have—I've got a very clear understanding with them, and they have been really just exemplary in the manner in which they've done that. So I have—we've all taken a part of the responsibility of this, and we all view this as a partnership challenge. This is not something which they say, you know, to the United States, "What are you going to do to help us out today?" No. They've been very forthcoming in terms of their approach and accepting their piece of the partnership responsibility in doing this. It's been commendable.

MR. MAHONE: We're going to go right here.

QUESTION: Mr. O'Keefe, Peter King, with CBS News Radio. Yesterday, we read the report, of course, and there were lines in there that expressed pessimism that NASA would be able to change, and in an interview after the report was issued, Admiral Gehman told my colleague, Bill Harwood, that some are or will be in denial about the changes needed and the flaws in the system. What message have you or will you send to those particular people at NASA?

MR. O'KEEFE: Well, again, and this is reminiscent of some of the earlier comments that we have shared here, this is tough stuff, and we shouldn't be a bit surprised when engineers, and technical folks and all of the rest of us as colleagues here in NASA act like all other human beings doing, which is, when you hear something, it really is tough, and it's hard to accept that it takes a little effort to work through it. And that's exactly what we've been really endeavoring to do in the last few months here is just kind of steeling ourselves for what we asked for, which was an unvarnished position, a very direct report, take off the gloves and let us know what's wrong. We didn't ask Admiral Gehman and his colleagues to tell us what's so right about this place. I mean, that's something that has, you know, again, been widely viewed as "over-thought" of. We got that point. The issue is we really wanted to know, in a very clear, distinctive way, exactly what they thought was flawed about the way we do business, what caused this accident, what were the contributing factors, all of the other things that may go to it, and they complied with that, and they did it with great skill, and it could—I can't imagine what the deliberations among the Board members must have been over these past several months. Trying to get 1[illegible] very, very smart, very thoughtful, very Type A people to come to closure on a set of views could not have been an easy task. And you can see that they really worked through some very differing approaches that ultimately came to a very crisp set of conclusions. So I think that's something we've got to work through, and this is part of the process we've been engaged in for the last few months is kind of strapping ourselves in for the fact this was going to be an unvarnished view and a very clinical, direct, straightforward position, and it has been. We got what we asked for and there's no question that we now need to go about the process of all of the steps that it takes in order to accept those findings and to comply with those recommendations, and that's a commitment we're not going to back off of.

QUESTION: Todd Halvorson of Florida Today. Now that you have had the CRIB report for 25 hours, and given the fact that you've gotten a good head start on your return-to flight activities, what are your thoughts now about your ability to make that March through April window for return to flight next year, and what are your thoughts about when you can get to core complete?

MR. O'KEEFE: Well, the answer to both is we'll see. From the technical hardware standpoint, all of the assessments we've gone through here in the last couple three months are there are a number of options that would certainly permit an opportunity after the new year to look at a return-to-flight set of objectives, and we've reviewed those with the Board. They're aware of that activity, and that's underway. The larger questions I think that are raised in this report too, that deal with some of the management systems, the processes, the procedures, the, again, the culture of how we do business, we really have to set this bar higher than what they did, what anybody would do. The standards that we are expecting of ourselves, we need to be our toughest critics on that. And so those are going to be a little more difficult to I think assess in terms of a calendar or a time line in terms of when they're done, and instead I think it's going to be a case where, when we've made the judgment that we are fit to fly, that's when it's going to occur.

Now, we're not going to just do this in isolation or a vacuum. We've asked a very impressive group of 27 folks who are part of the Tom Stafford and Dick Covey's Return-to-Flight Task Group to help us work through those options and assure that we're not just, you know, drinking our own bath water on this or singing ourselves to sleep on the options we love the most, you know. It's a case where we really want to lay out the full range of things we're going to do and have their assessment of whether they think that passes the sanity check. And that group of folks, I would suggest to you, if you haven't had the opportunity to do so, to look at the varied backgrounds that those 27 people bring, not only the technical and engineering and I think smart folks on the hard sciences side, but also a lot of management experts, a lot of folks who have dealt with large organizations, dealt with culture change. Walter Broadnax, who was the deputy secretary in the last administration for the Health and Human Services, is a member of that. He is now the president of Clark Atlanta University. This is a guy who has been through several different organizational shifts working for the State of New York, working for the last administration at HHS, and so forth, dealing with very large organizations, understanding management culture change requirements. Richard Danzig, who was the last Secretary of the Navy in the last administration, as well, was a member of this. Ron Fogleman, who was the Chief of Staff of the Air Force, who really set some standards in the Khobar Tower incident over what accountability standards should be adhered to within the Air Force, is a member of this group. So if you work through every one of those, what you find are folks that aren't just—or I shouldn't say "just"—it's not dominated by a group of folks looking strictly at the engineering-hardware kinds of challenges. It's also looking at these larger systems process changes, and those are the kind of folks that have been added to this, including a number of academics who have written about it, and thought about it, and worked through it like Dr. Vaughn and others, colleagues of hers, who have really looked at organization change issues and, in turn, are going the help us really think through this. And they've been there, done that, gotten several T-shirts and recognized lots of tendencies on the part of organizations or institutions to select options that may or may not be more or less convenient. They're going to be good sanity checkers as we work through that, and those are the kinds of people I think that their judgment will be invaluable as we work up to that inevitable return-to-flight milestone.

MR. MAHONE: And the complete list of those members are at www.nasa.gov. You can go to that and find their bios and so forth. A question right here.

QUESTION: Jim Oberg with NBC. I'd like to ask a question on culture and the issues of intellectual isolation of the NASA community from the outside world. The Board and other people have mentioned words, from the Board example, self-deception, introversion, diminished curiosity about the outside world, NASA's history of ignoring external recommendations. These are some pretty serious charges, and people have seen evidence of it. The Board did and other people have mentioned it, too. You have a situation where people who are here now are almost hunkering down into a siege mentality, where outside critics are cold and timid souls whose views should be ignored. How can you get the people to become what the Board wants you to be, a learning organization like that, when many of the same people who have been immersed in this culture for all of their working lives are the ones designing, developing and judging the success of your recovery process?

MR. O'KEEFE: Again, you have accurately recited what are the findings of the Board and their overarching view of what they have deemed or viewed to be the culture within the agency. The first step in any process is to accept the findings and to comply with those recommendations, and I think Admiral Gehman had been very fond of saying to the Board, "T equals zero," zero meaning anything that happened after February 1st is not something they're looking at. They're really focused on examining that. Well, to NASA today, T equals zero starts today, and we've really got to work our way through accepting those findings and complying with those recommendations and that will be the beginnings I think of sorting our way through these larger institutional challenges. I think the questions and comments and observations made by your colleagues here, as well as in my statements at the opening of this, suggest we've got to being that process and work with what is a very professional group of folks throughout this agency, who I think can step up and accept those responsibilities, and we all have, in working through this, and recognizing that this is a institutional set of failures that must be addressed. I don't see that the reticence on the part of any individual in this agency is going to be a setback in that regard. We've just got to work through that very methodically, very deliberately, very consistently, and employing a principle of the United States Marine Corps that I always found to be pretty pointed, which is "repeated rhythmic insult." If you always say the same thing, and you mean it, and you keep going at it, and you stick with that set of principles and values and discipline, it's going to resonate in time, and in time means sooner rather than later in order for us to really reconcile and come to grips with these findings, and accept them, and comply with those recommendations.

MR. MAHONE: Question, here.

QUESTION: Bill Glanz, Washington Times. I just want to find out what your gut reaction was while you were reading that part. For instance, were you appalled at some of the decisions that the program managers made, and also, when you were reading it, did you have any "holy crap" moments? [Laughter.]

MR. O'KEEFE: I've had so many of those since February 1st I can't count them all any more. Again, this was not a surprise. Among the emotions that I felt in reading through this, surprise was not among them, because again, they were very faithful in what they said they would do. Admiral Gehman and every member of that Board were very, very clear in the course of their proceedings of saying, "What we're telling you and what we're inquiring about in these public hearings is what you will read in this report." Very explicit about that. They never walked away from that point. Again, talking about repeated rhythmic insult, that was, a repetitive commentary that they followed through on and did precisely what they said they were going to do. So in reading through this, and again, our approach from day one, from the 1st of February on, was again to be as open as we possibly could conceive of being, release all of the information for everybody to see what was going on. So reading lots of the discourse and back and forthing and communication that went on that are now faithfully repeated in the report, was not the first time I'd read them, because we released them a lot here, and they talked about them a lot in the hearings, and so in the course of this, I think the terminology they used was very consistent with what I heard in the course of all those public hearings. And after 22, 23 hearings that lasted on average three-and-a-half to four hours each, that was a lot of volume. So really, distilling all of that and coming up with a report that was as succinct as this is, that it was only 248 pages by comparison to the thousands of pages of transcripts from all those hearings, was really the part that I found most impressive, was they were able to distill this into a very pointed set of findings and recommendations. But surprise was not among them, and there was nothing that I saw there that they had not previously talked about. They were very, very conscientious about following through on that commitment and they did what they said they were going to do.

QUESTION: [Off microphone] — appalled by some of the decisions that the program managers made, you know, being pressured by the long schedule, and all the missed opportunities that they mentioned in the report?

MR. O'KEEFE: Again, I mean the course of this. There have been countless hearings that I've been a witness at. There have been lots of different opportunities where we have gotten together among your colleagues in the press corps to discuss several of the events as we've walked through this in the last seven months. At each one of those there were plenty of cases in which you said, gosh, how could this have happened? But there's no question. None of it was a new revelation in that regard. It has been all by degrees over time in these last six, seven months, you know, rolling out and laying out in ways that we have really seen institutionally as well as with the hardware, as well as human failures were that led to this. By all means, they are a guidepost to figuring out exactly how you improve that communicate net, sharpen the decision-making process that informs, decision-making that includes all the information that's necessary to make those kinds of judgments at the time, and I think that's exactly what we saw come out of this.

QUESTION: Chris Stolnich from Bloomberg News. I was just wondering if you could describe what you believe the goals for manned space flight are in the wake of this report, and how or if they should change?

MR. O'KEEFE: We are, and have always been, dedicated to exploration objectives which in some instances require a multitude of different capabilities, to include human intervention. What we've laid out is a strategy, a stepping stone approach in which we conquer each of the technical and technology limitations as we pursue greater opportunities. Calls for a sequence of capabilities, which we see playing out right now. In early January we're going to see two Rovers land on the planet Mars, and it will follow, as it did several other missions that preceded this, in order to collect and gather the information and the knowledge necessary to inform the opportunity for human exploration at some point. And as we prepare those capabilities to proceed, we have a more complete knowledge of precisely what it is we're going to encounter, and what will be garnered and gathered from that set of missions and those that will follow, which are robotic, will inform that decision making and inform that understanding and judgment about exactly how human exploration thereafter could be permissible. The second phase of it though is an important one, because your question I think also speaks to the immediacy of instances and cases in which human involvement is imperative in order to preserve capacity.

Today there's a spirit of debate that's going on, that again I commend you all for having covered rather broadly, of exactly what is going to be the service life of the Hubble telescope. Just launched on Monday the SIRTF infrared telescope that will be a companion to Hubble, if you will, for all the infrared lower temperature observations and readings that could be observed by that imagery. But recall that the history of Hubble—which I have not seen very extensively discussed in all the coverage of the current debate about how long Hubble should be operational and what servicing missions are necessary—the history of that was your predecessors 10 years ago roundly viewed the deployment of that capability as a piece of $1 billion space junk because it couldn't see. The lens needed correction. It required a Lasik-equivalent surgery. And the only way that could be done was by human intervention. So in 1993 when that mission was launched to correct the Hubble, that was done successfully, and the only way it could be done was because a human being, several of them, spent many, many months training to be prepared for making those corrections on the spot, and for every contingency that could arise as you work through it. It was nothing we could do, adjust from the ground. The last round trip flight of the Columbia in March of 2002 was to the Hubble again to service it, to install new gyros, to install an infrared camera, to upgrade a number of different factors to it that improved its capacity by a factor of 10, according to all the astronomers who observed this, and they are just elated over the quality of what has come back from this. And yet it turned out that the primary human characteristic that was so important on that mission was embodied by a gent who will be joining us here in about a month, or a matter of fact, weeks—I'm losing track of days here—Dr. John Grunsfeld, who will be our Chief Scientist, and relieving Dr. Shannon Lucid, as she goes back to Johnson Space Center, as our Chief Scientist. He was on that mission. He's an astrophysicist, got all kinds of incredible scientific background. But his primary human characteristic trait that was most valuable proved to be that all the instruments for adjustment on the Hubble telescope are on the left-hand side.

So rather than having, like many of us—righties are stuck with the problem or reaching around the front of your face with a catcher's mitt equivalent capacity to adjust things and a big bubble over year head, trying to see what's going on—his primary human characteristic that was most valuable is he's a lefty. He's now referred to as "the southpaw savant." But it was a human characteristic that made those adjustments, that made that capacity work in a way that we never imagined possible, and that 10 years ago we were prepared to write off as garbage. And instead today, it's revolutionizing not only the field of astronomy, but also informing all of us as human beings of the origins of this universe, its progression over time. It has changed the way we look at everything. In the last 18 months it has been a remarkable set of discoveries that have emerged from that capability that would never have been possible were it not for human intervention. So those are the two areas we really have to focus on, is recognizing how we can advance the exploration opportunities by being informed as deeply as we can through a stepping-stone approach of always developing those capabilities and technologies that then permits the maximum opportunity for human involvement, and then in those cases in which nothing else will do than human intervention and cognitive judgment and determination, and making selections that only humans can do, where do you use those judiciously in order to avoid the unnecessary risk that's attendant to space flight for only those purposes and causes that are of greatest gain.

MR. MAHONE: Right here.

QUESTION: David Chandler with New Scientist Magazine. One thing that the Board explicitly avoided talking about, not because they didn't think it was important but because they didn't see it as their role to do, was issues of personal accountability. I'm wondering what your thoughts are on whether it is your role, and for example, people within the agency who failed to follow NASA's own rules. What kind of a message about the importance of safety will be sent if there is no personal accountability or personal consequences for people who didn't follow your own rules in this mission?

MR. O'KEEFE: Well, first and foremost, I am personally accountable, myself, for all the activities of this agency. I take that as a responsibility and I do not equivocate on that point. I think it is absolutely imperative that we all view our responsibilities, and that one is mine. The approach I think that is absolutely imperative to follow through with in this institutional change that we've talked about here, and had lots of different comments and observations about, that the report covers in depth, is that you must select folks in leadership and senior management capacities who understand exactly what that set of institutional change requirements are. So rather than saying I'm going to remove so-and-so, it's more a case of, I need to appoint folks who understand that. At this juncture of the four space flight centers that have any specific activity over Shuttle operations, International Space Station, et cetera, so among the 10 centers there are four that specifically and uniquely deal with space flight operations. The longest-serving tenure center director was appointed in April of 2002. He is now the elder statesman among them. The rest have been appointed since. And those are the folks who are, in my judgment, the kinds of leaders who very clearly understand, they get it, that this is about institutional change. Those are the folks that I fully anticipate are going to be the ones who will be the folks who will carry this out and accomplish the objectives we talked about here today, and they in turn select those managers, engineers, technical folks who share that same ethos. So as we work through this we've got to be very, very deliberate in relying on the judgment of individuals who have committed to those objectives. And I encourage you to just kind of scan through the senior leadership as well as the senior positions here throughout the agency that have been conducted, and you'll find a rather significant new management team in those capacities, new leadership team, and all of them share the view that I've just talked about here, which is this is an institutional challenge which is greater than any one of us individually or even collectively. It's about the longer-term values, discipline and principles that this agency should adhere to, and they share those goals and views.

MR. MAHONE: Last question.

QUESTION: Steven Young with SpaceFlightNow.com. I'm wondering if you've actually read the report cover to cover, or whether you intend to do that, and whether you would make it required reading for NASA employees and contractors?

MR. O'KEEFE: I think I don't need to direct that it be required reading. I haven't run into anybody in this agency, any colleague in the organization who have not felt that this is something they want to read in its fullness. So I think no amount of direction from me is going to make a difference. People are doing it because they view that as a responsibility, that we all need to view this is a responsibility that all of us must carry. I have read through it as of—again, it was a long day yesterday, but I started when Admiral Gehman dropped it off at 10 o'clock yesterday morning, so I had about a one hour head start from his press conference. And again, what I found in reading through it was that it remarkably patterns exactly what they said in all their public statements. So in many respects I was reading the same things I've been hearing, in listening to those public hearings, listening to their public comments. I've got to go back this weekend and read every single word for its content to do that right, but in reading through it briskly, as of yesterday morning and then last night after we left them, after a long session with them, had a chance for several hours to read through it again. But again, it struck me immediately as being remarkably close and right on to what it is they've been saying. So there were no surprises in that regard. But this weekend, you bet, word for word, from the first page to the last word on page 248 is what I intend to read. I don't need to instruct that anybody in the agency do that. I'll bet everybody is, because I think this is the sense of responsibility we all need to share, and I think that doesn't need to be directed by anybody.

MR. MAHONE: Mr. Administrator, thank you very much, and thank all of you for being here today. [Whereupon, at 12:31 p.m., the press briefing was concluded.]

The Mission Reports Series.

Freedom 7 ISBN 1896522807 America's first man in space Alan B Shepard Jr. puts the US firmly in the race against the Soviets. *Includes CD Rom*

Friendship 7 ISBN 1896522602 John Glenn pilots the US first manned orbital flight in 1962 and becomes a national hero. *Includes CD Rom*

Gemini 6 ISBN 1896522610 Piloted by Mercury astronaut Wally Schirra and Tom Stafford; the US achieves the first rendezvous in space with Gemini 7. *Includes CD Rom*

Gemini 7 ISBN 1896522823 Frank Borman and James Lovell not only achieve the first rendezvous with Gemini 6, but go on to an endurance record of two weeks in space. Includes *CD Rom*

Apollo 7 ISBN 1896522645. Wally Schirra takes the helm again to prove that America's Apollo moonship can perform flawlessly for a full 8 days. *CD Rom*

Apollo 8 ISBN 1896522661 Borman, Lovell and Anders pilot the Apollo spacecraft out of Earth orbit to the moon for the first time in human history. Genesis is read from Lunar orbit on Christmas eve 1968. *CD Rom*

Apollo 9 ISBN 1896522513 This mission to test the full Apollo configuration including the LM had so many firsts it was almost a space program in its own right. *CD Rom*

Apollo 10 ISBN 1896522688 The penultimate full up test of the CSM and LM in lunar orbit before the landing. The crew was probably the best ever to fly and they had a blast doing it. *CD Rom*

Apollo 11 Vol. One ISBN 189652253X Neil, Buzz and Mike take that "One small step.." and succeed in completing the greatest scientific achievement ever in the history of mankind. The first manned landing on the moon. *CD Rom of the EVA and over 1200 color photos.*

Apollo 11 Vol. Two ISBN 1896522491 The crew debrief of the mission, that was declassified specifically for this book. The story of the first moon landing in the crew's own words, including that pesky UFO sighting. *CD Rom with interactive lunar panoramas and exclusive interview with Buzz Aldrin.*

Apollo 11 Vol. Three ISBN 1896522858 The crew performed many scientific tests on the moon, here is what they found. *The DVD includes the landing video and EVA in a highly acclaimed format never seen before.*

Apollo 12 ISBN 1896522548 "It may have been a small one for Neil, but it was a big one for me!" Pete Conrad and Al Bean prove that a landing can be bang on target. *CD Rom includes the full EVA and exclusive interview with Dick Gordon.*

Apollo 13 ISBN 1896522556 Possibly the US space programs finest hour. Find out what went wrong and how they saved this dramatic mission that almost resulted in the deaths of the crew. *Exclusive video interview with Capt. Lovell.*

Apollo 14 ISBN 1896522564 It was up to America's first astronaut, Alan Shepard and his crew to get back to the moon and to...play a little golf! *CD Rom.*

Apollo 15 Vol. One ISBN 1896522572 Dave Scott and Jim Irwin land in the middle of a mountain range with their little moon buggy to go for a spin; while Al Worden conducts many experiments in lunar orbit. Includes *CD Rom*

Apollo 16 Vol. One ISBN 1896522580 Capt. John Young (The astronaut King) and Charlie Duke spend three days on the highlands at Descartes in typically flawless Young fashion. *CD Rom includes hours of EVA footage.*

Apollo 17 Vol. One ISBN 1896522599 The last men on the moon, Gene Cernan and Jack Schmitt (the first scientist to make the trip) almost make the whole thing look routine because they have so much fun. *CD Rom*

Space Shuttle STS 1-5 The NASA Mission Reports. ISBN 1896522696 The first amazing five missions of the Space Shuttle Columbia with crew debriefings and mission reports. *CD Rom hours of video and 100's of photos.*

X-15 ISBN 1896522653 Before and whilst the US space program was in full swing, there was another group of men becoming astronauts in a very promising precursor to the Shuttle program. Some wonderful film of rocket planes in full flight. *CD Rom*

Mars: The NASA Mission Reports. ISBN 1896522629 Every mission to Mars from Mariner 4 to Global Surveyor. See how our knowledge of the red planet has grown exponentially since the 60's. *CD Rom with 1,000's of photos.*

The Apogee Space Series

Rocket & Space Corporation Energia ISBN 1896522815 For the first time in English a pictorial history of the Russian Space program from the 1940's through Sealaunch.

Arrows To The Moon ISBN 1896522831 Did you know that in 1958 the most advanced fighter aircraft in the world was not US or Russian? It was Canadian! When their government cancelled that program many of the aerospace engineers went on to become an integral part of NASA.

The High Frontier ISBN 189652267X Gerard O'Neill writes about human colonies in space. The book that inspired the founders of the Space Frontier Foundation, among others. Includes 6 new contemporary chapters by leading visionaries.

The Unbroken Chain ISBN 189652284X by Guenter Wendt and Russell Still. The man who tucked all of the astronauts in their couches, shook their hands and closed the hatch from Alan Shepard through Shuttle. An amazing story! Hardcover with CD Rom

Creating Space ISBN 1896522866 by Mat Irvine. The story of the Space age told through model kits. Color book with foreword by Sir Arthur C Clarke.

Women Astronauts ISBN 1896522874 by Laura Woodmansee. Every woman who has ever flown in space, how they did it and more importantly WHY they did it. *Full color with CD Rom, which includes exclusive interviews with current astronauts.*

On To Mars ISBN 1896522904 Edited by Dr. Robert Zubrin. Many papers presented at the annual Mars Society Conferences, from the scientific to the moral and ethical questions posed by colonizing a new world. *CD Rom*

The Conquest of Space ISBN1896522920 by David Lasser. The first scientifically accurate book written about space travel in the English language. This book is the one that inspired people like Sir Arthur C. Clarke to become interested in space. It still holds up today and is a wonderful reference for the novice.

Lost Spacecraft ISBN1896522882 by Curt Newport. The search and recovery of Gus Grissom's Liberty Bell 7 space ship by the man who did it. This was the subject of a major Discovery Channel program. *CD Rom*

Virtual Apollo ISBN1896522947 by Scott Sullivan. If you ever wanted to know how to build a moonship here it is in awesome 3D color computer graphics.

A Vision Of Future Space Transportation ISBN1896522939 by Tim McElyea. The Official publication of SpaceDay 2003 by the man who produces the computer simulations of future spacecraft for NASA and the USAF. *Includes CD Rom with animations, videos and 3D renderings.*

Apollo EECOM ISBN1896522963 by Sy Liebergot. The amazing and heartrending story of one of the key figures in US space history. The first man in Mission Control to acknowledge that Apollo 13 really did have a problem. *CD Rom*

Interstellar Travel & Multi Generational Spacecraft ISBN1896522998 by Yoji Kondo and others from Goddard Space Flight Center. If you really wish to travel to the stars, these guys are the ones who could "make it so!"

Space Trivia ISBN189652298X by Bill Pogue. The latest book from the Skylab astronaut who brought us, "How Do You Go To The Bathroom In Space" a space bestseller. 1000's of irreverent facts about how and why we go to space.

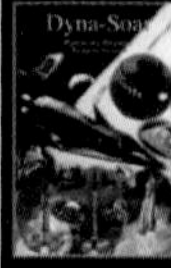

Dyna-Soar ISBN1896522955. Called "A treasure trove..." by IEEE Spectrum. The complete story of America's first military space shuttle and how it could have flown in the early 1960's. *DVD*

The Rocket Team ISBN1894959-00-0 by Frederick I Ordway III & Mitchell Sharpe. The historic tale of Germany's amazing rocket engineers from Peenemünde to the Moon. *Now with a DVD!*

Check us out on the Web -
http://www.apogeebooks.com

Return this completed form and become eligible to win free books!

One new space book almost every month! The world's number one space book publisher!

BUSINESS REPLY MAIL

FIRST CLASS MAIL PERMIT NO. 350 WHEATON IL

POSTAGE WILL BE PAID BY ADDRESSEE

COLLECTOR'S GUIDE
PUBLISHING INC
P.O. BOX 4588
WHEATON, IL 60189-9937

Check us out on the Web -
http://www.apogeebooks.com

Return this completed form and become eligible to win free books!

One new space book almost every month! The world's number one space book publisher!

NO POSTAGE NECESSARY IF MAILED IN THE UNITED STATES

BUSINESS REPLY MAIL

FIRST CLASS MAIL PERMIT NO. 350 WHEATON IL

POSTAGE WILL BE PAID BY ADDRESSEE

COLLECTOR'S GUIDE
PUBLISHING INC
P.O. BOX 4588
WHEATON, IL 60189-9937

THE SPACE BOOK COMPANY!

P.O. BOX 4588
WHEATON ILLINOIS 60189-9937
HTTP://WWW.APOGEEBOOKS.COM

Apogee Books is THE space book company, with almost one NEW space related book published every month.

☐ *To receive a catalog by mail check here and print your address below.*

☐ *Please update me by email (Print email address below)**

NAME: (Please PRINT) ______________________________

Company: ______________________________

Address: ______________________________

City: ______________ **State:** ______________ **ZIP:** ______________

E Mail: ______________________________

*CG Publishing will not disburse this email address to any other organizations for unrelated purposes.

Where did you buy this book?

Bookstore Chain ☐ Independent Bookstore ☐ Online Store ☐ Museum ☐ Other ☐

1

APOGEE BOOKS

THE SPACE BOOK COMPANY!

P.O. BOX 4588
WHEATON ILLINOIS 60189-9937
HTTP://WWW.APOGEEBOOKS.COM

Apogee Books is THE space book company, with almost one NEW space related book published every month.

☐ *To receive a catalog by mail check here and print your address below.*

☐ *Please update me by email (Print email address below)**

NAME: (Please PRINT) ______________________________

Company: ______________________________

Address: ______________________________

City: ______________ **State:** ______________ **ZIP:** ______________

E Mail: ______________________________

*CG Publishing will not disburse this email address to any other organizations for unrelated purposes.

Where did you buy this book?

Bookstore Chain ☐ Independent Bookstore ☐ Online Store ☐ Museum ☐ Other ☐

2